Engineering Design with SOLIDWORKS 2024

A Step-by-Step Project Based Approach
Utilizing 3D Solid Modeling

David C. Planchard
CSWP & SOLIDWORKS Accredited Educator

SDC
PUBLICATIONS

SDC Publications

P.O. Box 1334

Mission, KS 66222

913-262-2664

www.SDCpublications.com

Examination Copies

Books received as examination copies are for review purposes only and may not be made available for student use. Resale of examination copies is prohibited.

Electronic Files

Any electronic files associated with this book are licensed to the original user only. These files may not be transferred to any other party.

Trademarks

© Dassault Systèmes. All rights reserved. **3D**EXPERIENCE, the 3DS logo, the Compass icon, IFWE, 3DEXCITE, 3DVIA, BIOVIA, CATIA, CENTRIC PLM, DELMIA, ENOVIA, GEOVIA, MEDIDATA, NETVIBES, OUTSCALE, SIMULIA and SOLIDWORKS are commercial trademarks or registered trademarks of Dassault Systèmes, a European company (Societas Europaea) incorporated under French law, and registered with the Versailles trade and companies registry under number 322 306 440, or its subsidiaries in the United States and/or other countries.

All statements are strictly based on the author's opinion. Dassault Systèmes and its affiliates disclaim any liability, loss, or risk incurred as a result of the use of any information or advice contained in this book, either directly or indirectly.

SOLIDWORKS®, eDrawings®, SOLIDWORKS Simulation®, SOLIDWORKS Flow Simulation, and SOLIDWORKS Sustainability are a registered trademark of Dassault Systèmes SOLIDWORKS Corporation in the United States and other countries; certain images of the models in this publication courtesy of Dassault Systèmes SOLIDWORKS Corporation.

The publisher and the author make no representations or warranties with respect to the accuracy or completeness of the contents of this work and specifically disclaim all warranties, including without limitation warranties of fitness for a particular purpose. No warranty may be created or extended by sales or promotional materials. Dimensions of parts are modified for illustration purposes. Every effort is made to provide an accurate text. The authors and the manufacturers shall not be held liable for any parts, components, assemblies or drawings developed or designed with this book or any responsibility for inaccuracies that appear in the book. Web and company information was valid at the time of this printing.

The Y14 ASME Engineering Drawing and Related Documentation Publications utilized in this text are as follows: ASME Y14.1 1995, ASME Y14.2M-1992 (R1998), ASME Y14.3M-1994 (R1999), ASME Y14.41-2003, ASME Y14.5-1982, ASME Y14.5-1999, and ASME B4.2. Note: By permission of The American Society of Mechanical Engineers, Codes and Standards, New York, NY, USA. All rights reserved.

Download all needed model files from the SDC Publication website (www.SDCpublications.com/downloads/978-1-63057-630-1).

ISBN-13: 978-1-63057-630-1

ISBN-10: 1-63057-630-1

Printed and bound in the United States of America.

Introduction

Engineering Design with SOLIDWORKS® 2024 is written to assist students, designers, engineers and professionals.

The book provides a solid foundation in SOLIDWORKS by utilizing projects with step-by-step instructions for the beginner to intermediate SOLIDWORKS user featuring machined, plastic and sheet metal components.

It is a great starting point for those new to SOLIDWORKS or as a teaching aid in classroom training to become familiar with the software interface, basic commands and strategies as users complete a series of models while learning different ways to accomplish a particular task.

Desired outcomes and usage competencies are listed for each project.

The book is divided into six sections with 11 projects.

Project 1 - Project 6: Explore the SOLIDWORKS User Interface, CommandManager, Document and System properties. Address simple and complex parts, assemblies, design tables, configurations, multi-sheet, multi-view drawings, Bill of Material (BOM) and Revision table.

Formulate the skills to create, modify and edit sketches and solid features.

Understand and apply proper Design Intent.

Apply basic and advanced features along with Standard, Advanced and Mechanical mates.

Learn additional techniques to edit and reuse sketches, features and parts. Recover from Mate errors.

Utilize the SOLIDWORKS Toolbox.

Project 7: Understand Top-Down assembly modeling and Sheet Metal parts. Develop components In-Context with InPlace Mates, along with the ability to import parts using the Top-Down assembly method. Convert a solid part into a Sheet Metal part and insert and apply various Sheet Metal features.

Project 8: Recognize SOLIDWORKS Simulation. Understand a general overview of SOLIDWORKS Simulation and the type of questions that are on the SOLIDWORKS Associate - Simulation (CSWA-S) exam.

Details are provided on the login procedure; 3DEXPERIENCE® Certification Center at https://3dexperience.virtualtester.com/#home for students to register, examine their certifications, manage their account and take the (CSWA-S) certification.

The CSWA-S certification indicates a foundation in and apprentice knowledge of demonstrating an understanding in the principles of stress analysis and the Finite Element Method (FEM). The project covers and informs you on the required types of questions, layout and what to expect when taking the CSWA-S exam. The project examines and explains each area of the exam with sample questions, answers and tips.

Project 9: Explore SOLIDWORKS using the **3D**EXPERIENCE platform. Detailed steps are provided to obtain a **3D**EXPERIENCE ID (login credentials) to access the platform.

The **3D**EXPERIENCE platform provides downloadable content (DLC) from a secure cloud location within your SOLIDWORKS session. The platform enables you to collaborate with students and teams on and off campus through a browser. Think of the **3D**EXPERIENCE platform simply as a SOLIDWORKS Add-In on your desktop. Experience a single modeling environment. Access the platform within the SOLIDWORKS Task Pane. An e-Book on SOLIDWORKS and the **3D**EXPERIENCE platform is included.

Project 10: Additive Manufacturing - 3D Printing Fundamentals project provides a basic understanding between Additive vs. Subtractive manufacturing.

Discuss Fused Filament Fabrication (FFF), STereoLithography (SLA), and Selective Laser Sintering (SLS) printer technology.

Select suitable filament material. Comprehend 3D printer terminology. Knowledge of preparing, saving and printing a model on a Fused Filament Fabrication 3D printer.

Information on the Certified SOLIDWORKS Associate Additive Manufacturing (CSWA-AM) exam and general topic areas.

Information is provided on the login procedure at https://3dexperience.virtualtester.com/#home for students to register, examine their certifications, manage their account and to take the CSWA-AM certification.

Project 11: The Certified SOLIDWORKS Associate - Mechanical Design (CSWA) project provides detailed information and resources about the exam.

Information is provided on the login procedure at https://3dexperience.virtualtester.com/#home for students to register, examine their certifications, manage their account and take Segment 1 & 2 of the certification.

Take the sample online CSWA exam. **Note**: The sample online CSWA exam is provided in the book (PDF) format along with the SOLIDWORKS models, solutions, and additional practice exam problems.

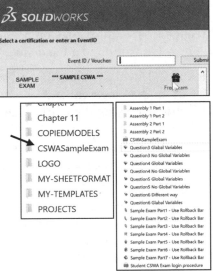

If your school is an academic certification provider, your instructor can allocate free exam credits for the CSWA (Segment 1 & 2), CSWA-SD, CSWA-S and CSWA-AM certifications. The instructor will require your .edu email address.

An instructor's folder is provided with log in and allocating student exam credit information.

About the Author

David Planchard is the founder of D&M Education LLC. Before starting D&M Education, he spent over 35 years in industry and academia holding various engineering, marketing, and teaching positions. He holds five U.S. patents. David has published and authored numerous papers and blogs on Machine Design, Product Design, Mechanics of Materials, Solid Modeling, SOLIDWORKS xDesign, SOLIDWORKS and the 3DEXPERIENCE Platform by Dassault Systèmes. David holds a BSME, MSM with the following professional certifications: CCAI, CCNP, CSWA-SD, CSWA-S, CSWA-AM, CSWP, CSWP-DRWT, SOLIDWORKS Accredited Educator, **3D**EXPERIENCE 3DSwymer, **3D**EXPERIENCE Collaborative Industry Innovator, **3D**EXPERIENCE Collaborative Designer for SOLIDWORKS and **3D**EXPERIENCE SOLIDWORKS Professional.

David is a SOLIDWORKS Solution Partner, **3D**EXPERIENCE Edu Champion and an Affiliate Professor of Emeritus in the Mechanical and Materials Engineering Department at WPI.

In 2012, David's senior Major Qualifying Project team (senior capstone) won first place in the Mechanical Engineering department at WPI.

In 2014, 2015 and 2016, David's senior Major Qualifying Project teams won the Provost award in the Mechanical Engineering department for design excellence.

In 2018, David's senior Major Qualifying Project team (Co-advisor) won the Provost award in the Electrical and Computer Engineering department. Subject area: Electrical System Implementation of Formula SAE Racing Platform.

In 2020, David was awarded Emeritus status at Worcester Polytechnic Institute (WPI) in the Department of Mechanical Engineering. David's ME design class achieved world recognition, featured in Compass published by Dassault Systèmes, in the technical article, "Open Innovation in a Pandemic."

In 2022, David's senior Major Qualifying Project team won second place in the Provost award in the Electrical and Computer Engineering department (ECE) at WPI. His FSAE Electric team won the IEEE Excellence in Electric Vehicle Engineering Award and achieved 4th place at the Formula Hybrid + Electric competition at the New Hampshire Motor Speedway in Loudon, NH. The competition is part of the Society of Automotive Engineers (SAE) Collegiate Design Series and is regarded as the most complex and dynamic of the series.

In 2023, David's senior Major Qualifying Project team won second place in the Provost award in the Electrical and Computer Engineering department (ECE) at WPI. His FSAE Electric team won the IEEE Excellence in Electric Vehicle Engineering Award, the TestEquity Electric Design award, and achieved 3rd place at the Formula Hybrid + Electric competition at the New Hampshire Motor Speedway in Loudon, NH.

In 2023, David was awarded the title of Affiliate Professor Emeritus in the Department of Mechanical and Materials Engineering at WPI.

David Planchard is the author of the following books:

- **Engineering Design with SOLIDWORKS® 2024**, 2023, 2022, 2021, 2020, 2019, 2018 and 2017

- **Engineering Graphics with SOLIDWORKS® 2024**, 2023, 2022, 2021, 2020, 2019, 2018 and 2017

- **SOLIDWORKS® 2024 Quick Start**, 2023, 2022, 2021, 2020, 2019 and 2018

- **SOLIDWORKS® 2024 Tutorial**, 2023, 2022, 2021, 2020, 2019 and 2017

- **Drawing and Detailing with SOLIDWORKS® 2022**, 2014 and 2012

- **Official Certified SOLIDWORKS® Professional (CSWP) Certification Guide 2020 - 2023**, 2019 - 2020 and 2015 - 2017

- **Official Guide to Certified SOLIDWORKS® Associate Exams: CSWA, CSWA-SD, CSWA-S, CSWA-AM 2020 - 2023**, 2019 - 2021, 2017 - 2019 and 2015 - 2017

Acknowledgements

Writing this book was a substantial effort that would not have been possible without the help and support of my loving family and of my professional colleagues. I would like to thank Professor John M. Sullivan Jr., Professor Jack Hall, and the community of scholars at Worcester Polytechnic Institute who have enhanced my life, my knowledge and helped to shape the approach and content to this text.

The author is greatly indebted to my colleagues from Dassault Systèmes SOLIDWORKS Corporation for their help and continuous support: Mike Puckett, Avelino Rochino, Yannick Chaigneau, Terry McCabe and the SOLIDWORKS Partner team.

Thanks also to Professor Richard L. Roberts of Wentworth Institute of Technology, Professor Dennis Hance of Wright State University, Professor Jason Durfess of Eastern Washington University and Professor Aaron Schellenberg of Brigham Young University - Idaho who provided vision and invaluable suggestions.

Contact the Author

We realize that keeping software application books current is imperative to our customers. We value the hundreds of professors, students, designers, and engineers that have provided us input to enhance the book. Please contact me directly with any comments, questions or suggestions on this book or any of our other SOLIDWORKS books at dplanchard@msn.com.

Note to Instructors

Download all needed model files (initial and final) from the SDC Publications website (www.SDCpublications.com/downloads/978-1-63057-630-1).

Instructors are provided PowerPoint presentations and project quizzes. Initial and final SOLIDWORKS models are included for each project and homework section.

Over 100 additional models are provided as classroom examples using various SOLIDWORKS features, sketches and mate types. Additional design projects are included.

The sample online CSWA exam is provided in the book (PDF) format along with the SOLIDWORKS models, solutions and additional practice exam problems.

An instructor's folder is provided with log in and allocating student exam credit information.

An eBook for SOLIDWORKS and the **3D**EXPERIENCE platform is included.

Trademarks, Disclaimer and Copyrighted Material

© Dassault Systèmes. All rights reserved. **3D**EXPERIENCE, the 3DS logo, the Compass icon, IFWE, 3DEXCITE, 3DVIA, BIOVIA, CATIA, CENTRIC PLM, DELMIA, ENOVIA, GEOVIA, MEDIDATA, NETVIBES, OUTSCALE, SIMULIA and SOLIDWORKS are commercial trademarks or registered trademarks of Dassault Systèmes, a European company (Societas Europaea) incorporated under French law, and registered with the Versailles trade and companies registry under number 322 306 440, or its subsidiaries in the United States and/or other countries.

All statements are strictly based on the author's opinion. Dassault Systèmes and its affiliates disclaim any liability, loss, or risk incurred as a result of the use of any information or advice contained in this book, either directly or indirectly.

SOLIDWORKS®, eDrawings®, SOLIDWORKS Simulation®, SOLIDWORKS Flow Simulation, and SOLIDWORKS Sustainability are a registered trademark of Dassault Systèmes SOLIDWORKS Corporation in the United States and other countries; certain images of the models in this publication courtesy of Dassault Systèmes SOLIDWORKS Corporation.

The publisher and the author make no representations or warranties with respect to the accuracy or completeness of the contents of this work and specifically disclaim all warranties, including without limitation warranties of fitness for a particular purpose. No warranty may be created or extended by sales or promotional materials. Dimensions of parts are modified for illustration purposes. Every effort is made to provide an accurate text. The authors and the manufacturers shall not be held liable for any parts, components, assemblies or drawings developed or designed with this book or any responsibility for inaccuracies that appear in the book. Web and company information was valid at the time of this printing.

References

- SOLIDWORKS Help Topics and What's New, SOLIDWORKS Corporation, 2024.

- 80/20 Product Manual, 80/20, Inc., Columbia City, IN, 2012.

- Ticona Designing with Plastics - The Fundamentals, Summit, NJ, 2009.

- Emerson-EPT Bearing Product Manuals and Gear Product Manuals, Emerson Power Transmission Corporation, Ithaca, NY, 2009.

- Emhart - A Black and Decker Company, On-line catalog, Hartford, CT, 2012.

During the initial SOLIDWORKS installation, you are requested to select either the ISO or ANSI drafting standard. ISO is typically a European drafting standard and uses First Angle Projection. The book is written using the ANSI (US) overall drafting standard and Third Angle Projection for drawings.

Table of Contents

Additional design projects are included. Use the web to locate a similar assembly and to understand the parts needed in the assembly. View the provided sample parts. Create your own parts and final assembly. View the provided motion file to create the proper movement in SOLIDWORKS.

Overview of Projects

Project 1: Overview of SOLIDWORKS and the SOLIDWORKS User Interface

SOLIDWORKS is a design software application used to create 2D and 3D sketches, 3D parts, 3D assemblies and 2D drawings.

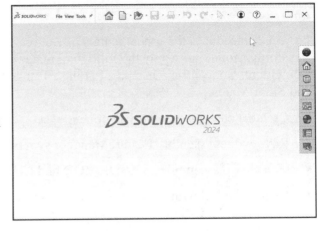

- Project 1 introduces the user to the SOLIDWORKS 2024 Welcome dialog box, User Interface (UI) and the CommandManager: Menu bar toolbar, Menu bar menu, Drop-down menus, Context toolbars, Consolidated drop-down toolbars, System feedback icons, Confirmation Corner, Heads-up View toolbar and Document Properties.

- Start a new SOLIDWORKS session. Create a new part. Open an existing part and view the created features and sketches using the Rollback bar. Design the part using proper design intent.

Project 2: Fundamentals of Part Modeling

Project 2 begins by creating file folders and sub-folders to manage projects. Apply System Options and Document Properties. Develop a Custom Part Template.

- Select the correct Sketch plane. Utilize the correct Sketch tool. Fully define all sketches.

- Create three parts: PLATE, ROD and GUIDE.

- Utilize the following features: Extruded Boss/Base, Instant3D, Extruded Cut, Fillet, Mirror, Chamfer, Hole Wizard and Linear Pattern.

- Apply material and appearance to a part.

- Reuse geometry.

- Copy and paste existing sketches and features.

Project 3: Fundamentals of Assembly Modeling (Bottom up)

Project 3 introduces the fundamentals of Bottom-up Assembly Modeling. Apply the following mate types: Coincident, Concentric, Distance, Parallel, Tangent, Slot and SmartMates.

- Create a custom Assembly template.

- Review the Assembly FeatureManager syntax.

- Create two assemblies: GUIDE-ROD and CUSTOMER.

- Utilize the following Assembly tools: Insert Component, Fix, Float, Hide/Show, Suppress, Move Component and Rotate Component.

- Edit component dimensions.

- Address tolerance and fit.

- Insert hardware from the SOLIDWORKS Toolbox.

Project 4: Fundamentals of Drawing

Project 4 provides an understanding of Drawing Templates, Part drawings, Assembly drawings, details and annotations.

- Apply the Edit Sheet Format and Edit Sheet modes.

- Develop and insert a Company logo and Third Angle Projection from a bitmap or picture file.

- Create the GUIDE SUPPORT drawing with Custom Properties, and various drawing views: (Front, Top, Right, Isometric, Section, Detailed and Auxiliary).

- Create a Revision table.

- Create the GUIDE-ROD drawing with Custom Properties.

- Create an Exploded Isometric view with balloons and a custom Bill of Materials (BOM).

Project 5: Extrude and Revolve Feature

Project 5 focuses on the customer's design requirements. Create four key FLASHLIGHT parts: BATTERY, BATTERYPLATE, LENS and BULB. Develop a custom ANSI - IPS Part Template.

- Create the BATTERY and BATTERYPLATE part with a custom part template.

- Apply the following design features: Extruded Boss/Base, Extruded Boss/Base with draft, Extruded Cut and Constant Size Fillet.

- Create the LENS and BULB part with a custom part template.

- Apply the following design features: Extruded Boss/Base, Extruded Cut, Revolved Base, Revolved Cut, Dome, Shell, Multi-radius Edge Fillet and Circular Pattern.

- Create Core and Cavity Tooling for the BATTERYPLATE.

- Utilize the following Mold tools: Scale, Parting Lines, Parting Surfaces, Shut-off Surfaces, Tooling Split and Draft.

Project 6: Swept, Lofted and Additional Features

Project 6 develops four additional components to complete the FLASHLIGHT assembly: O-RING, SWITCH, LENSCAP and HOUSING.

- Use the following: Extruded Boss/Base, Extruded Cut, Swept Boss/Base, Shell, Lofted Boss/Base, Draft, Rib, Linear Pattern, Circular Pattern, Mirror, Plane, Helix/Spiral, Dome and Revolved Cut.

- Create and edit a Design Table for an O-RING with three configurations.

- Utilize the following Assembly tools: Insert Component, Fix, Float, Hide/Show, Suppress, Move Component, Rotate Component, Exploded View and Interference Detection.

Project 7: Top-Down assembly and Sheet Metal parts

Project 7 focuses on the Top-Down assembly modeling approach. Develop a Layout Sketch.

- Create components and modify them In-Context of the assembly.

- Create Sheet metal features.

- Utilize the following features: Rip, Insert Sheet metal Bends, Base Flange, Edge Flange, Miter Flange, Break Corners, Hem and more.

- Utilize the Die Cut Feature and Louver Form tool.

- Add IGES format part files from the Internet.

- Replace fasteners in the assembly and redefine mates.

- Utilize equations, Global Variables and a Design Table to create multiple configurations of the BOX assembly.

The book is designed to expose the new SOLIDWORKS user to many tools, techniques and procedures. It may not always use the most direct tool or process.

Project 8: SOLIDWORKS Simulation

Project 8 provides a general overview of SOLIDWORKS Simulation and the type of questions that are on the SOLIDWORKS Associate - Simulation (CSWA-S) exam. On the completion of this project, you will be able to:

- Recognize the power of SOLIDWORKS Simulation.

- Utilize SOLIDWORKS Simulation to:

  - Define a Static Analysis Study.

  - Apply Material to a part model.

  - Work with a Solid and Sheet Metal model.

  - Define Solid, Shell and Beam elements.

  - Define Standard and Advanced Fixtures and External loads.

  - Define Local and Global coordinate systems.

  - Understand the axial forces, sheer forces, bending moments and factor of safety.

  - Define and apply component contacts (Local interactions) and contact sets (interactions) for assemblies and multi-body parts

  - Set and modify plots to display in the Results folder.

  - Work with Multi-body parts as different solid bodies.

  - Select different solvers as directed to optimize problems.

  - Determine if the result is valid.

  - Understand the type of problems and questions that are on the CSWA-S exam.

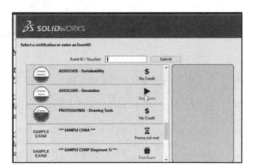

If your school is an academic certification provider, your instructor can allocate free exam credits for the CSWA (Segment 1 & 2), CSWA-SD, CSWA-S, and CSWA-AM certifications.

Perform step-by-step tutorials to practice and reinforce the subject matter and objectives.

Project 9: SOLIDWORKS and the 3DEXPERIENCE Platform

Project 9 offers a look at SOLIDWORKS and the **3D**EXPERIENCE platform. The project is written to answer a few keys questions for the new user. What is the **3D**EXPERIENCE platform? How does it work within my SOLIDWORKS session?

The platform enables you to collaborate with students and teams on and off campus through a browser in a secure cloud environment. Think of the **3D**EXPERIENCE platform simply as a SOLIDWORKS Add-In on your desktop. Through the Task Pane, students can manage, revise and share information. Experience a single modeling environment. Access the platform within the SOLIDWORKS Task Pane.

Detailed steps are provided to obtain a **3D**EXPERIENCE ID (login credentials) to access the platform. Nine step-by-step PDF tutorials and models are provided to practice and understand the tools and advantages of SOLIDWORKS and the platform. An e-Book on SOLIDWORKS and the **3D**EXPERIENCE platform is included.

On the completion of the project, you will be able to:

- Obtain a **3D**EXPERIENCE ID to access the platform.

- Understand **3D**EXPERIENCE platform terminology.

- Launch the SOLIDWORKS App directly from the **3D**EXPERIENCE platform.

- Utilize the new Save and Open tools.

- Identify and apply the Overwrite Protection tools.

- Create Bookmarks and a Collaborative space.

- Identify the auto generated embedded Product Lifecycle Management (PLM) model properties.

- Handle complex model relations in a Graph or Datagrid view.

- Apply the Revision and Lifecycle Maturity tools.

- Discuss and create a 3DDrive.

- Apply improved search tools (Searchable metadata and 6W Tags).

- Examine and utilize 3DPlay and 3DSwym.

Project 10: Additive Manufacturing - 3D Printing Fundamentals

Project 10 provides a basic understanding between Additive vs. Subtractive manufacturing. Discuss Fused Filament Fabrication (FFF), STereoLithography (SLA), and Selective Laser Sintering (SLS) printer technology.

Select suitable filament material. Comprehend 3D printer terminology.

Knowledge of preparing, saving, and printing a model on a Fused Filament Fabrication 3D printer.

Information on the Certified SOLIDWORKS Associate Additive Manufacturing (CSWA-AM) exam.

On the completion of the project, you will be able to:

- Discuss Additive vs Subtractive manufacturing.

- Review 3D printer technology: Fused Filament Fabrication (FFF), STereoLithography (SLA) and Selective Laser Sintering (SLS).

- Select the correct filament material: PLA (Polylactic acid), FPLA (Flexible Polylactic acid), ABS (Acrylonitrile butadiene styrene), PVA (Polyvinyl alcohol), Nylon 618 and Nylon 645.

- Create an STL (*.stl) file, Additive Manufacturing (*.amf) file and a 3D Manufacturing format (*.3mf) file.

- Prepare Geometric Code, "G-Code".

- Comprehend general 3D printer terminology.

- Understand optimum build orientation.

- Enter slicer parameters:
  - Raft, brim, skirt, layer height, percent infill, infill pattern, wall thickness, fan speed, print speed, bed temperature and extruder (hot end) temperature.

- Address tolerance for interlocking parts.

- Knowledge of the Certified SOLIDWORKS Associate Additive Manufacturing exam.

☀ If your school is an academic certification provider, your instructor can allocate free exam credits for the CSWA (Segment 1 & 2), CSWA-SD, CSWA-S, and CSWA-AM certifications. The instructor will require your .edu email address.

Login information and procedure is provided for students to register, examine their certifications, manage their account and take the exam.

Project 11: Certified SOLIDWORKS Associate - Mechanical Design Exam (CSWA)

Project 11 provides general information, resources, models and examples for instructors and students about the Certified SOLIDWORKS Associate - Mechanical Design exam (CSWA).

All categories of the CSWA Academic exam (Segment 1 & 2) are covered.

Information is provided on the login procedure at https://3dexperience.virtualtester.com/#home for students to register, examine their certifications, manage their account and take Segment 1 & 2 of the CSWA certification.

An instructor's folder is provided with log in and allocating student exam credit information.

Take the free sample online CSWA exam. There is a 30-day waiting period before retaking the sample online CSWA exam.

Note: The sample online CSWA exam is provided in the book (PDF) format along with the SOLIDWORKS models, solutions, and additional practice exam problems.

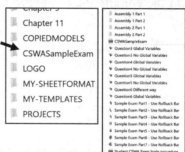

Additional references to help you prepare are:

- **Official Guide to Certified SOLIDWORKS Associate Exams: CSWA, CSWA-SD, CSWA-S, CSWA-AM; 2020 - 2023,** 2017 - 2019, 2015 - 2012. There are over 75 practice questions and examples (initial and final models) in the Official GUIDE to Certified SOLIDWORKS Exam book for the CSWA certification.

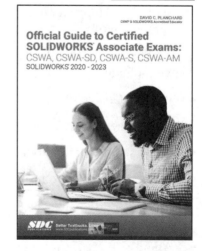

What is **SOLIDWORKS?**

SOLIDWORKS® is a design tool by Dassault Systèmes. The **3D**EXPERIENCE® platform is a technology platform that connects SOLIDWORKS to other design, simulation, intelligent information and collaboration Apps through a web browser.

SOLIDWORKS® is a mechanical design automation software package used to build parts, assemblies and drawings that takes advantage of the familiar Microsoft® Windows graphical user interface.

SOLIDWORKS is an easy to learn design and analysis tool (SOLIDWORKS Simulation, SOLIDWORKS Motion, SOLIDWORKS Flow Simulation, Sustainability, etc.), which makes it possible for designers to quickly sketch 2D and 3D concepts, create 3D parts and assemblies and detail 2D drawings.

The part, assembly and drawing documents are related. Additional information on SOLIDWORKS and its family of products can be obtained at their URL, www.SOLIDWORKS.com.

Features are the building blocks of parts. Use feature tools such as Extruded Boss/Base, Extruded Cut, Fillet, Revolved Boss/Base, Linear and Circular Patterns, Rib, Draft, etc. from the Features tab in the CommandManager to create 3D parts.

Extruded features begin with a 2D sketch created on a Sketch plane.

The 2D sketch is a profile or cross section. Use sketch tools such as Line, Center Rectangle, Slot, Circle Centerline, Mirror, etc. from the Sketch tab in the CommandManager to create a 2D sketch.

Sketch the general shape of the profile. Add geometric relationships and dimensions to control the exact size of the geometry and your Design Intent. Design for change.

Dimensions drive features. Change a dimension, and you change the size of the part.

Apply various Design Intent techniques using Geometric relationships: Horizontal, Vertical, Parallel, Collinear, Perpendicular, Parallel, Equal, Fix, etc.

Apply various Design Intent techniques using various End Conditions with Features: Blind, Through All, Up To Next, Up To Vertex, etc.

Create a hole that penetrates through a part (Through All). SOLIDWORKS maintains relationships through the change.

The step-by-step approach used in this text allows you to create, edit, modify and delete parts, assemblies and drawings. Change is an integral part of design. Learn how to recover for a design modification.

About the Book

You will find a wealth of information in this book. The book is a project based step-by-step text written for new and intermediate users.

The following conventions are used throughout this book:

- The term document refers to a SOLIDWORKS part, drawing or assembly file.

- The list of items across the top of the SOLIDWORKS interface is the Menu bar menu or the Menu bar toolbar. Each item in the Menu bar has a pull-down menu. When you need to select a series of commands from these menus, the following format is used: Click **Insert**, **Reference Geometry**, **Plane** from the Menu bar. The Plane PropertyManager is displayed.

- The book is organized into projects. Each project is focused on a specific subject or feature.

- The ANSI overall drafting standard and Third Angle projection is used as the default setting in this text. IPS (inch, pound, second) and MMGS (millimeter, gram, second) unit systems are used.

- The platform refers to the **3D**EXPERIENCE platform that provides downloadable content (DLC) from a secure cloud location within your SOLIDWORKS session.

- An e-Book on SOLIDWORKS and the **3D**EXPERIENCE platform is included.

- The following command syntax is used throughout the text. Commands that require you to perform an action are displayed in **Bold** text.

Format:	Convention:	Example:
Bold	• All commands actions. • Selected icon button. • Selected icon button. • Selected geometry: line, circle. • Value entries.	• Click **Options** ⚙ from the Menu bar toolbar. • Click **Corner Rectangle** ▱ from the Sketch toolbar. • Click **Sketch** 📝 from the Context toolbar. • Select the **centerpoint**. • Enter **3.0** for Radius.
Capitalized	• Filenames. • First letter in a feature name.	• Save the **FLATBAR** assembly. • Click the **Fillet** 🔲 feature.

Notes:

Project 1

Overview of SOLIDWORKS® 2024 and the User Interface

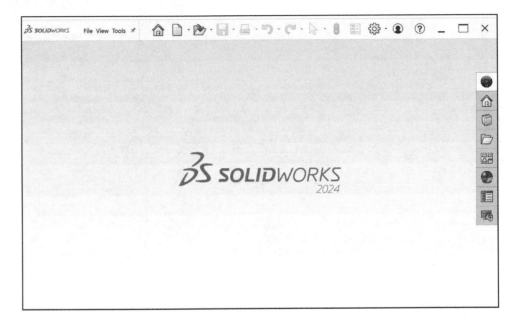

Below are the desired outcomes and usage competencies based on the completion of Project 1.

Desired Outcomes:	Usage Competencies:
• A comprehensive understanding of the SOLIDWORKS® 2024 User Interface (UI) and CommandManager.	• Ability to establish a SOLIDWORKS session. • Aptitude to utilize the following items: Menu bar toolbar, Menu bar menu, Drop-down menus, Context toolbars, Consolidated drop-down toolbars, System feedback icons, Confirmation Corner, Heads-up View toolbar and System and Document Properties. • Open a new and existing SOLIDWORKS part. • Knowledge to zoom, rotate and maneuver a three-button mouse in the SOLIDWORKS Graphics window.

Notes:

Project 1 - Overview of SOLIDWORKS® 2024 and the User Interface

Project Objective

Project 1 introduces the user to the SOLIDWORKS User Interface, CommandManager, Document and System properties. Review and address: Menu bar toolbar, Menu bar menu, Drop-down menus, Context toolbars, Consolidated drop-down toolbars, System feedback icons, Confirmation Corner and Heads-up View toolbar.

Start a new SOLIDWORKS session. Create a new part. Open an existing part and view the created features and sketches using the Rollback bar. Examine design intent in the part.

On the completion of this project, you will be able to:

- Utilize the Welcome - SOLIDWORKS dialog box.

- Establish a SOLIDWORKS session.

- Comprehend the SOLIDWORKS User Interface.

- Recognize the default Reference Planes in the FeatureManager.

- Open a new and existing SOLIDWORKS part.

- Utilize the Rollback bar.

- Locate Help and the SOLIDWORKS Tutorials.

- Zoom, rotate and maneuver a three-button mouse in the SOLIDWORKS Graphics window.

What is SOLIDWORKS®?

- SOLIDWORKS® is a design tool by Dassault Systèmes. The **3D**EXPERIENCE® platform is a technology platform that connects SOLIDWORKS to other design, simulation, intelligent information and collaboration Apps through a web browser.

- SOLIDWORKS® is a mechanical design automation software package used to build parts, assemblies and drawings that takes advantage of the familiar Microsoft® Windows graphical user interface.

- SOLIDWORKS is an easy to learn design and analysis tool (SOLIDWORKS Simulation, SOLIDWORKS Motion, SOLIDWORKS Flow Simulation, Sustainability, etc.), which makes it possible for designers to quickly sketch 2D and 3D concepts, create 3D parts and assemblies and detail 2D drawings.

- Model dimensions in SOLIDWORKS are associative between parts, assemblies and drawings. Reference dimensions are one-way associative from the part to the drawing or from the part to the assembly.

- The book is written for the beginner to intermediate user.

Start a SOLIDWORKS Session

Start a SOLIDWORKS session and familiarize yourself with the SOLIDWORKS User Interface. As you read and perform the tasks in this chapter, you will obtain a sense of how to use the book and the structure. Actual input commands or required actions in the chapter are displayed in bold.

The book does not cover starting a SOLIDWORKS session in detail for the first time. A default SOLIDWORKS installation presents you with several options. For additional information, visit http://www.SOLIDWORKS.com.

Activity: Start a SOLIDWORKS Session.

Start a SOLIDWORKS session.

1) Type **SOLIDWORKS 2024** in the Search window.

2) Click the **SOLIDWORKS 2024** application (or if available, **double-click** the SOLIDWORKS icon on the desktop). The Welcome - SOLIDWORKS dialog box is displayed by default.

The Welcome - SOLIDWORKS dialog box provides a convenient way to open recent documents (Parts, Assemblies and Drawings), view recent folders, access SOLIDWORKS resources, and stay updated on SOLIDWORKS news.

3) **View** your options. Do not open a document at this time. **Note:** The Task Pane icon order may be different.

Home Tab

The Home tab lets you open new and existing documents, view recent documents and folders and access SOLIDWORKS resources (*Part, Assembly, Drawing, Advanced mode, Open*).

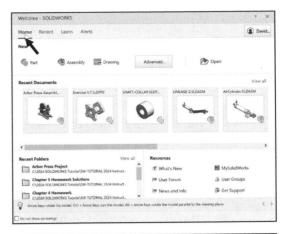

Recent Tab

The Recent tab lets you view a longer list of recent documents and folders. Sections in the Recent tab include *Documents* and *Folders*.

The Documents section includes thumbnails of documents that you have opened recently.

Click a thumbnail to open the document, or hover over a thumbnail to see the document location and access additional information about the document. When you hover over a thumbnail, the full path and last saved date of the document appears.

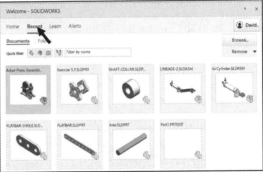

Learn Tab

The Learn tab lets you access instructional resources to help you learn more about the SOLIDWORKS software.

Sections in the Learn tab include:

- **Get Started with SOLIDWORKS CAD**.

- **Get Started with Collaborative Designer for SOLIDWORKS (Learning Path)**.

- **MySolidWorks Training (Learning Portal)**.

- **MySolidWorks CAD models.**

- **On my PC**.

- **Getting Started guide**.

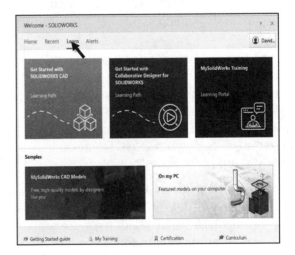

- **My Training**. Open the My Training section at MySolidWorks.com.

- **Certification**. Open the SOLIDWORKS Certification Program section at solidworks.com.

- **Curriculum**. Open the Curriculum section at solidworks.com.

🔅 When you install the software, if you do not install the Help Files or Example Files, the Tutorials and Samples links are unavailable.

Alerts Tab

The Alerts tab keeps you updated with SOLIDWORKS news.

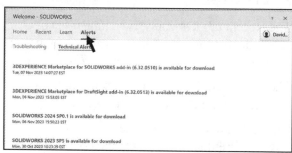

Sections in the Alerts tab include Critical, Troubleshooting, and Technical.

🔅 The Critical section does not appear if there are no critical alerts to display.

- **Troubleshooting**. Includes troubleshooting messages and recovered documents that used to be on the SOLIDWORKS Recovery tab in the Task Pane.

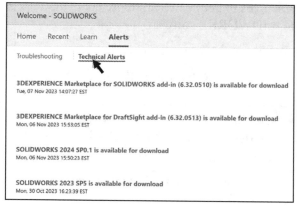

🔅 If the software has a technical problem and an associated troubleshooting message exists, the Welcome dialog box opens to the Troubleshooting section automatically on startup, even if you selected **Do not show at startup** in the dialog box.

- **Technical Alerts**. Open the contents of the SOLIDWORKS Support Bulletins RSS feed (Hot fixes, release news) at solidworks.com.

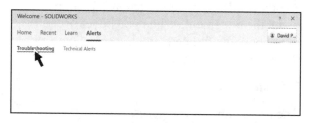

Close the Welcome - SOLIDWORKS dialog box.

4)   Click **Close** ✕ from the Welcome - SOLIDWORKS dialog box. The SOLIDWORKS Graphics window is displayed. **Note**: You can also click outside the Welcome - SOLIDWORKS dialog box, in the Graphics window.

5)   View the SOLIDWORKS Graphics window. **Note**: The Task Pane icon order may be different.

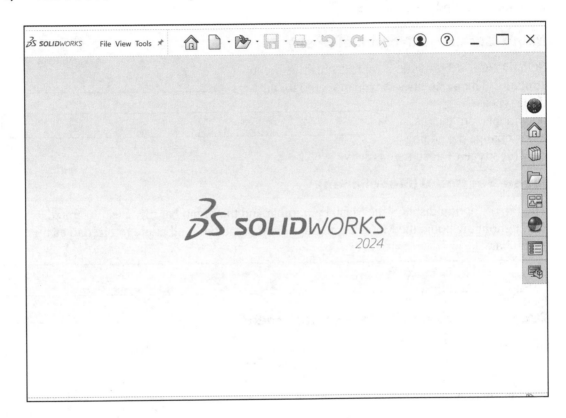

Menu Bar toolbar

The SOLIDWORKS (UI) is designed to make maximum use of the Graphics window. The Menu Bar toolbar contains a set of the most frequently used tool buttons from the Standard toolbar.

The following default tools are available:

Welcome to SOLIDWORKS ⌂ - Open the Welcome dialog box, **New** ▢ - Create a new document; **Open** 📂 - Open an existing document; **Save** 💾 - Save an active document;

Print 🖨 - Print an active document; **Undo** ↩ - Reverse the last action; **Redo** ↪ - Redoes the last action that you reverse; **Select** ⌖ - Select Sketch entities, components and more; **Rebuild** ▮ - Rebuild the active part, assembly or drawing; **File Properties** ▤ - Summary information on the active document; **Options** ⚙ - Change system options and Add-Ins for SOLIDWORKS; **Login** ⊕ - Login to SOLIDWORKS. **Help** ② - access to help, tutorials, updates and more.

Menu Bar menu (No model open)

SOLIDWORKS provides a context-sensitive menu structure. The menu titles remain the same for all three types of documents, but the menu items change depending on which type of document is active.

Menu Bar menu (Model open)

The Pin ✯ option displays the Menu bar toolbar and the Menu bar menu as illustrated. Throughout the book, the Menu bar menu and the Menu bar toolbar are referred to as the Menu bar.

Drop-down menu (Open part document)

SOLIDWORKS takes advantage of the familiar Microsoft® Windows user interface. Communicate with SOLIDWORKS through drop-down menus, Context sensitive toolbars, Consolidated toolbars or the CommandManager tabs.

To close a SOLIDWORKS drop-down menu, press the Esc key. You can also click any other part of the SOLIDWORKS Graphics window or click another drop-down menu.

Create a New Part Document

Activity: Create a new Part Document.

A part is a 3D model, which consists of features. What are features? Features are geometry building blocks. Most features either add or remove material. Some features do not affect material (Cosmetic Thread).

Features are created either from 2D or 3D sketched profiles or from edges and faces of existing geometry.

Features are individual shapes that combined with other features make up a part or assembly. Some features, such as bosses and cuts, originate as sketches. Other features, such as shells and fillets, modify a feature's geometry.

Features are displayed in the FeatureManager.

💡 The first sketch of a part is called the Base sketch. The Base sketch is the foundation for the 3D model. The book focuses on 2D sketches and 3D features.

💡 FeatureManager tabs, CommandManager tabs, Task Pane tabs and tree folders will vary depending on system setup, version and Add-ins.

There are two modes in the New SOLIDWORKS Document dialog box: Novice and Advanced. The Novice option is the default option with three templates. The Advanced mode contains access to additional templates and tabs that you create in system options. Use the Advanced mode in this book.

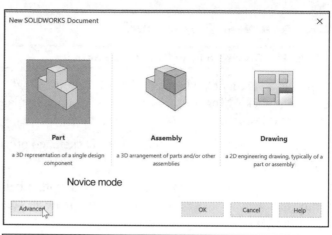

Create a new part.

6) Click **New** from the Menu bar. The New SOLIDWORKS Document dialog box is displayed.

Select the Advanced mode.

7) Click the **Advanced** button as illustrated. The Advanced mode is set.

8) Click the **Templates** tab.

9) Click **Part**. Part is the default template from the New SOLIDWORKS Document dialog box.

10) Click **OK** from the New SOLIDWORKS Document dialog box.

The Advanced mode remains selected for all new documents in the current SOLIDWORKS session. When you exit SOLIDWORKS, the Advanced mode setting is saved.

The default SOLIDWORKS installation contains two tabs in the New SOLIDWORKS Document dialog box: *Templates and MBD*. The *Templates* tab corresponds to the default SOLIDWORKS templates. The *MBD* tab corresponds to the templates utilized in the SOLIDWORKS (Model Based Definition). You may see a Tutorial tab depending on your version.

Note: **3D**EXPERIENCE users can create templates for the **3D**EXPERIENCE platform directly from SOLIDWORKS. Use the 3DEXPERIENCE tab in the New SOLIDWORKS Document dialog box. You will only see the 3DEXPERIENCE tab if you are logged into the **3D**EXPERIENCE Platform.

Part1 is displayed in the FeatureManager and is the name of the document. Part1 is the default part window name.

The Part Origin ⊥ is displayed in blue in the center of the Graphics window. The Origin represents the intersection of the three default reference planes: *Front Plane*, *Top Plane* and *Right Plane*. The positive X-axis is horizontal and points to the right of the Origin in the Front view. The positive Y-axis is vertical and points upward in the Front view. The FeatureManager contains a list of features, reference geometry, and settings utilized in the part.

💡 Edit the document units directly from the Graphics window as illustrated.

💡 FeatureManager tabs, CommandManager tabs, Task Pane tabs and tree folders will vary depending on system setup, version and Add-ins.

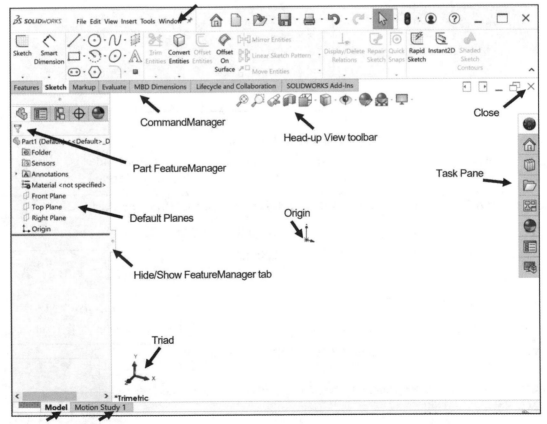

Main menu toolbar

CommandManager

Head-up View toolbar

Close

Part FeatureManager

Task Pane

Default Planes

Origin

Hide/Show FeatureManager tab

Triad

*Trimetric

Model Mode   Motion Mode

View the Default Sketch Planes.

11) Click the **Front Plane** from the FeatureManager.

12) Click the **Top Plane** from the FeatureManager.

13) Click the **Right Plane** from the FeatureManager.

14) Click the **Origin** from the FeatureManager. The Origin is the intersection of the Front, Top, and Right Planes.

15) Click **inside** the Graphics window.

Download all model files from the SDC Publications website (www.SDCpublications.com/downloads/978-1-63057-630-1). The top folder is ENGDESIGN-W-SOLIDWORKS. Work directly from a local hard drive.

Activity: Download the SOLIDWORKS folder. Open the Bracket Part.

Download the SOLIDWORKS folder. Open an existing SOLIDWORKS part.

16) **Download** the ENGDESIGN-W-SOLIDWORKS folder.

17) **Unzip** the folder.

18) **Work** from the unzip folder.

19) Click **Open** 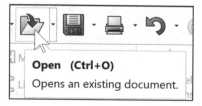 from the Main menu.

20) Browse to the **ENGDESIGN-W-SOLIDWORKS\Bracket** folder.

21) Double-click the **Bracket** part. The Bracket part is displayed in the Graphics window.

The FeatureManager design tree is located on the left side of the SOLIDWORKS Graphics window. The FeatureManager provides a summarized view of the active part, assembly, or drawing document. The tree displays the details on how the part, assembly or drawing document was created.

Use the FeatureManager rollback bar to temporarily roll back to an earlier state, to absorbed features, roll forward, roll to previous, or roll to the end of the FeatureManager design tree. You can add new features or edit existing features while the model is in the rolled-back state. You can save models with the rollback bar placed anywhere.

In the next section, review the features in the Bracket FeatureManager using the Rollback bar.

Activity: Use the FeatureManager Rollback Bar option.

Apply the FeatureManager Rollback Bar. Revert to an earlier state in the model.

22) Place the **mouse pointer** over the rollback bar in the FeatureManager design tree as illustrated. The pointer changes to a hand 👆. Note the provided information on the feature. This is called Dynamic Reference Visualization.

23) Drag the **rollback bar** up the FeatureManager design tree until it is above the features you want rolled back, in this case 10.0 (10) Diameter Hole1.

24) **Release** the mouse button.

View the first feature in the Bracket Part.

25) Drag the **rollback bar** up the FeatureManager above Fillet1. View the results in the Graphics window.

Return to the original Bracket Part FeatureManager.

26) Right-click **Extrude-Thin1** in the FeatureManager. The Pop-up Context toolbar is displayed.

27) Click **Roll to End**. View the results in the Graphics window.

Heads-up View toolbar

SOLIDWORKS provides the user with numerous view options. One of the most useful tools is the Heads-up View toolbar displayed in the Graphics window when a document is active.

💡 *Dynamic Annotation Views* : Only available with SOLIDWORKS MBD (Model Based Definition). Provides the ability to control how annotations are displayed when you rotate models.

In the next section, apply the following tools: Zoom to Fit, Zoom to Area, Zoom out, Rotate and select various view orientations from the Heads-up View toolbar.

Activity: Utilize the Heads-up View toolbar.

Zoom to Fit the model in the Graphics window.

28) Click the **Zoom to Fit** icon. The tool fits the model to the Graphics window.

Zoom to Area on the model in the Graphics window.

29) Click the **Zoom to Area** icon. The Zoom to Area icon is displayed.

Zoom in on the top left hole.

30) **Window-select** the top left corner as illustrated. View the results.

De-select the Zoom to Area tool.

31) Click the **Zoom to Area** icon.

Fit the model to the Graphics window.

32) Press the **f** key.

Rotate the model.

33) Hold the **middle mouse button** down. Drag **upward** ↻, **downward** ↻, to the **left** ↻ and to the **right** ↻ to rotate the model in the Graphics window.

Display a few Standard Views.

34) Click **inside** the Graphics window.

35) Click **Front** from the drop-down Heads-up view toolbar. The model is displayed in the Front view.

36) Click **Right** from the drop-down Heads-up view toolbar. The model is displayed in the Right view.

37) Click **Top** from the drop-down Heads-up view toolbar. The model is displayed in the Top view.

Display a Trimetric view of the Bracket model.

38) Click **Trimetric** from the drop-down Heads-up view toolbar as illustrated. Note your options. View the results in the Graphics window.

SOLIDWORKS Help

Help in SOLIDWORKS is context-sensitive and in HTML format. Help is accessed in many ways, including Help buttons in all dialog boxes and PropertyManager and

Help ? on the Standard toolbar for SOLIDWORKS Help.

39) Click ⑦ from the Standard toolbar.

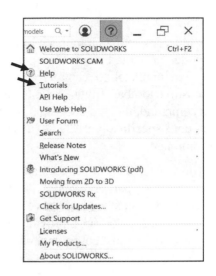

40) Click ⑦ **Help** from the drop-down menu. The SOLIDWORKS Home Page is displayed by default. View your options.

🔆 SOLIDWORKS Web Help is active by default under Help in the Main menu.

Close Help. Return to the SOLIDWORKS Graphics window.

41) **Close** ✖ SOLIDWORKS Home.

SOLIDWORKS Tutorials

Display and explore the SOLIDWORKS tutorials.

42) Click ⑦ from the Standard toolbar.

43) Click **Tutorials**. The SOLIDWORKS Tutorials are displayed. The SOLIDWORKS Tutorials are presented by category.

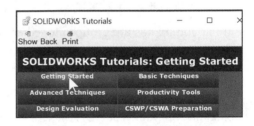

44) Click the **Getting Started** category. The Getting Started category provides lessons on parts, assemblies, and drawings.

In the next section, close all models, tutorials and view the additional User Interface tools.

Activity: Close all Tutorials and Models.

Close SOLIDWORKS Tutorials and models.

45) **Close** ✖ SOLIDWORKS Tutorials.

46) Click **Window**, **Close All** from the Menu bar menu.

User Interface Tools

The book utilizes additional areas of the SOLIDWORKS User Interface. Explore an overview of these tools in the next section.

Right-click

Right-click in the Graphics window on a model, or in the FeatureManager on a feature or sketch to display the Context-sensitive toolbar. If you are in the middle of a command, this toolbar displays a list of options specifically related to that command.

Right-click an empty space in the Graphics window of a part or assembly, and a selection context toolbar above the shortcut menu is displayed. This provides easy access to the most commonly used selection tools.

Consolidated toolbar

Similar commands are grouped together in the CommandManager. For example, variations of the Rectangle sketch tool are grouped in a single fly-out button as illustrated.

If you select the Consolidated toolbar button without expanding:

For some commands such as Sketch, the most commonly used command is performed. This command is the first listed and the command shown on the button.

For commands such as rectangle, where you may want to repeatedly create the same variant of the rectangle, the last used command is performed. This is the highlighted command when the Consolidated toolbar is expanded.

System feedback icon

SOLIDWORKS provides system feedback by attaching a symbol to the mouse pointer cursor.

The system feedback symbol indicates what you are selecting or what the system is expecting you to select.

As you move the mouse pointer across your model, system feedback is displayed in the form of a symbol, riding next to the cursor as illustrated. This is a valuable feature in SOLIDWORKS.

Confirmation Corner

When numerous SOLIDWORKS commands are active, a symbol or a set of symbols is displayed in the upper right-hand corner of the Graphics window. This area is called the Confirmation Corner.

When a sketch is active, the confirmation corner box displays two symbols. The first symbol is the sketch tool icon. The second symbol is a large red X. These two symbols supply a visual reminder that you are in an active sketch. Click the sketch symbol icon to exit the sketch and to save any changes that you made.

When other commands are active, the confirmation corner box provides a green check mark and a large red X. Use the green check mark to execute the current command. Use the large red X to cancel the command.

Confirm changes you make in sketches and tools by using the D keyboard shortcut to move the OK and Cancel buttons to the pointer location in the Graphics window.

Heads-up View toolbar

SOLIDWORKS provides the user with numerous view options from the Standard Views, View and Heads-up View toolbar.

The Heads-up View toolbar is a transparent toolbar that is displayed in the Graphics window when a document is active.

For a drawing document

You can hide, move or modify the Heads-up View toolbar. To modify the Heads-up View toolbar, right-click on a tool and select or deselect the tools that you want to display.

The following views are available.
Note: available views are document dependent.

- *Zoom to Fit* : Fit the model to the Graphics window.

- *Zoom to Area* : Zoom to the areas you select with a bounding box.

- *Previous View* : Display the previous view.

- *Section View* : Display a cutaway of a part or assembly, using one or more cross section planes.

- *Dynamic Annotation Views* : Only available with SOLIDWORKS MBD. Control how annotations are displayed when you rotate a model.

The Orientation dialog has an option to display a view cube (in-context View Selector) with a live model preview. This helps the user to understand how each standard view orientates the model. With the view cube, you can access additional standard views. The views are easy to understand and they can be accessed simply by selecting a face on the cube.

To activate the Orientation dialog box, press (Ctrl + spacebar) or click the View Orientation icon from the Heads-up View toolbar. The active model is displayed in the View Selector in an Isometric orientation (default view).

Click the View Selector icon in the Orientation dialog box to show or hide the in-context View Selector.

 Press **Ctrl + spacebar** to activate the View Selector.

 Press the **spacebar** to activate the Orientation dialog box.

- *View Orientation box* : Select a view orientation or the number of viewports. The options are: *Top, Left, Front, Right, Back, Bottom, Single view, Two view - Horizontal, Two view - Vertical, Four view*. Click the drop-down arrow to access Axonometric views: *Isometric, Dimetric* and *Trimetric*.

- *Display Style* : Display the style for the active view. The options are: *Wireframe, Hidden Lines Visible, Hidden Lines Removed, Shaded, Shaded With Edges*.

- *Hide/Show Items* : Select items to hide or show in the Graphics window. The available items are document dependent. Note the View Center of Mass ✛ icon.

- *Edit Appearance* : Edit the appearance of entities of the model.

- *Apply Scene* : Apply a scene to an active part or assembly document. View the available options.

- *View Setting* : Select the following settings: *RealView Graphics, Shadows In Shaded Mode, Ambient Occlusion, Perspective* and *Cartoon.*

- *Rotate view* : Rotate a drawing view. Input Drawing view angle and select the ability to update and rotate center marks with view.

- *3D Drawing View* : Dynamically manipulate the drawing view in 3D to make a selection.

To display a grid for a part, click Options , Document Properties tab. Click Grid/Snaps, check the Display grid box.

Add a custom view to the Heads-up View toolbar. Press the space key. The Orientation dialog box is displayed. Click the New View tool. The Name View dialog box is displayed. Enter a new named view. Click OK.

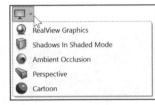

Use commands to display information about the triad or to change the position and orientation of the triad. Available commands depend on the triad's context.

Save space in the CommandManager, limit your CommandManager tabs. **Right-click** on a CommandManager tab. Click **Tabs**. View your options to display CommandManager tabs.

SOLIDWORKS CommandManager

The SOLIDWORKS CommandManager is a Context-sensitive toolbar. By default, it has toolbars embedded in it based on your active document type. When you click a tab below the CommandManager, it updates to display that toolbar. For example, if you click the Sketch tab, the Sketch toolbar is displayed.

The Lifecycle and Collaboration tab is displayed with all active tools if you have purchased 3DEXPERIENCE education roles and are logged in.

Below is an illustrated CommandManager for a **Part** document. Tabs will vary depending on system setup and Add-ins.

The SOLIDWORKS CommandManager is a Context-sensitive toolbar that automatically updates based on the toolbar you want to access. By default, it has toolbars embedded in it based on your active document type. The available tools are feature and document dependent.

Below is an illustrated CommandManager for a **Drawing** document. Tabs will vary depending on system setup and Add-ins.

To add a custom tab, right-click on a tab in the CommandManager and click Customize. You can also select to add a blank tab and populate it with custom tools from the Customize dialog box.

The SOLIDWORKS CommandManager is a Context-sensitive toolbar that automatically updates based on the toolbar you want to access. By default, it has toolbars embedded in it based on your active document type. The available tools are feature and document dependent.

Below is an illustrated CommandManager for an **Assembly** document. Tabs will vary depending on system setup and Add-ins.

The Markup tab is displayed by default in the CommandManager on some systems. The Share as Picture option is available if you purchased 3DEXPERIENCE education roles and are logged in.

Float the CommandManager. Drag the Features, Sketch or any CommandManager tab. Drag the CommandManager anywhere on or outside the SOLIDWORKS window.

To dock the CommandManager, perform one of the following:

While dragging the CommandManager in the SOLIDWORKS window, move the pointer over a docking icon -

 and click the needed command.

Double-click the floating CommandManager to revert the CommandManager to the last docking position.

Collapse the CommandManager

Collapse the CommandManager to only display the available tabs. No tools are shown.

Click the Collapsed CommandManager arrow to the right of the active CommandManager tab as illustrated. Only the tabs are displayed.

To display the CommandManager tools back, click on a CommandManager tab. Click the Pin icon in the CommandManager as illustrated.

Part FeatureManager Design Tree

The Part FeatureManager consists of various tabs:

- *FeatureManager design tree* tab.

- *PropertyManager* tab.

- *ConfigurationManager* tab.

- *DimXpertManager* tab.

- *DisplayManager* tab.

- *CAM FeatureManager tree* tab.

- *CAM Operation tree* tab.

- *CAM Tools tree* tab.

Click the direction arrows to expand or collapse the FeatureManager design tree.

CommandManager and FeatureManager tabs and folder files will vary depending on your SOLIDWORKS applications and Add-ins.

Select the Hide/Show FeatureManager Area tab

as illustrated to enlarge the Graphics window for modeling.

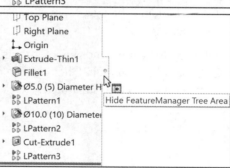

The Sensors tool located in the FeatureManager monitors selected properties in a part or assembly and alerts you when values deviate from the specified limits. There are five sensor types: Simulation Data, Mass properties, Dimensions, Measurement and Costing Data.

Various commands provide the ability to control what is displayed in the FeatureManager design tree.

1. Show or Hide FeatureManager items.

Click **Options** ⚙ ˙ from the Menu bar. Click **FeatureManager** from the System Options tab. Customize your FeatureManager from the Hide/Show tree Items dialog box.

2. Filter the FeatureManager design tree. Enter information in the filter field. You can filter by *Type of features, Feature names, Sketches, Folders*, *Mates, User-defined tags* and *Custom properties*.

Tags are keywords you can add to a SOLIDWORKS document to make them easier to filter and to search. The Tags 🏷 icon is located in the bottom right corner of the Graphics window.

Collapse all items in the FeatureManager, **right-click** and select **Collapse items**, or press the **Shift** + **C** keys.

The FeatureManager design tree and the Graphics window are dynamically linked. Select sketches, features, drawing views, and construction geometry in either pane.

Split the FeatureManager design tree and either display two FeatureManager instances, or combine the FeatureManager design tree with the ConfigurationManager or PropertyManager.

Move between the FeatureManager design tree, PropertyManager, ConfigurationManager, DimXpertManager, DisplayManager and others by selecting the tab at the top of the menu.

Split

The ConfigurationManager tab is located to the right of the PropertyManager tab. Use the ConfigurationManager to create, select and view multiple configurations of parts and assemblies.

The icons in the ConfigurationManager denote whether the configuration was created manually or with a design table.

The DimXpertManager ⊕ tab provides the ability to insert dimensions and tolerances manually or automatically. The options are: **Auto Dimension Scheme** , **Auto Pair Tolerance** , **Basic, Location Dimension** , **Basic Size Dimension** , **General Profile Tolerance** , **Show Tolerance Status** , **Copy Scheme** , **Import Scheme** , **TolAnalyst Study** and **Datum Target** .

TolAnalyst is available in SOLIDWORKS Premium.

Fly-out FeatureManager

The fly-out FeatureManager design tree provides the ability to view and select items in the PropertyManager and the FeatureManager design tree at the same time.

Throughout the book, you will select commands and command options from the drop-down menu, fly-out FeatureManager, Context toolbar, or from a SOLIDWORKS toolbar.

Another method for accessing a command is to use the accelerator key. Accelerator keys are special key strokes, which activate the drop-down menu options. Some commands in the menu bar and items in the drop-down menus have an underlined character.

Task Pane

The Task Pane is displayed when a SOLIDWORKS session starts. You can show, hide, and reorder tabs in the Task Pane. You can also set a tab as the default so it appears when you open the Task Pane, pin or unpin to the default location. **Note**: The Task Pane icon order may be different.

The Task Pane contains the following default tabs:

- *SOLIDWORKS Resources* .

- *Design Library* .

- *File Explorer* .

- *View Palette* .

- *Appearances, Scenes and Decals* .

- *Custom Properties* .

- *3DEXPERIENCE files on This PC* .

- *3DEXPERIENCE* .

Additional tabs are displayed with Add-Ins.

SOLIDWORKS Resources

The SOLIDWORKS Resources icon displays the following default selections:

- *Welcome to SOLIDWORKS.*

- *SOLIDWORKS Tools.*

- *Online Resources.*

- *Subscription Services.*

Other user interfaces are available during the initial software installation selection: *Machine Design, Mold Design, Consumer Products Design, etc.*

Design Library

The Design Library 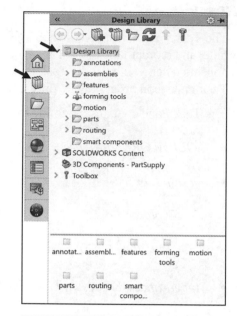 contains reusable parts, assemblies, and other elements including library features.

The Design Library tab contains default selections. Each default selection contains additional sub categories.

The default selections are:

- *Design Library.*

- *SOLIDWORKS Content (Internet access required).*

- *3D Components - PartSupply.*

- *Toolbox.*

Activate the SOLIDWORKS Toolbox. Click Tools, Add-Ins.., from the Main menu. Check the SOLIDWORKS Toolbox Library and SOLIDWORKS Toolbox Utilities box from the Add-ins dialog box or click SOLIDWORKS Toolbox from the SOLIDWORKS Add-Ins tab.

To access the Design Library folders in a non-network environment, click Add File Location and browse to the needed path. Paths may vary depending on your SOLIDWORKS version and window setup. In a network environment, contact your IT department for system details.

File Explorer

File Explorer ⌷ tab duplicates Windows Explorer from your local computer and displays:

- *Open in SOLIDWORKS.*

- *Desktop.*

View Palette

The View Palette ⬚⬚ tab provides the ability to insert drawing views of an active document, or click the Browse button to locate the desired document.

Click and drag the view from the View Palette into an active drawing sheet to create a drawing view.

Appearances, Scenes, and Decals

Appearances, Scenes, and Decals ● tab provides a simplified way to display models in a photo-realistic setting using a library of Appearances, Scenes, and Decals.

An appearance defines the visual properties of a model, including color and texture. Appearances do not affect physical properties, which are defined by materials.

Scenes provide a visual backdrop behind a model. In SOLIDWORKS they provide reflections on the model. PhotoView 360 is an Add-in. Drag and drop a selected appearance, scene or decal on a feature, surface, part or assembly.

Custom Properties

The Custom Properties 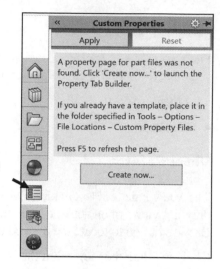 tab provides the ability to enter custom and configuration specific properties directly into SOLIDWORKS files.

3DEXPERIENCE files on This PC

The 3DEXPERIENCE files on This PC tab displays the models on your PC which were saved to the **3D**EXPERIENCE platform.

3DEXPERIENCE

The 3DEXPERIENCE tab provides access to the **3D**EXPERIENCE platform. The **3D**EXPERIENCE platform provides downloadable content (DLC) from a secure cloud location within your SOLIDWORKS session. The platform enables you to collaborate with students and teams on and off campus through a browser.

Think of the **3D**EXPERIENCE platform simply as a SOLIDWORKS Add-In on your desktop. Experience a single modeling environment. Access the platform within the SOLIDWORKS Task Pane.

Note: If you purchased 3DEXPERIENCE education roles, follow the below procedure.

Create and register your **3D**EXPERIENCE ID (credentials) to the platform. Go to https://solidworks.com/academic-cloud to create and register a new ID. **Note**: In an academic environment, it is strongly recommended to use your school's assigned email address.

Input the requested information.

Click **Register**.

You will receive the **3D**EXPERIENCE platform Invitation link via email. The email you use must be the same as your **3D**EXPERIENCE ID.

Click **Launch your 3DEXPERIENCE platform** from the Invitation email.

Bookmark the address.

Store the Invitation email. It's your unique link to the **3D**EXPERIENCE platform.

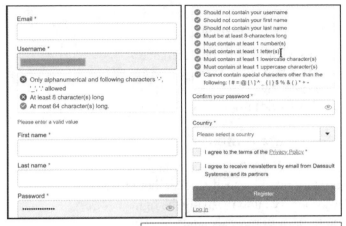

Log in with your **3D**EXPERIENCE ID.

Dynamic Reference Visualization (Parent/Child)

Dynamic Reference Visualization provides the ability to view the parent/child relationships between items in the FeatureManager design tree.

When you hover over a feature with references in the FeatureManager design tree, arrows display showing the relationships.

If a reference cannot be shown because a feature is not expanded, the arrow points to the feature that contains the reference and the actual reference appears in a text box to the right of the arrow.

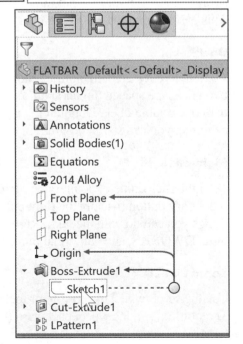

Use Dynamic reference visualization for a part, assembly and mates.

To display the Dynamic Reference Visualization, click **View**, **User Interface**, **Dynamic Reference Visualization Parent/Child)** from the Main menu bar.

Mouse Movements

A mouse typically has two buttons: a primary button (usually the left button) and a secondary button (usually the right button). Most mice also include a scroll wheel between the buttons to help you scroll through documents and to Zoom in, Zoom out and rotate models in SOLIDWORKS. It is highly recommended that you use a mouse with at least a Primary, Scroll and Secondary button.

① Primary button   ② Scroll wheel   ③ Secondary button

Single-click

To click an item, point to the item on the screen, and then press and release the primary button (usually the left button). Clicking is most often used to select (mark) an item or open a menu. This is sometimes called single-clicking or left-clicking.

Double-click

To double-click an item, point to the item on the screen, and then click twice quickly. If the two clicks are spaced too far apart, they might be interpreted as two individual clicks rather than one double-click. Double-clicking is most often used to open items on your desktop. For example, you can start a program or open a folder by double-clicking its icon on the desktop.

Right-click

To right-click an item, point to the item on the screen, and then press and release the secondary button (usually the right button). Right-clicking an item usually displays a list of things you can do with the item. Right-click in the open Graphics window or on a command in SOLIDWORKS, and additional pop-up context is displayed.

Scroll wheel

Use the scroll wheel to zoom-in or to zoom-out of the Graphics window in SOLIDWORKS. To zoom-in, roll the wheel backward (toward you). To zoom-out, roll the wheel forward (away from you).

Saving SOLIDWORKS Documents as Previous version

Beginning with SOLIDWORKS 2024, you can save SOLIDWORKS parts, assemblies and drawings created or saved in the latest version of SOLIDWORKS as fully functional documents in a previous version of SOLIDWORKS.

You can save documents back to the previous two releases. Pack and Go also supports this functionality.

You can save SOLIDWORKS 2024 files as SOLIDWORKS 2023 or SOLIDWORKS 2022 versions.

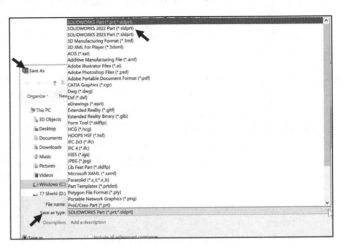

This previous release compatibility lets you share files with others who use one of the two previous versions of SOLIDWORKS. You cannot extend the previous release compatibility beyond those two releases.

Note: SOLIDWORKS users must have an active subscription license to access this functionality. 3DEXPERIENCE users are active subscribers by default. The SOLIDWORKS (UI) is designed to make maximum use of the Graphics window for your model. Displayed toolbars and commands are kept to a minimum.

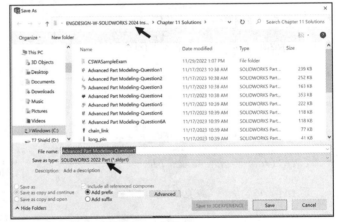

To open the Previous Release Check dialog box, click Tools, Evaluate, from the Main menu.

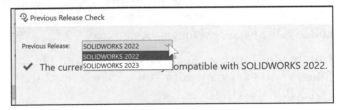

Summary

The SOLIDWORKS (UI) is designed to make maximum use of the Graphics window for your model. Displayed toolbars and commands are kept to a minimum.

The SOLIDWORKS User Interface and CommandManager consist of the following main options: Menu bar toolbar, Menu bar menu, Drop-down menus, Context toolbars, Consolidated fly-out menus, System feedback icons, Confirmation Corner and Heads-up View toolbar.

The Part CommandManager controls the display of tabs: *Features, Sketch*, Markup, *Evaluate, MBD Dimensions, Lifecycle and Collaboration* and various additional *SOLIDWORKS Add-Ins*.

The FeatureManager consists of various tabs:

- *FeatureManager design tree* tab.

- *PropertyManager* tab.

- *ConfigurationManager* tab.

- *DimXpertManager* tab.

- *DisplayManager* tab.

- *CAM FeatureManager tree* tab.

- *CAM Operation tree* tab.

- *CAM Tools tree* tab.

Click the direction arrows to expand or collapse the FeatureManager design tree.

CommandManager, FeatureManager, and file folders will vary depending on system set-up, Add-ins and 3DEXPERIENCE Roles.

You learned about creating a new SOLIDWORKS part and opening an existing SOLIDWORKS part along with using the Rollback bar to view the sketches and features.

If you modify a document property from an Overall drafting standard, a modify message is displayed as illustrated.

Overall drafting standard
ANSI-MODIFIED
Derived from: ANSI

Use the Search box, in the upper left corner of the Materials dialog box, to search through the entire materials library.

Templates are part, drawing and assembly documents which include user-defined parameters. Open a new part, drawing or assembly. Select a template for the new document.

In Project 2, create three folders: PROJECTS, MY-TEMPLATES, and VENDOR COMPONENTS.

Create the PART-MM-ANSI Part template utilized for the parts in this project.

Create three parts:

- PLATE

- ROD

- GUIDE

Apply and edit sketch dimensions.

Insert the following Geometric relations: Horizontal, Vertical, Symmetric, Equal and Coradial.

Create and modify 3D parts.

Create and modify the following SOLIDWORKS features: Extruded Boss\Base, Extruded Cut, Fillet, Mirror, Hole Wizard (Slot hole), Chamfer, and Linear Pattern.

The book provides information on creating and storing special Part templates, Assembly templates and Drawing templates in the MY-TEMPLATES folder. The MY-TEMPLATES folder is added to the New SOLIDWORKS Document dialog box. Talk to your IT department *before you set* any new locations on a network system. The procedure in the book is designed for your personal computer.

If you do not create the MY-TEMPLATE tab or the special part, drawing or assembly templates, use the standard SOLIDWORKS default template and apply all of the needed document properties and custom properties.

Notes:

Project 2

Fundamentals of Part Modeling

Below are the desired outcomes and usage competencies based on the completion of Project 2.

Project Desired Outcomes:	Usage Competencies:
• Address File Management.	• Use folders for various Project Models and Document Templates.
• PART-ANSI-MM Part Template.	• Apply and modify System Options and Document Properties.

• PART-ANSI-MM Part Template.
   • PLATE
   • ROD
   • GUIDE
	• Create 2D Sketch profiles on the correct Sketch plane.
	
	• Understand and apply proper design intent.
	
	• Knowledge to create and modify the following SOLIDWORKS 3D features: Extruded Boss/Base, Extruded Cut, Instant3D, Fillet, Mirror, Chamfer, Hole Wizard (Slot hole) and Linear Pattern.

Notes:

Project 2 - Fundamentals of Part Modeling

Project Objective

Use folders: PROJECTS, MY-TEMPLATES, MY-SHEETFORMAT, VENDOR COMPONENTS, etc. to address file management in the book.

	LOGO
	MY-SHEETFORMAT
	MY-TEMPLATES
	PROJECTS
	VENDOR COMPONENTS

Create the PART-MM-ANSI Part template utilized for the parts in this project.

Create three parts:

- PLATE

- ROD

- GUIDE

On the completion of this project, you will be able to:

- Manage and organize files and folders.

- Apply and understand System Options and Document Properties.

- Create a Part template.

- Open, Save and Close Part documents and templates.

- Create 2D sketch profiles on the proper Sketch plane.

- Utilize the following Sketch tools: Line, Corner Rectangle, Circle, Arc, Trim Entities, Convert Entities, Straight Slot and Mirror Entities.

- Apply and edit sketch dimensions.

- Insert the following Geometric relations: Horizontal, Vertical, Symmetric, Equal and Coradial.

- Create and modify 2D sketches.

- Create and modify the following SOLIDWORKS 3D features: Extruded Boss\Base, Extruded Cut, Fillet, Mirror, Hole Wizard (Slot hole), Chamfer and Linear Pattern.

Project Situation

You receive an email or a concept sketch from a customer. The customer is retooling an existing assembly line.

Customer Concept

You are required to design and manufacture a ROD. The ROD part is 10mm in diameter x 100mm in length.

One end of the ROD connects to an existing customer GUIDE CYLINDER assembly.

The other end of the ROD connects to the customer's tool. The ROD contains a 3mm hole and a key-way to attach the tool.

The ROD requires a support GUIDE. The ROD travels through the support GUIDE. The GUIDE-ROD assembly is the finished customer product.

The GUIDE-ROD assembly is a component used in a low volume manufacturing environment. Investigate a few key design issues:

- How will the customer use the GUIDE-ROD assembly?

- How are the parts: PLATE, ROD and GUIDE used in the GUIDE-ROD assembly?

- Does the GUIDE-ROD assembly affect other components?

- Identify design requirements for load, structural integrity or other engineering properties.

- Identify cost effective materials.

- How are parts manufactured?

- What are their critical design features?

- How will each part behave when modified?

You may not have access to all of the required design information. Placed in a concurrent engineering situation, you are dependent on others and are ultimately responsible for the final design.

Dimensions for this project are in millimeters. Design information is provided from various sources. Ask questions. Part of the learning experience is to know which questions to ask and who to ask.

The ROD part requires support. During the manufacturing operations, the ROD exhibits unwanted deflection.

The engineering group calculates working loads on test samples. Material test samples include Plain Carbon Steel, AISI 304 steel and a machinable plastic Nylon 101.

The engineering group recommends AISI 304 steel for the ROD and GUIDE parts.

In the real world, there are numerous time constraints. The customer requires a quote, design sketches and a delivery schedule, YESTERDAY. If you wait for all of the required design information, you will miss the project deadline.

You create a rough concept sketch with notes in a design review meeting. Your colleagues review and comment on the concept sketch.

The ROD cannot mount directly to the customer's GUIDE CYLINDER assembly without a mounting PLATE part.

The PLATE part mounts to the customer's PISTON PLATE part in the GUIDE CYLINDER assembly.

☀ CommandManager and FeatureManager tabs and tree folders will vary depending on system setup and Add-ins.

Concept Sketch and Design
Review Notes

Project Overview

A key goal in this project is to create three individual parts: PLATE, ROD and GUIDE for the requested GUIDE-ROD assembly.

Incorporate the three individual parts into the GUIDE-ROD assembly in Project 3 using Standard mates.

File management organizes documents and folders. The GUIDE-ROD assembly consists of numerous documents.

Use file folders to organize parts, assemblies, drawings, templates, sheet formats, and vendor components.

Drafting standards such as ANSI or ISO and unit systems such as IPS (inch, pound, second) or MMGS (millimeter, gram, second) are defined in Document Properties and are stored in the Part, Assembly or Drawing template.

Plan file organization and templates before you create parts, assemblies or drawings.

Create a PART-MM-ANSI Part template for the parts required for the GUIDE-ROD assembly.

Note: If you do not create the templates or the MY-TEMPLATES tab in the New SOLIDWORKS Document dialog box, use a standard SOLIDWORKS template and address the needed document and custom properties.

File Management

File management organizes parts, assemblies and drawings. Why do you need file management? A large assembly contains hundreds or even thousands of parts.

Parts and assemblies are distributed between team members to save time. Design changes occur frequently in the development process. How do you manage and control changes? Answer: Through file management. File management is a very important tool in the development process.

The GUIDE-ROD assembly consists of many files. Utilize file folders to organize projects, vendor parts and assemblies, templates and various libraries.

Folders exist on the local hard drive, example C:\. Folders also exist on a network drive, example Z:\. The letters C:\ and Z:\ are used as examples for a local drive and a network drive respectively. The following example utilizes the folder "Documents" to contain the folders for your projects.

Download all model files from the SDC Publications website (www.SDCpublications.com/downloads/978-1-63057-630-1). The top folder is ENGDESIGN-W-SOLIDWORKS. Work directly from a local hard drive

File Management - View the sub-folders.

1) If needed, download the **ENGDESIGN-W-SOLIDWORKS** folder.

2) **Unzip** the ENGDESIGN-W-SOLIDWORKS folder.

3) **View** the file folders.

Bracket
Chapter 2 - Homework
Chapter 3 - Homework
Chapter 4 - Homework
Chapter 5 - Homework
Chapter 6 - Homework
Chapter 7 - Homework
Chapter 8
Chapter 9
Chapter 11
Chapter 11 Solutions
CSWASampleExam
LOGO
MY-SHEETFORMAT
MY-TEMPLATES
PROJECTS
VENDOR COMPONENTS

Activity: Start a SOLIDWORKS 2024 Session. Open a New Part Document.

Start a SOLIDWORKS session.

4) **Start** a SOLIDWORKS 2024 session. The SOLIDWORKS 2024 Welcome dialog box is displayed.

5) **Close** the Welcome dialog box. The SOLIDWORKS Graphics window is displayed.

Create a new part document.

6) Click **New** ▢ from the Main menu. The New SOLIDWORKS Document dialog box is displayed. If needed, click the **Advance** button.

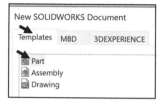

7) Double-click **Part** from the Templates tab.

8) **View** the default Part FeatureManager and empty Graphics window.

System Options

System Options are stored in the registry of the computer. System Options are not part of the document. Changes to the System Options affect current and future documents.

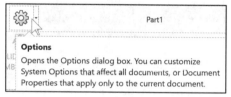

System Options, File Locations, Document Templates option is only valid for the current session of SOLIDWORKS on most *network locations*.

Talk to your IT department *before you set* any new locations on a network system. The procedure on the next page is designed for your personal computer.

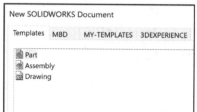

Set the System Options (on your own computer) before you start a project. The File Locations Option contains a list of folders referenced during a SOLIDWORKS session.

Add the ENGDESIGN-W-SOLIDWORKS\MY-TEMPLATES folder path name to the Document Templates File Location.

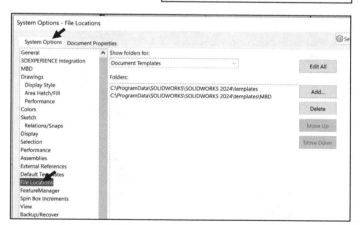

Note: If you do not create the templates or the MY-TEMPLATES tab, use a standard SOLIDWORKS template and address the needed document and custom properties.

Activity: Create a New Document Templates Tab.

Create a New Document Tempates tab. Set the folder location.

9)   Click **Options** ⁻ from the Main menu. The System Options General dialog box is displayed.

10)   Click the **File Locations** folder from the System Options tab.

11)   Select **Document Templates** from Show folders for.

Add a new tab.

12)   Click the **Add** button.

13)   Browse to the **ENGDESIGN-W-SOLIDWORKS\MY-TEMPLATES** folder. **Note**: The folder and files were downloaded from the SDC Publications website (www.SDCpublications.com/downloads/978-1-63057-630-1).

14)   Click **Select Folder**. View the new folder.

15)   Position the folder at the bottom of the column. Click the **Move Down** button.

16)   Click **OK** from the System Options, File Locations dialog box.

17)   Click **Yes** to the SOLIDWORKS dialog box. The SOLIDWORKS dialog box displays the new Folder location.

18)   Click **Yes**. Click **Yes**.

☀ Talk to your IT department *before you set* any new locations on a network system.

☀ The document templates and tutorial location will vary depending on your network system setup.

Each folder listed in the System Options, File Locations, Document Templates, Show Folders For option produces a corresponding tab in the New SOLIDWORKS Document dialog box.

The new tab, MY-TEMPLATES is not displayed in the New SOLIDWORKS Document dialog box, **until a template** is added to the folder.

Part Document Template and Document Properties.

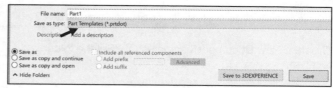

The Part template is the foundation for a SOLIDWORKS part. Part1 displayed in the FeatureManager utilizes the Part #.prtdot default template.

Document Properties contain the default settings for the Part template. The Document Properties include the drafting/dimensioning standard, units, dimension decimal display, grids, note font and line styles.

There are hundreds of document properties. Modify the following Document Properties: Drafting standard, Units and Decimal Places.

The Drafting standard determines the display of dimension text, arrows, symbols and spacing. Units are the measurement of physical quantities. Millimeter (MMGS) dimensioning and decimal inch (IPS) dimensioning are the two most common unit types specified for engineering parts and drawings.

Document Properties are stored with the document. Apply the Document Properties to the Part template.

Create a Part template named PART-MM-ANSI from the default Part template. Save the Custom Part template in the ENGDESIGN-W-SOLIDWORKS\MY-TEMPLATES folder. Utilize the PART-MM-ANSI Part template for all metric parts.

Conserve modeling time. Set the Document Properties and create the required templates before starting a project.

The Overall Drafting standard options are ANSI, ISO, DIN, JIS, BSI, GOST and GB.

Overall Drafting standard options:	Abbreviation:	Description:
ANSI ISO DIN JIS BSI GOST GB	ANSI	American National Standards Institute
	ISO	International Standards Organization
	DIN	Deutsche Institute für Normumg (German)
	JIS	Japanese Industry Standard
	BSI	British Standards Institution
	GOST	Gosndarstuennye State Standard (Russian)
	GB	Guo Biao (Chinese)

Display the parts, assemblies and drawings created in the GUIDE-ROD assembly in the ANSI drafting standard.

The Units Document Property assists the designer by defining the Unit System, Length unit, Angular unit, Density unit and Force unit of measurement for the Part, Assembly or Drawing document.

The Decimals option displays the number of decimal places for the Length and Angle unit of measurement.

Activity: Create a Part Document Template.

Set Document Properties. Set overall drafting standard, units and precision.

19) Click **Options** ⚙️ ˘ from the Main menu. The System Options General dialog box is displayed

20) Click the **Document Properties** tab.

21) Click the **Drafting Standard** folder.

22) Select **ANSI** from the Overall drafting standard drop-down menu. Various Detailing options are available depending on the selected standard.

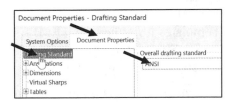

Set document units. Set length and angle precision.

23) Click **Units**.

24) Click **MMGS (millimeter, grams, second)** for Unit system.

25) Select **.12** for length decimal place.

26) Select **None** for angle decimal place.

27) Click **OK** from the Document Properties - Units dialog box.

Save the part template.

28) Click **Save As** 📇 .

29) Select **Part Templates *.prtdot** from the Save as type box. The default Part Templates folder is displayed.

30) Click the **drop-down arrow** in the Save in box.

31) Select the **ENGDESIGN-W-SOLIDWORKS\MY-TEMPLATES** folder.

32) Enter **PART-MM-ANSI** in the File name box.

33) Click **Save**. PART-MM-ANSI is displayed in the FeatureManager.

Close the PART-MM-ANSI template.

34) Click **File**, **Close** from the Main menu. All SOLIDWORKS documents are closed. If required, click **Window**, **Close All** from the Main menu to close all open documents.

CommandManager and FeatureManager tabs and tree folders will vary depending on system setup and Add-ins.

Review File Folders and a Custom Part Template.

Use folders: PROJECTS, MY-TEMPLATES, MY-SHEETFORMAT, VENDOR COMPONENTS, etc. to address file management in the book.

System Options, File Locations directed SOLIDWORKS to open Document Templates from the MY-TEMPLATES folder. System Options are stored in the registry of the computer.

You created a custom Part template, PART-MM-ANSI from the default Part document. Document Properties control the drafting standard, units, tolerance and other properties. Document Properties are stored in the current document. The Part template file extension is .prtdot.

Note: If you do not create the templates or the MY-TEMPLATES tab, use a standard SOLIDWORKS template and address the needed document and custom properties.

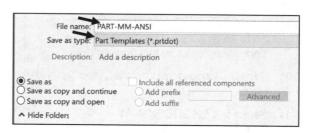

PLATE Part Overview

Determine the functional and geometric requirements of the PLATE.

Functional:

The PLATE part fastens to the customer's PISTON PLATE part in the GUIDE-CYLINDER assembly.

Fasten the PLATE part to the ROD part with a countersunk screw.

Geometric:

The dimensions of the PISTON PLATE part are 56mm x 22mm.

The dimensions of the PLATE part are 56mm x 22mm.

Locate the 4mm mounting holes with respect to the PISTON PLATE mounting holes. The mounting holes are 23mm apart on center.

The PLATE part requires a Countersink Hole to fasten the ROD part to the PLATE part.

Review the mating part dimensions before creating the PLATE part.

The GUIDE-CYLINDER assembly dimensions referenced in this project are derived from the SMC Corporation of America (www.smcusa.com).

🔆 Click **View**, **Hide/Show**, **Origins** from the Main menu to display the Origin in the Graphics window.

PLATE Part   GUIDE-CYLINDER PISTON

Start the translation of the initial design functional and geometric requirements into SOLIDWORKS features.

What are features?

- Features are geometry building blocks.

- Features can add or remove material.

- Features are created from sketched profiles or from edges and faces of existing geometry.

Utilize the following features to create the PLATE part:

- ***Extruded Boss/Base*** : The Extruded Boss/Base feature adds material to the part. The Boss-Extrude1 (Base) feature is the first feature of the PLATE. An extrusion extends a profile along a path normal to the profile plane for a specified distance. The movement along that path becomes the solid 3D model. Sketch the 2D rectangle (close profile) on the Front plane. Fully define the Base Sketch with Geometric relations and dimensions.

- ***Extruded Cut*** : The Extruded Cut feature removes material. The Extruded Cut starts with a 2D circle sketched on the front face. The front face is your Sketch plane. Copy the sketched circle to create the second circle. Utilize the Through All Depth End Condition (Design Intent). The holes extend through the entire Boss-Extrude1 feature. Note: Use the Hole Wizard feature when you need to incorporate complex or blind holes in a part. The Hole Wizard (Call out) information can be imported directly into a drawing and provides the correct hole annotations.

- ***Fillet*** : The Fillet feature removes sharp edges of the PLATE. Add Fillets to a solid, not the sketch. Group small corner edge fillets together. In this exercise, group Tangent edge fillets together.

- ***Hole Wizard*** : The Hole Wizard feature creates the Countersink hole at the center of the PLATE. The Hole Wizard requires a Sketch plane. Select the back face of the PLATE for your Sketch plane. Note: In this book, we focus on 2D Sketches vs. 3D Sketches.

A goal of this book is to expose the new user to various SOLIDWORKS design tools and features.

Activity: Create the PLATE Part. Use a Custom Part Template.

Create a new part.

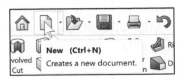

35) Click **New** ⬜ from the Main menu. The New
SOLIDWORKS Document dialog box is displayed.

36) Click the **MY-TEMPLATES** tab.

37) Double click **PART-MM-ANSI**. The Part FeatureManager is
displayed. **Note**: There are other templates displayed.

🔅 The first system default Part filename is Part1. The
system attaches the .sldprt suffix to the created part. The
second created part in the same session increments to the
filename Part2.

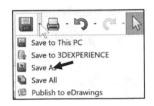

Save the part. Enter a part name. Enter part description.

38) Click **Save As** 📋 from the drop-down menu.

39) Select the **ENGDESIGN-W-SOLIDWORKS\PROJECTS** folder.

40) Enter **PLATE** for File name.

41) Enter **PLATE 56MM x 22MM**
for description.

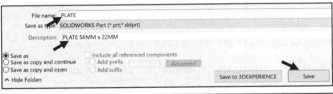

42) Click **Save** from the Save As
dialog box.

Note: <u>Save to This PC</u>: Saves
the document to your local hard
drive in the folder which was
last opened from. <u>Save to
3DEXPERIENCE</u>: Opens the
Save as New dialog box to save
the file to a Collaborative space
on the **3D**EXPERIENCE
platform. The Save to
3DEXPERIENCE option is
displayed if you purchased
3DEXPERIENCE education
roles and are logged into the
3DEXPERIENCE platform.
<u>Save As</u>: Saves the document to
a new file name that becomes the active document without
saving the original document. <u>Save All</u>: Saves all the open files
that have been modified since they were last saved.

PLATE is displayed in the FeatureManager design tree. The Part icon is displayed at the top of the FeatureManager.

CommandManager and FeatureManager tabs and tree folders will vary depending on system setup and Add-ins.

The origin ⵜ represents the intersection of the Front, Top and Right planes.

Base Feature

What is a Base feature? The Base feature (Boss-Extrude1) is the first feature that is created. The Base feature is the foundation of the part. Keep the Base feature <u>simple!</u>

The Base feature geometry for the PLATE is an extrusion. Rename the base feature Extruded-Base. Note: The default name is Boss-Extrude1.

How do you create a 3D Extruded Boss/Base feature?

- Select a Sketch plane.

- Sketch a 2D profile on the Sketch plane.

- Apply all needed Geometric relations and dimensions to fully define the sketch.

- Apply the Extruded Boss/Base feature tool. Extend the profile perpendicular (⊥) to the Sketch plane.

It is considered very poor design practice not to fully define a sketch. You will be tempted in order to save design time, but keep in mind that the extra couple of minutes you take to do something right the first time will save you additional time in the end. Fully defined sketches are required to manufacture the part.

Review the part manufacturing, materials, reference planes and orthographic projection before creating a Base feature.

Machined Part

In earlier conversations with manufacturing, a decision was made that the part would be machined.

Block, Angle and Rod Stock

Your material supplier stocks raw material in rod, sheet, plate, angle and block forms.

You decide to start with a standard plate form. A standard plate form will save time and money.

Select the best profile for the extrusion. The best profile is a simple 2D rectangle.

The boss and square internal cuts are very costly to manufacture from machined bar stock.

Boss is costly for machine stock.

As a designer, review your manufacturing options.

Utilize standard material thickness and hole sizes in the PLATE design. Utilize standard stock in the ROD design.

Utilize slot cuts in the GUIDE design. Machined parts require dimensions to determine the overall size and shape. Datum planes determine the location of referenced dimensions. Add geometric relations and dimensions to fully define the geometry Extruded Boss/Base feature.

Internal Square Cut (expensive)

Holes and Slots (less expensive)

Understanding reference planes and views is important as you design machined parts. Before you create the PLATE, review the next topic on Reference Planes and Orthographic Projection.

Reference Planes and Orthographic Projection

The three default ⊥ Reference planes represent infinite 2D planes in 3D space:

- Front
- Top
- Right

Planes have no thickness or mass. Orthographic projection is the process of projecting views onto parallel planes with ⊥ projectors.

Default ⊥ datum Planes are:

- Primary
- Secondary
- Tertiary

Use the following planes in manufacturing:

- Primary datum plane: Contacts the part at a minimum of three points.

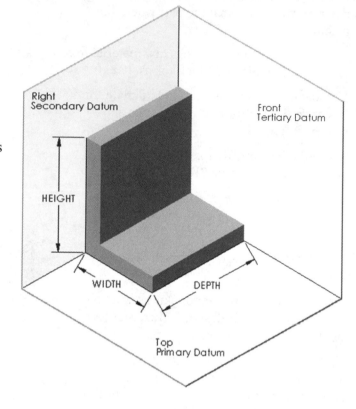

- Secondary datum plane: Contacts the part at a minimum of two points.

- Tertiary datum plane: Contacts the part at a minimum of one point.

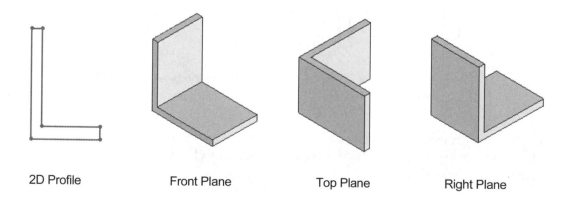

2D Profile   Front Plane   Top Plane   Right Plane

The part view orientation is dependent on the first feature Sketch plane. Compare the available default Sketch planes in the FeatureManager: Front Plane, Top Plane and Right Plane.

Each Extruded feature above was created with an L-shaped 2D Sketch profile. There are six principle views of Orthographic projection listed in the ASME Y14.3M standards:

- Top

- Front

- Right side

- Bottom

- Rear (Back)

- Left side

SOLIDWORKS Standard view names correspond to these Orthographic projection view names.

ASME Y14.3 Principle View Name:	SOLIDWORKS Standard View:
Front	Front
Top	Top
Right side	Right
Bottom	Bottom
Rear	Back
Left side	Left

The most common standard drawing views in Third angle Orthographic projection are:

- Front

- Top

- Right

- Isometric

There are two Orthographic projection drawing systems. The first Orthographic projection system is called Third Angle Projection. The second Orthographic projection system is called First Angle Projection. The two drawing systems are derived from positioning a 3D object in the Third or First quadrant as illustrated.

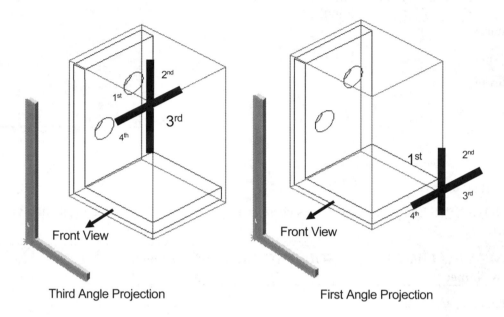

Third Angle Projection   First Angle Projection

Third Angle Projection

Third angle projection is a method of creating a 2D drawing of a 3D object. The 3D object is positioned in the third quadrant. The 2D projection planes are located between the viewer and the part. The projected views are placed on a drawing.

First Angle Projection

First-angle projection is a method of creating a 2D drawing of a 3D object. It is mainly used in Europe and Asia. The 3D object is positioned in the first quadrant. Views are projected onto the planes located behind the part. The projected views are placed on a drawing.

Third Angle Projection is primarily used in the U.S. & Canada and is based on the ASME Y14.3M multi and sectional view drawings standard. Designers should have knowledge and understanding of both systems.

There are numerous multi-national companies. Example: A part is designed in the U.S., manufactured in Japan and destined for a European market.

Third Angle Projection is used in this text. A truncated cone symbol should appear on the drawing to indicate the Projection system.

Third Angle Projection Symbol

First Angle Projection Symbol

Utilize the Heads-up View toolbar or the Orientation dialog box to orient or set view style for your model.

Before incorporating your design intent into the Sketch plane, ask a question: How will the part be oriented in the assembly? Answer: Orient or align the part to assist in the final assembly. Utilize the Front Plane to create the Extruded Base feature for the PLATE part.

PLATE Part - Extruded Boss/Base Feature

An Extruded Boss/Base ⬚ feature is a feature in SOLIDWORKS that utilizes a sketched profile and extends the profile perpendicular (⊥) to the Sketch plane.

In SOLIDWORKS, a 2D profile is called a sketch. A sketch requires a Sketch plane and a 2D profile. The sketch in this example uses the Front Plane. The 2D profile is a rectangle. Geometric relationships and dimensions define the exact size of the rectangle. The rectangle is extruded perpendicular to the Sketch plane.

The Front Plane's imaginary boundary is represented on the screen with four visible edges. Planes are flat and infinite. Planes are used as the primary sketch surface for creating Extruded Boss/Base and Extruded Cut features.

Activity: PLATE Part - Extruded-Base Feature.

Select the Sketch plane. Create a 2D sketch.

43) Right-click **Front Plane** from the FeatureManager. The Context toolbar is displayed.

44) Click **Sketch** ⬚ from the Context toolbar. The Sketch toolbar is displayed. Front Plane is your Sketch Plane.

The plane or face you select for the Base sketch determines the orientation of the part.

☼ Click **View**, **Hide/Show**, **Origins** from the Main menu to display the Origin in the Graphics window.

You can also click the
Front Plane from the
FeatureManager and
click the Sketch tab
from the

CommandManager. Front Plane is the Sketch plane and the Sketch toolbar is displayed.

The Origin represents the intersection of the Front, Top and Right Planes. The left corner point of the rectangle is Coincident with the origin. The Origin is displayed in red.

Sketch a Corner rectangle.

45) Click the **Corner Rectangle** tool from the Sketch toolbar. The sketch opens in the Front view. The mouse pointer displays the Corner Rectangle feedback symbol . The Front Plane feedback indicates the current Sketch plane.

The Rectangle-based tool uses a Consolidated Rectangle PropertyManager. The SOLIDWORKS application defaults to the last used rectangle type.

46) Click the **Origin** ⊥. This is the first point of the rectangle. It is very important that you always reference the sketch to the origin. This helps to fully define the sketch.

47) Drag the **mouse pointer** up and to the right.

48) Release the **mouse button** to create the second point and the rectangle. The first point is Coincident to the Origin.

View the illustrated sketch relations in the Graphics window.

49) Click **View**, **Hide/Show**, **Sketch Relations** from the Main menu. The sketch relations are displayed in the Graphics window.

To deactivate the illustrated sketch relations, click **View**, **Hide/Show**, and uncheck **Sketch Relations** from the Main menu.

The X-Y coordinates of the rectangle are displayed above the mouse pointer as you drag the mouse pointer up and to the right. The X-Y coordinates display different values. Define the exact width and height with the Smart dimension tool.

The Corner Rectangle tool ⬚ remains selected. The rectangle sketch contains predefined geometric sketch relations: two Horizontal ⁻ relations and two Vertical ❘ relations. The first point of the rectangle contains a Coincident ⬠ relation at the Origin.

If you make a mistake, select the **Undo** ↺ icon in the Main menu.

SOLIDWORKS uses color and cursor feedback to aid in the sketching process. The Corner Rectangle tool remains active until you select another tool or right-click the Select. The right mouse button contains additional tools.

Deactivate the Corner Rectangle tool.

50) Right-click a **position** in the Graphics window. The Context toolbar is displayed.

51) Click **Select**. The mouse pointer displays the Select icon and the Corner Rectangle Sketch tool is deactivated.

The phrase "Right-click Select" means **right-click** the mouse in the Graphics window. Click **Select**. This will de-select the current tool that is active.

Size the geometry of the rectangle.

52) Click and drag the **top horizontal line** of the rectangle upward. View the Line Properties PropertyManager.

53) **Right-click** and view your options.

54) Release the **mouse button**.

The Line Properties PropertyManager is displayed to the left of the Graphics window.

The selected line displays a Horizontal relation in the Existing Relations box.

The geometry consists of four lines.

Context menus and toolbars save time. Commands and tools vary depending on the mouse position in the SOLIDWORKS window and the active Sketch tool and history.

💡 Rename a feature or sketch. Slowly click the feature or sketch name twice and enter the new name when the old one is highlighted.

💡 Design Intent is how your part reacts as parameters are modified. Example: If you have a hole in a part that must always be .125≤ from an edge, you would dimension to the edge rather than to another point on the sketch. As the part size is modified, the hole location remains .125≤ from the edge.

Dimension the bottom horizontal line.

55) Click the **Smart Dimension** ✎ tool from the Sketch toolbar. The pointer displays the dimension feedback symbol ✎.

56) Click the **bottom horizontal line** of the rectangle. **Note**: A dimension value is displayed.

57) Click a **position** below the bottom horizontal line. The Modify dialog box is displayed. The Smart Dimension tool uses the Smart Dimension PropertyManager. The PropertyManager provides the ability to select three tabs. Each tab provides a separate menu.

58) Enter **56**mm in the Modify dialog box. **Note**: The Dimension Modify dialog box provides the ability to select a unit drop-down menu to modify units in a sketch or feature from the document properties.

59) Click the **Green Check mark** ✔ in the Modify dialog box.

You can either press the enter key or click the Green Check mark to accept the dimension value.

Dimension the vertical line.

60) Click the **left vertical line**.

61) Click a **position** to the left of the vertical line.

62) Enter **22**mm.

63) Click the **Green Check mark** ✔ in the Modify dialog box. The sketch is fully defined. All lines and vertices are displayed in black.

☼ To flip the arrow direction, click the control point as illustrated.

Deactivate the sketch relations and select the dimension value.

64) Click **View**, **Hide/Show**, and uncheck **Sketch Relations** from the Main menu.

65) Right-click **Select** in the Graphics window.

66) Position the **mouse pointer** over the dimension text. The pointer changes to a linear dimension symbol with a displayed text box. D2@Sketch1 represents the second linear dimension created in Sketch1.

Modify the dimension text.

67) Double-click the **22** dimension text in the Graphics window.

68) Enter **42**mm.

🔆 Click the **Spin Box Arrows** ⬍ to increase or decrease dimensional values.

Return to the original vertical dimension.

69) Double-click the **42**mm dimension text in the Graphics window.

70) Enter **22**mm.

71) Click the **Green Check mark** ✔ from the Modify dialog box.

🔆 Click the **Undo** ↺ tool to return to the original dimension.

Options in the Modify dialog box:

- Green Check mark ✔ - Saves the current value and exits the Modify dialog box.

- Restore ✖ - Restores the original value and exits.

- Rebuild ● - Rebuilds the model with the current value.

- Reverse - Reverse the sense of the direction.

- Reset - Reset the spin increment value.

- Mark for Drawing - Set by default. Inserts part dimensions into the drawing.

- Set Functions and File Properties. Create on the fly Global Variables. Enter an = sign as illustrated to start an equation in the Modify dialog box. View the available drop-down menu for functions, file properties and any created variables.

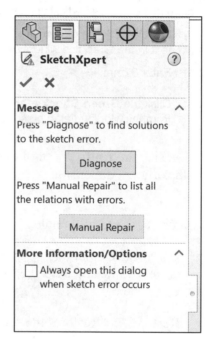

The System displays Under Defined in the Status bar located in the lower right corner of the Graphics window. In an under defined Sketch, the entities that require position, dimensions or Sketch relations are displayed in blue.

In a fully defined sketch, all entities are displayed in black. The Status bar displays Fully Defined. In machining practices, parts require fully defined sketches. *A fully defined sketch has defined positions, dimensions and/or relationships.*

In an over defined sketch, there is geometry conflict with the dimensions and or relationships. In an over defined sketch, entities are displayed in red.

SOLIDWORKS provides the SketchXpert tool to correct an over defined sketch. The SketchXpert PropertyManager is displayed by default for an over defined sketch. Click the Diagnose button and accept an option to correct the over defined sketch.

Insert an Extruded Boss/Base feature. This is the first (Base) feature of the model. Utilize the default Blind End Condition.

72) Click the **Features** tab from the CommandManager.

73) Click the **Extruded Boss/Base** tool from the Features toolbar. The Boss-Extrude PropertyManager is displayed. The extruded sketch is previewed in a Trimetric view. The preview displays the direction of the extrude feature.

Reverse the direction of the extruded depth.

74) Click the **Reverse Direction** box. Note the location of the Origin in the Graphics window. Blind is the default End Condition for Direction 1.

75) Enter **10**mm for Depth in Direction1.

76) Click **OK** from the Boss-Extrude PropertyManager. The Boss-Extrude1 feature is displayed in the Graphics window. The name Boss-Extrude1 is displayed in the PLATE FeatureManager.

Origin

The OK ✔ button accepts and completes the Feature process.

If you exit the sketch before selecting the Extruded Boss/Base tool from the Features toolbar, Sketch1 is displayed in the FeatureManager. To create the Extruded Boss/Base feature, click Sketch1 from the FeatureManager. Click Extruded Boss/Base to create the feature.

For many features (Extruded Boss/Base, Extruded Cut, Simple Hole, Revolved Boss/Base, Revolved Cut, Fillet, Chamfer, Scale, Shell, Rib, Circular Pattern, Linear Pattern, Curve Driven Pattern, Revolved Surface, Extruded Surface, Fillet Surface, Edge Flange and Base Flange) you can enter and modify equations directly in the PropertyManager fields that allow numerical inputs. You can create equations with Global Variables, functions, and file properties without accessing the Equations, Global Variables and Dimensions dialog box.

For example, in the PropertyManager for the Extruded Boss/Base feature, you can enter equations in:

- Depth fields for Direction 1 and Direction 2.

- Draft fields for Direction 1 and Direction 2.

- Thickness fields for a Thin Feature with two direction types.

- Offset Distance field.

To create an equation in a numeric input field, start by entering an = (equal sign). A drop-down list displays options for Global Variables, functions and file properties. Numeric input fields that contain equations can display either the equation itself or its evaluated value. You can toggle between the equation and the value by clicking the Equations or Global Variable button that appears at the beginning of the field.

In edit mode, right-click anywhere on an extruded feature and modify the end condition from a pop-up shortcut menu. Click in empty space, on geometry, or on the handle. The pop-up shortcut menu provides the document dependent options for Direction 1 and Direction 2.

💡 Display an Isometric view of the model. Press the **space bar** to display the Orientation dialog box. Click the **Isometric view** 📦 icon.

PLATE Part - Modify Dimensions and Rename.

Incorporate a design change into the PLATE. Modify dimension values in the Modify dialog box or directly in the Graphics window. Utilize the Rebuild 🔘 tool to update the Extruded Boss/Base feature.

Either double-click the Boss-Extrude1 folder from the FeatureManager or double-click on the feature in the Graphics window to display the model dimensions. Rename entries by selecting on the text (slowly double-click) in the FeatureManager.

Activity: PLATE Part - Modify Dimensions and Rename.

Modify the PLATE part.

77) Double-click on the **front face** of Boss-Extrude1 in the Graphics window as illustrated. Boss-Extrude1 is highlighted in the FeatureManager.

Fit the model to the Graphics window.

78) Press the **f** key.

Modify the width dimension.

79) Click the **10**mm dimension in the Graphics window.

80) Enter **5**mm.

Display an Isometric view.

81) Press the **space bar** to display the Orientation dialog box.

82) Click **Isometric view** 📦. You can also access the Isometric view tool from the Heads-up View toolbar.

The Instant3D tool provides the ability to drag geometry and dimension manipulator points to resize or create new features in the Graphics window. Use the on-screen ruler to measure and apply modifications. In this book, you will primarily use the PropertyManager and dialog box to modify model dimensions and to create new features. It is very difficult to apply design intent (End Conditions) using the Instant 3D tool. See SOLIDWORKS Help for additional information.

Incorporate the machining process into the PLATE design. Reference dimensions from the three datum planes with the machined Origin in the lower left hand corner of the PLATE.

Maintain the Origin on the front lower left hand corner of the Extruded Boss/Base feature and co-planar with the Front Plane.

Rename Sketch1 and Boss-Extrude1. The first Extruded Boss/Base feature is named Boss-Extrude1 in the FeatureManager. Sketch1 is the name of the sketch utilized to create the Boss-Extrude1 feature.

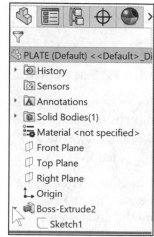

The Arrow ▸ icon indicates that additional feature information is available.

Expand the Boss-Extrude1 entry from the FeatureManager.

83) Click the **Arrow** ▸ icon of the Boss-Extrude1 feature in the FeatureManager.

Rename Sketch1.

84) Right-click **Sketch1** in the FeatureManager. Click **Rename tree item**. A white box is displayed around the selected item. Enter **Sketch-Base** for the sketch name. The Down arrow ▾ icon indicates that the feature information is expanded.

85) Click the **Down arrow** ▾ icon to collapse the Boss-Extrude1 feature.

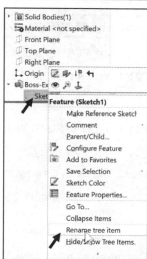

Rename the Boss-Extrude1 feature.

86) Right-click **Boss-Extrude1** in the FeatureManager. Click **Rename tree item**. A white box is displayed around the selected item.

87) Enter **Base-Extrude** for the feature name as illustrated.

Save the PLATE.

88) Click **Save** 🖫.

Display Modes, View Modes, View tools and Appearances.

Access the display modes, view modes, view tools and appearances from the Standard Views toolbar and the Heads-up View toolbar. Apply these tools to display the required modes in your document.

The Apply scene ![scene icon] tool adds the likeness of a material to a model in the Graphics window without adding the physical properties of the material. Select a scene from the drop-down menu.

Activity: Display Modes, View Modes, View tools and Appearances.

View the display modes from the Heads-up View toolbar.

89) Click **Wireframe** ![icon]. View the results.

90) Click **Hidden Lines Visible** ![icon]. View the results.

91) Click **Hidden Lines Removed** ![icon]. View the results.

92) Click **Shaded** ![icon]. View the results.

93) Click **Shaded With Edges** 🔲. View the results.

💡 The Hidden Lines Removed option may take longer to display than the Shaded option depending on your computer configuration and the size of the part and file.

Display modes and View modes remain active until deactivated.

Display the View modes.

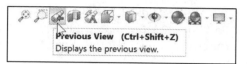

94) Click **Previous view** ⬦ to display the previous view of the part in the current window.

Previous View   (Ctrl+Shift+Z)
Displays the previous view.

95) Click **Zoom to Fit** 🔍 to display the full size of the part in the current window.

96) Click **Zoom to Area** 🔍.

97) **Zoom in** on a corner of the PLATE.

98) Click **Zoom to Area** 🔍 to deactivate the tool.

99) Press the **f** key to fit the model to the Graphics window.

💡 Press the lower-case z key to zoom out. Press the upper-case Z key to zoom in.

100) Right-click **Select** in the Graphics window. View the available view tools.

101) Click **inside** the Graphics window.

Return to an Isometric Shaded With Edges view.

102) Press the **space bar** to display the Orientation dialog box.

103) Click **Isometric view** 🔲. View the results.

104) Click **Shaded With Edges** 🔲 from the Heads-up View toolbar.

💡 The Normal To view ⬥ tool displays the part ⊥ to the selected plane. The View Orientation tool ⬦ from the Standard Views toolbar creates and displays a custom named view.

💡 To rotate your model, use the middle mouse button. The rotate icon ⬤ is displayed.

PLATE Part - Extruded Cut Feature.

An Extruded Cut feature removes material perpendicular ⊥ to the sketch for a specified depth. The Through All End Condition option creates the holes through the entire depth of the PLATE.

Utilize the Circle Sketch tool to create the hole profile on the front face. The first circle is defined with a center point and a point on its circumference. The second circle utilizes the Ctrl-C (Copy) key to create a copy of the first circle.

Geometric relations are constraints between one or more entities. The holes may appear aligned and equal, but they are not! Add two relations: Insert an Equal relation between the two circumferences and a Horizontal relation between the two center points.

The machinist references all dimensions from the reference datum planes to produce the PLATE holes. Reference all three linear dimensions from the Right datum plane (left vertical edge) and the Top datum plane (bottom horizontal edge). Utilize a dimension to define the diameter of the circle.

The linear dimensions are positioned to the left and below the Boss-Extrude1 feature. You will learn more about dimensioning to a drawing standard in Project 3.

☀ Position the dimensions off the sketch profile. Place smaller dimensions to the inside. These dimensioning techniques in the part sketch will save time in dimensioning the drawing.

Design intent is not a static concept that controls changing geometry. As the designer, think about how to maintain the design intent with geometric changes to the model. Dimension and geometric relation schemes vary depending on the design intent of the part. In the next section there are alternate dimension schemes to define the position of two circles.

The following two examples are left as an exercise.

Example A: The center points of the holes reference a horizontal centerline created at the Midpoint of the two vertical lines. A dimension is inserted between the two center points.

Example B: The center points of the circles reference the horizontal line. Insert a vertical line at the Midpoint of the two horizontal lines. The center points of the circles utilize a Symmetric relation with the vertical centerline.

Dimension schemes defined in the part can also be redefined in the future drawing.

Activity: PLATE Part - Extruded Cut Feature.

View the mouse pointer feedback.

105) Position the mouse pointer on the **front face**.

106) View the Select Face icon ⬚. The mouse pointer provides feedback depending on the selected geometry.

107) Position the mouse pointer on the **right edge**. View the Select Edge icon ⬚.

Select the Sketch plane for the holes.

108) Right-click the **front face** of the Base-Extrude feature in the Graphics window.

Create a sketch.

109) Click **Sketch** ⬚ from the Context toolbar. The Sketch toolbar is displayed.

Display the Front view.

110) Click **Front view** ⬚. The Origin is located in the lower left corner of the part.

111) Click **Hidden Lines Visible** ⬚ from the Heads-up View toolbar.

Sketch the first circle.

112) Click the **Circle** ⬚ Sketch tool. The cursor displays the Circle feedback symbol ⬚. The Circle PropertyManager is displayed.

113) Click a **center point** on the front face diagonally to the right of the Origin as illustrated.

114) Click a **position** to the right of the center point. The circumference and the center point are selected.

Copy the first circle.

115) Right-click **Select** in the Graphics window to deselect the circle Sketch tool.

116) Hold the **Ctrl** key down.

117) Click and drag the **circumference** of the first circle to the right.

118) Release the **mouse button**.

119) Release the **Ctrl** key. The second circle is displayed.

120) Click a position in the **Graphics window**, off the sketch. The first and second circles are displayed in blue.

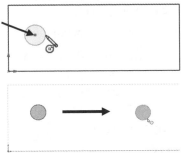

Add an Equal relation between the circles.

121) Click the **circumference** of the first circle.

122) Hold the **Ctrl** key down.

123) Click the **circumference** of the second circle.

124) Release the **Ctrl** key. Arc1 and Arc2 are displayed in the Selected Entities text box in the Properties PropertyManager.

125) Click **Equal** ⬛ from the Add Relations box. The Equal radius/length is displayed in the Existing Relations text box.

126) Click **OK** ✅ from the Properties PropertyManager.

💡 You can right-click the **Make Equal** tool from the Context toolbar.

The SOLIDWORKS default name for curve entities is Arc#. There is an Arc# for each circle. The default name for any point is Point#. The two center points are entities named Point2 and Point4.

💡 To remove unwanted entities from the Selected Entities box, right-click in the **Selected Entities** window. Click **Clear Selections**. The Selected Entities # differs if geometry was deleted and recreated. The Existing Relations box lists Equal radius/length0.

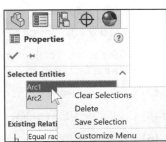

Add a Horizontal relation between the first circle center point and the second circle center point.

127) Click the first circle **center point**.

128) Hold the **Ctrl** key down.

129) Click the second circle **center point**. The Properties PropertyManager is displayed. The selected entities are displayed in the Selected Entities box.

130) Release the **Ctrl** key.

131) Click **Horizontal** from the Add Relations box.

132) Click **OK** ✔ from the Properties PropertyManager.

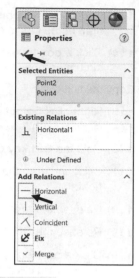

Add a diameter dimension using the Smart Dimension tool.

133) Click the **Smart Dimension** ↖ Sketch tool.

134) Click the **circumference** of the first circle.

135) Click a **position** diagonally upward.

136) Enter **4**mm.

137) Click the **Green Check mark** ✔ from the Modify dialog box.

💡 If required, click the dimension arrowhead to toggle the dimension arrow as illustrated.

Add a vertical dimension.

138) Click the **bottom horizontal** line of the Base-Extrude feature.

139) Click the **center point** of the first circle.

140) Click a **position** to the left of the profile.

141) Enter **11**mm.

142) Click the **Green Check mark** ✔ from the Modify dialog box.

Add a horizontal dimension.

143) Click the **left vertical** line of the Base-Extrude feature.

144) Click the **center point** of the first circle.

145) Click a **position** below the profile.

146) Enter **16.5**mm.

147) Click the **Green Check mark** ✔ from the Modify dialog box.

Add a second horizontal dimension.

148) Click the **left vertical** line of the Base-Extrude feature.

149) Click the **center point** of the second circle.

150) Click a **position** below the 16.50 dimension in the Graphics window.

151) Enter **39.5**mm.

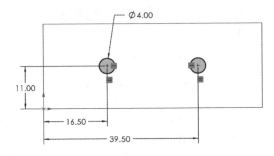

152) Click the **Green Check mark** ✔ from the Modify dialog box.

Display an Isometric view. Shaded With Edges.

153) Press the **f** key.

154) Press the **space bar** to display the Orientation dialog box.

155) Click **Isometric view** 🔲. View the results.

156) Click **Shaded With Edges** 🔲 from the Heads-up View toolbar.

Create an Extruded Cut feature.

157) Click the **Features** tab from the CommandManager.

158) Click the **Extruded Cut** 🔲 tool from the Features toolbar. The Cut-Extrude PropertyManager is displayed. The direction arrow points into the Extruded feature as illustrated.

159) Select **Through All** for End Condition in Direction 1 as illustrated. View the End Condition options.

160) Click **OK** ✔ from the Cut-Extrude PropertyManager. Cut-Extrude1 is displayed in the FeatureManager.

Fit the PLATE to the Graphics window.

161) Press the **f** key.

Save the PLATE.

162) Click **Save**. View the results.

The Through All End Condition option in Direction 1 creates the Mounting Holes feature through the Base-Extrude feature. As you modify the depth dimension of the Base-Extrude, the Mounting Holes update to reflect the change.

Rename the Cut-Extrude1 feature.

163) Right-click **Cut-Extrude1** from the FeatureManager.

164) Click **Rename tree item**. The feature name is highlighted.

165) Enter **Mounting Holes** for new name.

Save the PLATE.

166) Click **Save**. View the results.

A goal of this book is to expose the new user to various SOLIDWORKS design tools and features.

Review the Extruded Boss/Base and Extruded Cut Features.

An Extruded Boss/Base feature adds material. An Extruded Cut feature removes material. Extrude features require:

- Sketch Plane

- Sketch

- Geometric relations and dimensions

- Depth and End Condition

PLATE Part - Fillet Feature.

The Fillet feature removes sharp edges, strengthens corners and or cosmetically improves appearance. Fillets blend inside and outside surfaces. Fillet features are applied features. Applied features require edges or faces from existing features.

There are many fillet options. The Fillet PropertyManager provides the ability to select either the Manual or FilletXpert tab. Each tab has a separate menu. The Fillet PropertyManager displays the appropriate selections based on the type of fillet you create.

In the next activity, select the Manual tab and single edges for the first Fillet. Select the Tangent propagation option for the second Fillet.

Activity: PLATE Part - Fillet Feature.

Insert a Constant radius edge Fillet.

167) Click **Hidden Lines Visible** from the Heads-up View toolbar.

168) Click the **Fillet** Features tool. The Fillet PropertyManager is displayed. Click the **Manual** tab from the Fillet PropertyManager.

169) Click the **Constant Size Fillet** type icon.

170) Enter **1.0**mm in the Fillet Parameters Radius box.

171) Click the **top left corner edge** as illustrated. The fillet option pop-up toolbar is displayed. Options are model dependent.

172) Select the **Connected to start loop, 3 Edges** icon. The four selected edges are displayed in the Edges, Faces, Features and Loop box.

173) Click **OK** ✔ from the Fillet PropertyManager. Fillet1 is displayed in the FeatureManager.

Rename Fillet1.

174) Rename **Fillet1** to **Small Edge Fillet**.

Save the PLATE.

175) Click **Save** 💾. View the results.

💡 Click the FilletXpert tab in the Fillet PropertyManager to display the FilletXpert PropertyManager. The FilletXpert can only create and edit a Constant radius fillet type. See SOLIDWORKS Help for additional information.

Insert a Multi Radius Constant size Fillet.

176) Click the **Fillet** 🔲 Features tool. The Fillet PropertyManager is displayed. The Tangent propagation check box is checked.

177) Click the **Manual** tab.

178) Click the **Multi Radius Fillet** box.

179) Enter **1.0mm** in the Fillet Radius box.

180) Click the **front top horizontal edge**.

181) Click the **back top horizontal edge**. The selected edges Edge<1>, Edge<2> are displayed in the Items To Fillet box.

182) Click **OK** ✔ from the Fillet PropertyManager. Fillet2 is displayed in the FeatureManager.

Rename Fillet2.

183) Rename **Fillet2** to **Front-Back Edge Fillet**.

Save the PLATE.

184) Click **Save** 💾.

PLATE Part - Hole Wizard Feature.

The PLATE part requires a Countersink hole. Apply the Hole Wizard feature. The Hole Wizard creates simple and complex Hole features by stepping through a series of options (Wizard) to define the hole type and hole placement. The Hole Wizard requires a face or Sketch plane to position the Hole feature. Select the back face.

Activity: PLATE Part - Hole Wizard Feature.

Rotate the view.

185) Click and drag the **mouse pointer** using the middle mouse key to the left to rotate the part as illustrated. The Rotate icon ↻ is displayed.

Display the Back view.

186) Click **Back view** 🗘 from the Heads-up View toolbar. The Origin is displayed at the lower right corner.

Insert the Countersink hole.

187) Click the **Hole Wizard** 🗘 Feature tool. The Hole Specification PropertyManager is displayed.

188) Click the **Type** tab.

189) Click the **Countersink** icon.

190) Select **ANSI Metric** for Standard.

191) Select **Flat Head Screw - ANSI B18.6.7M** for Type.

192) Select **M4** for Size.

193) Select **Through All** End Condition. Accept the default settings.

194) Click the **Positions** tab.

Select the Sketch plane for the Hole Wizard.

195) Click the **middle back face** of Base-Extrude as illustrated. The Point ▱ tool is selected.

196) Click **again to place** the center point of the hole.

197) Right-click **Select** in the Graphics window to de-select the Point tool.

Add a horizontal dimension.

198) Click the **Smart Dimension** Sketch tool. The Point tool is de-selected and the Smart Dimension icon is displayed.

199) Click the **center point** of the Countersink hole.

200) Click the **far right vertical line**.

201) Click a **position** below the bottom horizontal edge.

202) Enter **56/2**. The dimension value 28mm is calculated.

203) Click the **Green Check mark** from the Modify dialog box.

Add a vertical dimension.

204) Click the **center point** of the Countersink hole.

205) Click the **bottom horizontal line**.

206) Click a **position** to the right of the profile.

207) Enter **11**mm.

208) Click the **Green Check mark** from the Modify dialog box.

209) Click **OK** from the Hole Position PropertyManager. View the results.

Fit the PLATE to the Graphics window.

210) Press the **f** key.

The Countersink hole is named CSK for M4 Flat Head in the FeatureManager.

💡 Enter dimensions as a formula in the Modify Dialog Box. Example 56/2 calculates 28. Note you can also enter an equation.

Enter dimensions for automatic unit conversion. Example 2.0in calculates 50.8mm when primary units are set to millimeters (1in. = 25.4mm).

Display an Isometric view of the PLATE.

211) Press the **space bar** to display the Orientation dialog box.

212) Click **Isometric view** . You can also access the Isometric view tool from the Heads-up View toolbar.

213) Click the **Shaded With Edges** icon.

Save the PLATE.

214) Click **Save** .

Close the PLATE.

215) Click **File**, **Close** from the Main menu.

 Review the PLATE Part.

The PLATE part utilized an Extruded Boss/Base feature. An Extruded Boss/Base feature consisted of a rectangular profile sketched on the Front Plane. You added linear dimensions to correspond to the overall size of the PLATE based on the GUIDE-CYLINDER assembly, Piston Plate.

The two holes utilized an Extruded Cut feature. You sketched two circles on the front face and added Equal and Horizontal relations. You utilized Geometric relations in the sketch to reduce dimensions and to maintain the geometric requirements of the part. Linear dimension defined the diameter and position of the circles.

You created fully defined sketches in both the Extruded Base and Extruded Cut feature to prevent future rebuild problems and obtain faster rebuild times. The Fillet feature inserted the edge Fillets and tangent edge Fillets to round the corners of the Base-Extrude feature.

The Hole Wizard feature created the M4 Flat Head Countersink on the back face of the PLATE. You renamed all features in the PLATE FeatureManager.

Note: Hole Wizard was updated for 2024. Sketching with the Hole Wizard is enhanced when you use the Positions tab of the PropertyManager. Under Hole Positions, you can click Existing 2D Sketch and select an existing 2D sketch to position and automatically create the holes at all endpoints, vertices and points of the sketch geometry.

CommandManager and FeatureManager tabs and tree folders will vary depending on system setup and SOLIDWORKS Add-ins.

ROD Part Overview

Recall the functional requirements of the customer:

- The ROD is part of a sub-assembly that positions materials onto a conveyor belt.

- The back end of the ROD fastens to the PLATE.

- The front end of the ROD mounts to the customer's components.

- The customer supplies the geometric requirements for the keyway cut and hole.

The ROD utilizes an Extruded Boss/Base feature with a circular profile sketched on the Front Plane. The ROD also utilizes the Hole Wizard feature for a simple hole and the Extruded Cut feature. Explore new features and techniques with the ROD.

Utilize the following features to create and modify the ROD part:

- **Extruded Boss/Base** : Create the Extruded Boss/Base feature (Boss-Extrude1) on the Front Plane with a circular sketched profile. Use the Circle Sketch tool.

The plane or face you select for the Base sketch determines the orientation of the part.

- **Hole Wizard** : Utilize the Hole Wizard feature to create the first hole on the front face of the ROD. Later, copy the first hole to the back face of the ROD.

A goal of this text is to expose the new user to various SOLIDWORKS design tools and features. For manufacturing, always apply the Hole Wizard feature to create a complex hole.

Press **Ctrl + spacebar** to activate the View Selector.

- **Chamfer** : Create the Chamfer feature from the circular edge of the front face. The Chamfer feature removes material along an edge or face. The Chamfer feature assists the ROD by creating beveled edges for ease of movement in the GUIDE-ROD assembly.

- **Extruded Cut** : Create the Extruded Cut feature from a 2D sketch. Convert the edge of the Boss-Extrude1 (Base Extrude) feature to form a Keyway. Utilize the Trim Entities sketch tool to delete sketched geometry.

- **Hole Wizard** : Create the back hole with the Copy/Paste tool. Copy the front hole to the back face. Modify the Hole Wizard hole diameter from 3mm to 4mm.

- **Extruded Cut** : Apply the Rollback bar and Edit Feature function to implement a customer design change. Add a new Extruded Cut feature to the front face.

- **Redefine** the Chamfer feature and Keyway cut feature. Modify the Sketch Plane.

ROD Part - Extruded Boss/Base Feature

The geometry of the Boss-Extrude1 (Base) feature is a cylindrical extrusion. The Extruded Boss/Base feature is the foundation for the ROD. What is the Sketch plane? Answer: The Front Plane is the Sketch plane for this model. What is the shape of the sketched 2D profile? Answer: A circle.

Activity: ROD Part - Extruded Base Feature

Create a new part.

216) Click **New** ⬜ from the Menu bar.

217) Click the **MY-TEMPLATES** tab.

218) Double-click **PART-MM-ANSI**. Note: Additional templates are displayed.

💡 If the MY-TEMPLATES tab is not displayed, click **Options** ⚙, **File Locations** from the Menu bar. Add the full pathname to the **MY-TEMPLATES** folder in the Document Templates box.

Save the part. Enter name. Enter description.

219) Click **Save As** 🖫. The Save As dialog box is displayed.

220) Select **ENGDESIGN-W-SOLIDWORKS\PROJECTS**.

221) Enter **ROD** for File name.

222) Enter **ROD 10MM DIA x 100MM** for description.

223) Click **Save** from the Save As dialog box.

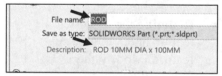

Select the Sketch plane. Create Sketch1.

224) Right-click **Front Plane** from the FeatureManager. Front Plane is your Sketch plane. The Base Sketch plane determines the orientation of the part.

225) Click **Sketch** 🖊 from the Context toolbar. The Sketch toolbar is displayed.

💡 To de-activate the display Reference planes, right-click on the **selected plane** in the FeatureManager, click **Hide** 🖉 from the Context toolbar or click **View**, **Hide/Show** and uncheck **Planes** from the Menu bar.

Sketch a circle Coincident at the Origin. Only a single dimension is required to fully define the Base Sketch.

226) Click the **Circle** Sketch tool. The Circle PropertyManager is displayed. Use the Origin in the sketch.

227) Click the **Origin** ⊥ . The Origin is Coincident with the center point of the circle.

Origin

228) Click a **position** to the right of the Origin.

Add a dimension. Use the Smart Dimension tool.

229) Right-click the **Smart Dimension** ⟋ Sketch tool. The Smart Dimension ⟋ icon is displayed.

230) Click the **circumference** of the circle.

231) Click a **position** diagonally off the profile.

232) Enter **10**mm.

233) Click the **Green Check mark** ✅ from the Modify dialog box.

🔆 If required, click the arrowhead dot to toggle the direction of the dimension arrow.

Insert an Extruded Boss/Base feature.

234) Click the **Extruded Boss/Base** 🗗 feature tool from the Features tab in the CommandManager. Blind is the default End Condition in Direction 1.

235) Enter **100**mm for Depth. Accept the default settings. Note the direction of the extrude feature.

236) Click **OK** ✅ from the Boss-Extrude PropertyManager. Boss-Extrude1 is displayed in the FeatureManager. Note the location of the Origin.

Fit the model to the Graphics window.

237) Press the **f** key.

Rename Boss-Extrude1.

238) Rename **Boss-Extrude1** to **Base Extrude**.

Save the ROD.

239) Click **Save** 💾. View the results.

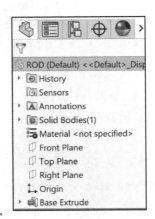

ROD Part - HOLE Wizard Feature

Utilize the Hole Wizard feature to create the first hole on the front face of the ROD. Later, copy the first hole to the back face of the ROD. What is the Sketch Plane for the first Hole Wizard feature? Answer: Front face of the ROD.

You can define the end condition of a hole to the depth of the drill tip or to the depth of the shoulder. The options are available for all Hole Wizard features (including the Hole Wizard Assembly feature) and Advanced Hole types with the following end conditions: Blind, Up to Vertex, Up to Surface, Offset to Surface. Previously, the end condition of a hole was calculated only up to the full diameter of the shoulder.

Activity: ROD Part - Hole Wizard Feature

Select the Front view.

240) Press the **space bar** to display the Orientation dialog box.

241) Click **Front view** .

Create a simple ANSI Metric hole using the Hole Wizard feature.

242) Click the **Hole Wizard** feature tool. The Hole Specification PropertyManager is displayed. Type is the default tab.

243) Click **Hole** for Hole Type.

244) Select **ANSI Metric** for Standard.

245) Select **Drill sizes** for Type. Select **3.0** for Hole Size. Select **Blind** for End Condition.

246) Enter **10mm** for Depth. Click the **Positions** tab.

247) Click a position to the right of the **Origin** as illustrated. Do not click the Origin. The Point tool icon is displayed.

248) Click the **Origin** to place the center point of the hole at the Origin. Note the Geometric relation.

De-select the Point tool.

249) Right-click **Select** in the Graphics area. Note: If you Right-click on the Origin, it will exit directly out of the PropertyManager.

Exit the Dimension/Hole Position.

250) Click **OK** from the Dimension/Hole Position PropertyManager.

251) Expand Diameter Hole1 in the FeatureManager. View the two sketches. Both sketches are fully defined.

Display an Isometric view with Hidden Lines Visible.

252) Press the **space bar** to display the Orientation dialog box.

253) Click Isometric view .

254) Click Hidden Lines Visible from the Heads-up View toolbar.

Rename Diameter Hole1.

255) Rename Diameter Hole1 to **Front Hole**.

Save the ROD.

256) Click Save .

The Hole Wizard feature creates either a 2D or 3D sketch for the placement of the hole in the FeatureManager. You can consecutively place multiple holes of the same type. The Hole Wizard creates 2D sketches for holes unless you select a non-planar face or click the 3D Sketch button in the Hole Position PropertyManager.

ROD Part - Chamfer Feature

The Chamfer feature removes material along an edge or face. The Chamfer feature assists the ROD by creating beveled edges for ease of movement in the GUIDE-ROD assembly. The various Chamfer features are:

- *Angle Distance.* Selected by default. Set the Distance and Angle.

- *Distance Distance.* Enter values for both distances on either side of the selected chamfer edges or Equal Distance and specify a single value.

- *Offset Face.* Are solved by offsetting the faces adjacent to selected edges. The software calculates the intersection point of the offset faces, then calculates the normal from that point to each face to create the chamfer.

- *Face Face.* Creates symmetric, asymmetric, hold line, and chord width chamfers.

- *Vertex.* Enter values for the three distances on each side of the selected vertex or click Equal Distance and specify a single value.

Chamfer the front circular edge and utilize the Angle distance option in the next activity.

Activity: ROD Part - Chamfer Feature

Insert an Angle distance Chamfer feature.

257) Click the **front outer circular edge** of the ROD.
Note the mouse icon feedback symbol for an edge.
The edge turns blue.

258) Click the **Chamfer** feature tool from the
Consolidated toolbar. The Chamfer
PropertyManager is displayed. The arrow indicates
the chamfer direction.

259) Select **Angle Distance** for Chamfer type.

260) Enter **1.00**mm for Distance. Accept the
default 45deg Angle.

261) Click **OK** from the Chamfer
PropertyManager. Chamfer1 is displayed in
the FeatureManager.

Save the ROD.

262) Click **Save**.

To deselect geometry inside the Chamfer Parameters box, right-
click **Clear Selections**. Enter values for Angle distance in the
Chamfer Property Manager or inside the Angle distance Pop-up
box.

Design Intent is how your part reacts as parameters are
modified. Example: If you have a hole in a part that must always go through, apply the
Through All End Condition. As the part thickness changes, the hole will always go
through the part.

ROD Part-Extruded Cut Feature & Convert Entities Sketch tool

The Extruded Cut feature removes
material from the front of the ROD to
create a keyway. A keyway locates the
orientation of the ROD into a mating part in the assembly.

Utilize the Convert Entities Sketch tool in the Sketch toolbar to extract existing geometry
to the selected Sketch plane. The Trim Entities Sketch tool deletes sketched geometry.
Utilize the Power Trim option from the Trim PropertyManager.

Activity: ROD Part-Extruded Cut Feature and Convert Entities Sketch tool

Select the Sketch plane for the keyway. The Sketch plane is the front circular face of the ROD.

263) Click **Shaded With Edges** 🗔 from the Heads-up View toolbar.

264) Right-click the **front circular face** of the ROD. Base Extrude is highlighted in the FeatureManager.

Create a sketch.

265) Click **Sketch** 🖫 from the Context toolbar.

Zoom in on the front face.

266) Click **Zoom to Area** 🔍 from the Heads-up View toolbar.

267) Zoom in on the front face. Click **Zoom to Area** 🔍 to deactivate from the Heads-up View toolbar.

Convert the outside edge to the Sketch plane.

268) Click the **outside front circular edge** of the Base Extrude ▷❘ as illustrated.

269) Click the **Convert Entities** 🗇 Sketch tool from the Sketch toolbar. The system extracts the outside edge and positions it on the Sketch plane.

Display a Front view.

270) Press the **space bar** to display the Orientation dialog box.

271) Click **Front view** 🗔. The front face of the ROD is displayed.

Sketch a vertical line.

272) Click the **Line** ✏ Sketch tool. The Insert Line PropertyManager is displayed.

273) Sketch a **vertical line** as illustrated. The end points of the line extend above and below the diameter of the circle. Insert a Vertical relation if needed.

De-select the Line Sketch tool.

274) Right-click **Select** in the Graphics window.

🔆 Model about the Origin; this provides a point of reference for your dimensions to fully define the sketch.

Trim the sketch.

275) Click the **Trim Entities** ✂ Sketch tool. The Trim PropertyManager is displayed. Note the available options.

276) Click **Power trim** 🔲 from the Trim PropertyManager Options box.

277) Click a **position** to the right of the vertical line, above the circular edge.

278) Drag the **mouse pointer** to intersect the vertical line above the circular edge.

279) Release the **mouse button** when you intersect the vertical line. The line is removed.

The Power trim option trims the vertical line to the converted circular edge and adds a Coincident relation between the endpoint of the vertical line and the circle.

280) Click a **position** to the left of the circular edge.

281) Click and drag the **mouse pointer** to intersect the left circular edge.

282) Release the **mouse button**. The left circular edge displays the original profile.

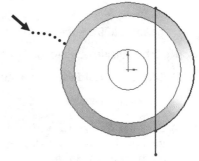

283) Click a **position** to the right of the vertical line, below the circular edge.

284) Drag the **mouse pointer** to intersect the vertical line below the circular edge.

285) Release the **mouse pointer**.

286) Click **OK** ✅ from the Trim PropertyManager.

A selection box allows selection of multiple entities at one time.

To Box-select, click the upper left corner of the box in the Graphics window. Drag the mouse pointer to the lower right corner. A blue box is displayed. Release the mouse pointer. If you create the box from left to right, the entities completely inside the box are selected.

Box-select from left to right.

287) Click a **position** to the upper left corner of the profile as illustrated.

288) Drag the **mouse pointer** diagonally to the lower right corner.

289) Release the **mouse pointer**. The Properties PropertyManager is displayed. The Arc and Line sketch entity are displayed in the Selected Entities dialog box.

290) Click **OK** from the Properties PropertyManager.

Insert a vertical dimension using the Smart Dimension tool.

291) Click the **Smart Dimension** Sketch tool. The Smart Dimension icon is displayed.

292) Click the **vertical line** as illustrated.

293) Click a **position** to the right of the profile.

294) Enter **6.0**mm.

295) Click the **Green Check mark** from the Modify dialog box. The vertical line endpoints are coincident with the circle outside edge. The profile is a single closed loop.

Fit the sketch to the Graphics window.

296) Press the **f** key.

Although dimensions are not required to create features in SOLIDWORKS, dimensions provide location and size information. Models require dimensions for manufacturing.

Insert an Extruded Cut feature.

297) Click the **Extruded Cut** feature tool. The Cut-Extrude PropertyManager is displayed.

298) Enter **15**mm for the Depth in Direction 1. Accept the default settings. Blind is the default End Condition.

299) Click **OK** ✔️ from the Cut-Extrude PropertyManager. Cut-Extrude1 is displayed in the FeatureManager.

Rename the feature and sketch.

300) Rename **Cut-Extrude1** to **Keyway Cut**.

301) Expand Keyway Cut from the FeatureManager.

302) Rename **Sketch4** to **Keyway**.

Exit the sketch.

303) Exit the sketch.

Fit the model to the Graphics window.

304) Press the **f** key.

Save the ROD.

305) Click **Save** 💾.

Rename features or sketches in the FeatureManager by slowly clicking the name twice and entering the new name when the old one is highlighted.

Utilize the Save as copy and continue command to save the document to a new file name without replacing the active document. Utilize the Save as copy and open command to save the document to a new file name that becomes the active document. The original document remains open. References to the original document are not automatically assigned to the copy.

The Extruded Cut feature created the Keyway Cut by converting existing geometry. The Hole Wizard feature created the Front Simple Hole on the front face.

The Convert Entities tool extracts feature geometry. Copy features utilizing the Copy/Paste Edit option.

🔅 To remove Tangent edges, click **Display**, from the System Options menu, and check the **Removed** box.

ROD Part - View Orientation, Named View & Viewport Option

The View Orientation defines the preset position of the ROD in the Graphics window. It is helpful to display various views when creating and editing features. The View Orientation options provide the ability to perform the following tasks:

- Create a named view

- Select a standard view

- Combine views by using the Viewport option

Pressing the **Shift key** + **Up arrow** rotates the model in 90deg increments. Start in an Isometric view. Hold the Shift key down and press the Up arrow key two times to create a Back Isometric view.

Activity: ROD Part - View Orientation and Named View

Rotate the ROD.

306) Hold the **Shift** key down.

307) Press the **Up arrow** key twice.

308) Release the **Shift** key.

Insert a new view.

309) Press the **space bar** on your keyboard or click **View Orientation** ✐ from the Standard Views toolbar. The Orientation dialog box is displayed.

310) Click the **Push Pin** 📌 icon from the Orientation dialog box to maintain the displayed menu.

311) Click the **New View** ✐ icon in the Orientation dialog box.

312) Enter **back iso** in the View name text box.

313) Click **OK** from the Named View dialog box.

Close the Orientation menu.

314) Close ☒ the dialog box.

Save the ROD in the new view.

315) Click **Save** 💾. View the results.

ROD Part - Copy/Paste Feature

The Copy and Paste feature allows you to copy selected sketches and features from one face to another face or from one model to different models.

The ROD requires an additional hole on the back face. Copy the Front Hole feature to the back face. Edit the sketch and modify the dimensions and geometric relations.

Activity: ROD Part - Copy/Paste Feature

Copy the Front Hole to the Back Face.

316) Click **Back view** either from the Heads-up View toolbar or from the Orientation dialog box.

317) Click **Front Hole** from the FeatureManager.

318) Click **Edit**, **Copy** from the Menu bar.

319) Click the **back face** of the ROD as illustrated. DO NOT select the center point.

320) Click **Edit**, **Paste** from the Menu bar. View the results.

The Copy Confirmation dialog box is displayed. The box states that there are external constraints in the feature being copied. The external constraint is the Coincident geometric relation used to place the Front Hole.

Delete the old constraint.

321) Click **Delete** from the Copy Confirmation dialog box. The back face of the ROD contains a copy of the Front Hole feature.

Rename the feature.

322) Rename Ø3.0 (3) **Diameter Hole2** to Ø4.0 (4) **Back Hole**.

Edit the Sketch. Locate the Back Hole. Insert a Coincident relation to the Origin.

323) Expand Ø4.0 (4) **Back Hole** for the FeatureManager.

324) Right-click **(-) Sketch5**.

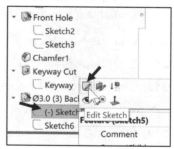

325) Click **Edit Sketch** from the Context toolbar.

326) Click the **center point** of the circle.

327) Hold the **Ctrl** key down.

328) Click the **Origin** of the ROD as illustrated.

329) Release the **Ctrl** key. The Properties PropertyManager is displayed. The selected sketch entities are displayed in the Selected Entities box.

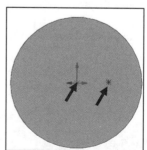

330) Click **Coincident** ⌐ from the Add Relations box. The center point of the back hole and the origin are Coincident.

331) Click **OK** ✔ from the Properties PropertyManager.

Display an Isometric view. Modify the diameter dimension of the hole. Edit the Hole Wizard feature.

332) Click **Isometric view** 🔲.

333) Right-click Ø**4.0 (4) Back Hole** from the FeatureManager.

334) Click **Edit Feature** 🔲 from the Context toolbar. The Hole Specification PropertyManager is displayed.

335) Select Ø**4.0** for Hole Size.

336) Click **OK** ✔ from the Hole Specification PropertyManager.

Save the ROD.

337) Click **Save** 🔲.

💡 The Confirmation Corner in the sketch Graphics window indicates that a sketch is active.

ROD Part - Design Changes with Rollback Bar

You are finished for the day. The phone rings. The customer voices concern about the GUIDE-ROD assembly. The customer provided incorrect dimensions for the mating assembly to the ROD. The ROD must fit into a 7mm hole. The ROD fastens to the customer's assembly with a 4mm Socket head cap screw.

You agree to make the changes at no additional cost since the ROD has not been machined. You confirm the customer's change in writing.

The customer agrees but wants to view a copy of the GUIDE-ROD assembly design by tomorrow. You are required to implement the design change and to incorporate it into the existing part.

Use the Rollback bar and Edit Feature function to implement a design change. The Rollback bar provides the ability to redefine a feature in any state or order. Reposition the Rollback bar in the FeatureManager.

The Edit Feature function provides the ability to redefine feature parameters. Implement the design change.

- Modify the ROD with the Extruded Cut feature to address the customer's new requirements.

- Edit the Chamfer feature to include the new edge.

- Redefine a new Sketch plane for the Keyway Cut feature.

In the next section, develop rebuild errors and correct the rebuild errors.

Activity: ROD Part - Design Changes with Rollback Bar in the FeatureManager

---

Position the Rollback bar in the FeatureManager.

338) Place the **mouse pointer** over the Rollback bar at the bottom of the FeatureManager. The mouse pointer displays a symbol of a hand.

339) Drag the **Rollback** bar upward below the Base Extrude feature. The Base Extrude feature is displayed.

Select the Sketch plane and display the Sketch toolbar.

340) Click **Front view** .

341) Right-click the **Front face** of the Base Extrude feature. This is your Sketch plane. Base Extrude is highlighted in the FeatureManager.

342) Click **Sketch** from the Context toolbar. The Sketch toolbar is displayed.

Sketch a circle Coincident about the Origin.

343) Click the **Circle** Sketch tool. The Circle PropertyManager is displayed.

344) Click the **Origin** of the Base Extrude feature.

345) Click a **position** to the right of the Origin as illustrated.

Add a diameter dimension.

346) Right-click the **Smart Dimension** tool.

347) Click the **circumference** of the circle. Click a **position** diagonally off the profile.

348) Enter **7**mm.

349) Click the **Green Check mark** from the Modify dialog box.

Fit the ROD to the Graphics window.

350) Press the **f** key.

Insert an Extruded Cut feature.

351) Click the **Extruded Cut** feature tool. The Cut-Extrude PropertyManager is displayed.

352) Click the **Flip side to cut** box. The direction arrow points outward. Blind is the default End Condition.

353) Enter **10**mm for Depth. Accept the default settings.

354) Click **OK** ✔ from the Cut-Extrude PropertyManager.

Display an Isometric view.

355) Press the **space bar** to display the Orientation dialog box.

356) Click **Isometric view** 🗊.

Rename the feature.

357) Rename the **Cut-Extrude2** feature to **Front Cut**.

Save the ROD.

358) Click **Save** 💾.

ROD Part - Recover from Rebuild Errors

Rebuild errors can occur when using the Rollback function. A common error occurs when an edge or face is missing. Redefine the edge for the Chamfer feature. When you delete sketch references, Dangling dimensions, sketches and plane errors occur. Redefine the face for the Keyway Sketch plane and the Keyway Sketch.

Activity: ROD Part - Recover from Rebuild Errors

Redefine the Chamfer feature.

359) Drag the **Rollback** bar downward below the Chamfer1 feature. A red x 🗊 ❌ Chamfer1 is displayed next to the name of the Chamfer1 feature. A red arrow is displayed 🗊 ⬇ ROD (Default next to the ROD part icon. Red indicates a model rebuild error. When you created the Front Cut feature, you deleted the original edge from the Base Extrude feature.

Insert the Chamfer feature on the Front Cut edge.

360) Right-click **Chamfer1** from the FeatureManager.

361) Click **Edit Feature** 🗊. The Chamfer1 PropertyManager is displayed. **Missing**Edge<1> is displayed in the Chamfer Parameters box.

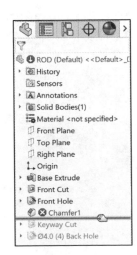

362) Right-click **inside** the Edges and Faces or Vectors box.

363) Click **Clear Selections**.

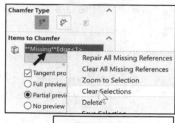

364) Click the **front edge** of the Base Extrude feature as illustrated. Edge<2> is displayed in the Items to Chamfer box.

365) Enter **1.0**mm for Distance. Accept the default settings.

Update the Chamfer feature.

366) Click **OK** ✔ from the Chamfer1 PropertyManager.

Redefine the face for the Keyway Sketch plane.

367) Drag the **Rollback** bar downward below the Keyway Sketch. The Rod Rebuild Error box is displayed. The ⚠ Keyway sketch indicates a dangling sketch plane. The Front Cut deleted the original sketch plane.

Redefine the sketch plane and sketch relations to correct the rebuild errors.

368) Right-click the **Keyway** sketch from the FeatureManager.

369) Click **Edit Sketch Plane** 🖉 from the Context toolbar. The Sketch Plane PropertyManager is displayed.

370) **Delete** any selected entities.

371) Click the **large circular face** of the Front Cut as illustrated. Face<2> is displayed in the Sketch Plane/Face box.

372) Click **OK** ✔ from the Sketch Plane PropertyManager. The SOLIDWORKS dialog box is displayed.

373) Click the **Stop and Repair** button. The What's Wrong dialog box is displayed. View the description.

374) Click **Close** from the What's Wrong dialog box.

The Keyway sketch moved to the new sketch plane. To access the Edit Sketch Plane option, right-click the Sketch entry in the FeatureManager.

Redefine the sketch relations and dimensions to correct the two build errors. Dangling sketch entities result from a dimension or relation that is unresolved. An entity becomes dangling when you delete previously defined geometry.

The Display/Delete Relations sketch tool lists dangling geometry.

Edit the Keyway Cut sketch.

375) Right-click **Keyway Cut** from the FeatureManager.

376) Click **Edit Sketch** from the Context toolbar.

Display the Sketch relations. Delete the On Edge 0 relation.

377) Click the **Display/Delete Relations** Sketch tool. The Display/Delete Relations PropertyManager is displayed.

378) Right-click the **On Edge 0** relation.

379) Click **Delete**. The Dangling status is replaced by Satisfied. The sketch is under defined.

380) Click **OK** from the Display/Delete Relations PropertyManager.

When you select a relation from the Relations list box, the appropriate sketch entities are highlighted in the Graphics window. If Sketch Relations (**View**, **Hide/Show**, **Sketch Relations**) is selected, all icons are displayed, but the icons for the highlighted relations are displayed in a different color.

Display a Front view.

381) Press the **space bar** to display the Orientation dialog box.

382) Click **Front view**.

Add a Coradial relation.

383) Click the **right blue arc** as illustrated.

384) Hold the **Ctrl** key down.

385) Click the **outside circular edge**.

386) Release the **Ctrl** key. The Properties PropertyManager is displayed.

387) Click **Coradial** from the Add Relations box.

388) Click **OK** ✔ from the Properties PropertyManager.

389) Exit the Sketch. The sketch is fully defined and is displayed in black.

Display an Isometric view. Roll back the Rollback bar in the FeatureManager.

390) Click **Isometric view** 🔲.

391) Drag the **Rollback** bar downward below the Back Hole feature in the FeatureManager.

Save the ROD.

392) Click **Save** 💾.

💡 There are three ways to add geometric relations. Utilize the Add Relations Sketch tool ⊥ , use the Properties PropertyManager, or use the Context toolbar.

Utilize the Properties PropertyManager 🔲 or the Context toolbar when you select multiple sketch entities in the Graphics window. Select the first entity. Hold the Ctrl key down and select the remaining geometry to create Sketch relations. The Properties and Context toolbar technique requires fewer steps.

💡 View the mouse pointer icon to confirm selection of the correct point, edge, face, etc.

Modify the depth of the Back Hole.

393) Right-click Ø**4.0 (4) Back Hole** from the FeatureManager.

394) Click **Edit Feature** from the Context toolbar. The Hole Specification PropertyManager is displayed.

395) Select **Blind** for End Condition. Enter **20.00**mm for Blind Hole Depth.

396) Click **OK** from the Hole Specification PropertyManager.

Save the ROD part.

397) Click **Save** .

A machinist rule of thumb states that the depth of a hole does not exceed 10 times the diameter of the hole. The purpose of this book is to expose various SOLIDWORKS design tools and features.

A Countersunk screw fastens the PLATE to the ROD. The depth of the PLATE and the depth of the Back Hole determine the length of the Countersunk screw. Utilize standard length screws.

ROD Part - Edit Part Appearance

Parts are shaded gray by system default. The Edit Part Color option provides the ability to modify part, feature and face color. In the next activity, modify the part color and the machined faces.

Activity: ROD Part - Edit Part Appearance

Edit the part color.

398) Right-click the **ROD Part** icon at the top of the FeatureManager.

399) Click the **Appearances** drop-down arrow as illustrated.

400) Click **ROD** as illustrated. The Color PropertyManager is displayed.

401) Right-click **ROD** inside the Selection Geometry box.

402) Click **Clear Selections**.

Modify machined surfaces on the ROD part.

403) Click **Select Features** in the Selection Geometry box.

404) Click **Keyway Cut** from the fly-out FeatureManager.

405) Click **Front Hole** from the fly-out FeatureManager.

406) Click **Back Hole** from the fly-out FeatureManager. The selected entities are displayed in the Selected Entities box.

407) Click a **color swatch** for the selected faces. Example: Blue. The ROD faces are displayed in the selected color.

408) Click **OK** ✔ from the Color PropertyManager.

Expand the Show Display Plane.

409) Click the **Show Display Pane arrow**. View the results.

410) Click **Show Display Plane** icon. View the results.

Collapse the Hide Display Plane.

411) Click the **Hide Display Pane arrow** to collapse the FeatureManager.

Display an Isometric view.

412) Click **Isometric view** .

The ROD part is complete. Save the ROD.

413) Click **Save** .

Close the ROD.

414) Click **File**, **Close** from the Menu bar. View the results.

💡 Standardize the company part colors. Example: Color red indicates a cast surface. Color blue indicates a machined surface.

Review of the ROD Part.

The ROD part utilized an Extruded Boss/Base feature with a circular profile sketched on the Front Plane. Linear and diameter dimensions corresponded to the customer's requirements.

The Chamfer feature removed the front circular edge. The Chamfer feature required an edge, distance and angle parameters. The Extruded Cut feature utilized a converted edge of the Base Extrude feature to form the Keyway Cut. Reuse geometry with the Convert Entity Sketch tool.

You utilized the Hole Wizard feature for a simple hole on the front face to create the Front Hole. You reused geometry. The Front Hole was copied and pasted to the back face to create the Back Hole. The Back Hole dimensions were modified. The Display\Delete Sketch Tool deleted dangling dimensions of the copied feature. You inserted an Extruded Cut feature by moving the Rollback bar to a new position in the FeatureManager. You recovered from the errors that occurred with Edit Feature, Edit Sketch, and Edit Sketch plane tools by redefining geometry.

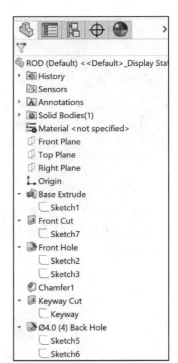

GUIDE Part Overview

The GUIDE part supports the ROD. The ROD moves linearly through the GUIDE. Two slot cuts provide flexibility in locating the GUIDE in the final assembly.

The GUIDE supports a small sensor mounted on the angled right side. You do not have the information on the exact location of the sensor. During the field installation, you receive additional instructions.

Create a pattern of holes on the angled right side of the GUIDE to address the functional requirements.

Address the geometric requirements with the GUIDE part:

- **Extruded Boss/Base** : Use the Extruded Boss/Base feature to create a symmetrical sketched profile for the GUIDE part.

Utilize a centerline and the Dynamic Mirror Sketch tool. The centerline acts as the mirror axis. The copied entity becomes a mirror image of the original across the centerline.

A centerline is a line with a property that makes it exempt from the normal rules that govern sketches. When the system validates a sketch, it does not include centerlines when determining if the profile is disjointed or self-intersecting.

- **Extruded Cut** : Use the Extruded Cut feature with the Through All End Condition to create the first slot. The Extruded Cut feature utilizes the Straight Slot Sketch tool on the top face to form the slot.

- **Mirror** : Use the Mirror feature to create a second slot about the Right Plane. Reuse geometry with the Mirror feature.

- **Extruded Cut** : Use the Extruded Cut feature with the Through All End Condition to create the Guide Hole. The ROD glides through the Guide Hole. Note: The purpose of this book is to expose users to various SOLIDWORKS design tools and features.

- **Hole Wizard** : Use the Hole Wizard feature to create an M3 tapped hole, seed feature for the Linear Pattern.

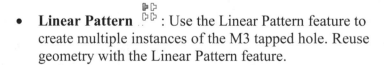

- **Linear Pattern** : Use the Linear Pattern feature to create multiple instances of the M3 tapped hole. Reuse geometry with the Linear Pattern feature.

GUIDE Part - Extruded Boss/Base Feature and Dynamic Mirror

The GUIDE utilizes an Extruded Boss/Base feature. What is the Sketch plane? Answer: The Front Plane is the Sketch plane. The Mid Plane End Condition extrudes the sketch symmetric about the Front Plane.

How do you sketch a symmetrical 2D profile? Answer: Use a sketched centerline and the Dynamic Mirror Sketch tool.

Sketching lines and centerlines produce different behavior depending on how you sketch. Review sketching line techniques before you begin the GUIDE part.

There are two methods to create a line:

- Method 1: Click the first point of the line. Drag the mouse pointer to the end of the line and release. Utilize this technique to create an individual line segment.

- Method 2: Click the first point of the line. Release the left mouse button. Move to the end of the line and click again. The endpoint of the first line segment becomes the start point of the second line segment. Move to the end of the second line and click again.

Utilize these methods to create a chain of line segments. To end a chain of line segments, double-click on the endpoint or right-click and select End chain.

Utilize Method 1 to create the individual centerline.

Utilize Method 2 to create the chained profile lines.

Activity: GUIDE Part - Extruded Base Feature and Dynamic Mirror

Create a new part.

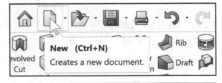

415) Click **New** from the Menu bar.

416) Click the **MY-TEMPLATES** tab.

417) Double-click **PART-MM-ANSI**.

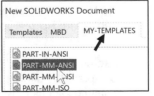

Save the part. Enter name. Enter description.

418) Click **Save As** .

419) Select **ENGDESIGN-W-SOLIDWORKS\PROJECTS**.

420) Enter **GUIDE** for the File name.

421) Enter **GUIDE SUPPORT** for the Description.

422) Click **Save**. The GUIDE FeatureManager is displayed.

Create the first sketch. Select the Front Plane.

423) Right-click **Front Plane** from the GUIDE FeatureManager. This is your Sketch Plane.

424) Click **Sketch** ⌙ from the Context toolbar. The Sketch toolbar is displayed.

Sketch a centerline.

425) Click the **Centerline** ✎ Sketch tool from the Consolidate toolbar. The Insert Line PropertyManager is displayed.

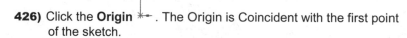

💡 A centerline is displayed with a series of long and short dashes.

426) Click the **Origin** ↧. The Origin is Coincident with the first point of the sketch.

427) Drag and click a **point** above the Origin as illustrated. Right-click **Select**.

Activate the Mirror Sketch tool.

428) Click **Tools, Sketch Tools, Dynamic Mirror** from the Menu bar.

429) Click the **vertical centerline**. The Line Properties PropertyManager is displayed.

Sketch the profile.

430) Click the **Line** ✎ Sketch tool. The Insert Line PropertyManager is displayed.

431) Sketch the first horizontal line from the **Origin** to the right as illustrated. The mirror of the sketch is displayed on the left side of the centerline. **Note**: Click the endpoint of the line.

432) Sketch the **four additional line segments** to complete the sketch as illustrated. The last point is coincident with the centerline. The profile is continuous; there should be no gaps or overlaps.

433) Right-click **Select** to end the sketch.

Add angular, vertical and horizontal dimensions.

434) Click the **Smart Dimension** Sketch tool.

Create an angular dimension.

435) Click the **right horizontal line**.

436) Click the **right-angled line**.

437) Click a **position** between the two lines.

438) Enter **110**.

439) Click the **Green Check mark** .

Create two vertical dimensions.

440) Click the **left small vertical line**.

441) Click a **position** to the left of the profile.

442) Enter **10**mm. Click the **Green Check mark** .

443) Click the **bottom horizontal line**.

444) Click the **top horizontal line**.

445) Click a **position** to the left of the profile.

446) Enter **50**mm.

447) Click the **Green Check mark** .

Create two linear horizontal dimensions.

448) Click the **top horizontal line**.

449) Click a **position** above the profile.

450) Enter **10**mm.

451) Click the **Green Check mark** .

452) Click the **bottom horizontal line**.

453) Click a **position** below the profile.

454) Enter **80**mm.

455) Click the **Green Check mark** .

💡 Select edges instead of points to create linear dimensions. Points are removed when Fillet and Chamfer features are added.

💡 Dimension angles, smaller line segments, and then larger line segments to maintain the profile shape.

Insert an Extruded Boss/Base feature.

456) Click the **Extruded Boss/Base** feature tool from the Features tab in the CommandManager. The Boss-Extrude PropertyManager is displayed. The direction arrow points to the front.

457) Enter **30**mm for Depth.

458) Select **Mid Plane** for End Condition in Direction1. Think Symmetry with the Mid Plane End Condition option.

459) Click **OK** ✔ from the Boss-Extrude PropertyManager. Boss-Extrude1 is displayed in the FeatureManager.

Display an Isometric view.

460) Click **Isometric view** either from the Heads-up View toolbar or from the Orientation dialog box.

Rename the feature.

461) Rename **Boss-Extrude1** to **Base Extrude**.

Save the GUIDE.

462) Click **Save** 🖫.

💡 There are two Mirror Sketch tools in SOLIDWORKS: *Mirror* and *Dynamic Mirror*. Mirror requires a predefined profile and a centerline. Select **Mirror**, select the **profile** and the **centerline**. Dynamic Mirror requires a centerline. As you sketch, entities are created about the selected centerline.

GUIDE Part - Extruded Cut Slot Profile

Create an Extruded Cut 📄 feature. Utilize the Straight Slot Sketch tool to create the 2D sketch profile for the slot.

Activity: GUIDE Part - Extruded Cut Slot Profile

Display a Top view. Create the Slot sketch on the top right face of the model.

463) Click **Top view** ⬜.

464) Right-click the **top right flat face** of the GUIDE in the Graphics window.

465) Click **Sketch** ⬜ from the Context toolbar. The Sketch toolbar is displayed.

Apply the Straight Slot Sketch tool. The Straight Slot Sketch tool requires three points,

466) Click the **Straight Slot** Sketch ⊙⊙ tool from the Consolidated Slot toolbar as illustrated. The Slot PropertyManager is displayed.

467) Click the **first point** as illustrated.

468) Click the **second point** directly above the first point.

469) Click the **third point** directly to the right of the second point. The Slot Sketch is created.

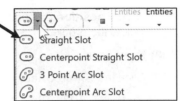

💡 Create a slot without using the Slot Sketch tool by sketching two lines, two 180-degree tangent arcs and apply the needed geometric relations.

Add dimensions.

470) Click the **Smart Dimension** ✏ Sketch tool.

471) Click the **Origin** ↳ as illustrated.

472) Click the **center point** of the top arc. Note: When the mouse pointer is positioned over the center point.

473) Click a **position** above the profile.

474) Enter **30**mm.

475) Click the **Green Check mark** ✔.

Create a vertical dimension.

476) Click the **top horizontal edge** of the GUIDE.

477) Click the **center point** of the top arc.

478) Click a **position** to the right of the profile.

479) Enter **10**mm.

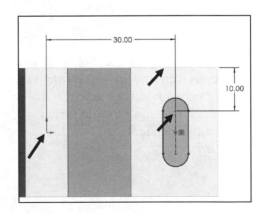

480) Click the **Green Check mark** .

Create a dimension between the two arc center points.

481) Click the top arc **center point**.

482) Click the second arc **center point**.

483) Click a **position** to the right of the profile.

484) Enter **10**mm.

485) Click the **Green Check mark** .

Create a radial dimension.

486) Click the **bottom arc** radius.

487) Click a **position** diagonally below the profile.

488) Enter **3**mm.

489) Click the **Green Check mark** . The Sketch is fully defined and is displayed in black.

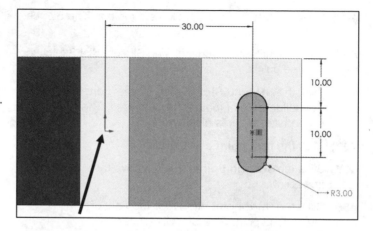

Insert an Extruded Cut feature.

490) Click the **Extruded Cut** feature tool. The Cut-Extrude PropertyManager is displayed.

491) Select **Through All** (Think Design Intent) for End Condition in Direction1. Accept all default settings.

492) Click **OK** from the Cut-Extrude PropertyManager. Cut-Extrude1 is displayed in the FeatureManager.

Right-click anywhere on an extruded feature; view the end conditions from the pop-up shortcut menu. Click in empty space, on geometry, or on the handle. The shortcut menu provides the options for Direction 1 and Direction 2. Note: Options are document dependent.

Display an Isometric view.

493) Click **Isometric view** 🔲 either from the Heads-up View toolbar or the Orientation dialog box.

Rename the feature.

494) Rename **Cut-Extrude1** to **Slot Cut**.

Save the GUIDE.

495) Click **Save** 💾.

SOLIDWORKS contains additional methods to create sketch slots. Here are three examples that you can explore as an exercise:

- Sketch a rectangle. Delete two opposite line segments. Insert two arcs. Extrude the profile.

- Sketch a single vertical centerline. Utilize Offset Entities, Bidirectional option and Cap Ends option. Extrude the profile.

- Utilize a Design Library Feature.

The machined, metric, straight slot feature provides both the Slot sketch and the Extruded Cut feature. You define the references for dimensions, overall length and width of the Slot. Explore the SOLIDWORKS Design Library later in the book.

GUIDE Part - Mirror Feature

The Mirror feature mirrors a selected feature about a mirror plane. Utilize the Mirror feature to create the Slot Cut on the left side of the GUIDE. The mirror plane is the Right Plane.

🔆 Use the Mirror Feature PropertyManager to mirror features. In parts, you can mirror faces, features, and bodies. In assemblies, you can mirror assembly features.

Mirrors features, faces, and bodies about a plane or a planar face. In assemblies, you can mirror assembly features.

Activity: GUIDE Part - Mirror Feature with Geometric Pattern Option.

Insert a Mirror feature with the Geometric Pattern options.

496) Click the **Mirror** feature tool. The Mirror PropertyManager is displayed.

497) **Expand** the GUIDE fly-out FeatureManager in the Graphics window.

498) Click **Right Plane** from the fly-out FeatureManager. Right Plane is displayed in the Mirror Faces/Plane box.

499) Click inside the **Features to Mirror** box.

500) Click **Slot Cut** from the fly-out FeatureManager for the Features to Mirror. Slot Cut is displayed in the Features to Mirror box.

501) Check the **Geometry Pattern** box.

502) Click **OK** ✔ from the Mirror PropertyManager. Mirror1 is displayed in the FeatureManager.

Save the GUIDE part.

503) Click **Save** 💾.

Fit the GUIDE part to the Graphics window.

504) Press the **f** key.

💡 Save rebuild time, check the Geometry pattern option in the Mirror Feature PropertyManager. Each instance is an exact copy of the faces and edges of the original feature.

GUIDE Part - Holes

The Extruded Cut 🔲 feature removes material from the front face of the GUIDE to create a Guide Hole. The ROD moves through the Guide Hole.

Create the tapped holes with the Hole Wizard feature and the Linear Pattern feature. Apply dimensions to the tapped holes relative to the Guide Hole.

Activity: GUIDE Part - Holes

Create a sketch on the front face.

505) Right-click the **front face** of the GUIDE in the Graphics window. This is your Sketch Plane.

506) Click **Sketch** 🔳 from the Context toolbar.

507) Click **Normal To view** ⬇ from the Heads-up View toolbar. The front face of the GUIDE is displayed.

508) Click the **Circle** ⊙ Sketch tool. The Circle PropertyManager is displayed.

509) Sketch a **circle** in the middle of the GUIDE as illustrated.

510) Right-click **Select** to deselect the Circle sketch tool.

511) If needed, add a **Vertical Relation** between the center of the circle and the Origin.

Add dimensions.

512) Click the **Smart Dimension** Sketch tool.

513) Click the circle **circumference**.

514) Click a **position** below the horizontal line.

515) Enter **10**mm.

516) Click the **Green Check mark** .

517) Click the **center point** of the circle.

518) Click the **Origin**.

519) Click a **position** to the right of the profile.

520) Enter **29**mm.

521) Click the **Green Check mark** . The sketch is fully
defined.

Create an Extruded Cut feature.

522) Click the **Extruded Cut** feature tool. The Cut-Extrude
PropertyManager is displayed.

523) Select **Through All** for End Condition in Direction1. Note the
direction of the cut. Accept the default conditions.

524) Click **OK** from the Cut-Extrude PropertyManager. Cut-
Extrude1 is displayed in the FeatureManager.

525) Press the **space bar** to display the Orientation dialog box.

Display an Isometric view with Hidden Lines Removed.

526) Click **Isometric view** either from the Heads-up View toolbar or
from the Orientation dialog box.

527) Click **Hidden Lines Removed** from the Heads-up View
toolbar.

Rename the feature.

528) Rename **Cut-Extrude1** to **Guide Hole**.

Display the Temporary Axes.

529) Click **View**, **Hide/Show**, check **Temporary
Axes** from the Menu bar. The Temporary
Axes are displayed in the Graphics window.

Create the Tapped Hole. Utilize the Hole Wizard.

530) Click the **Hole Wizard** feature tool. The
Hole Specification PropertyManager is
displayed.

531) Click the **Type** tab.

532) Click **Straight Tap** for Hole Specification.

533) Select **ANSI Metric** for Standard.

534) Select **Bottoming Tapped Hole** for Type.

535) Select **M3x0.5** for Size.

536) Select **Blind** for End Conditions. Accept the default conditions.

537) Click the **Positions** tab.

538) Click the **right angled face** below the Guide Hole Temporary Axis as illustrated. The Point icon is displayed.

539) Click a **position** below the **Center hole Temporary Axes** as illustrated.

Dimension the tapped hole relative to the Guide Hole axis. Add the horizontal dimension.

Insert Dimensions.

540) Click the **Smart Dimension** ✏ Sketch tool. The Point tool is de-selected. The Smart Dimension icon is displayed.

541) Click the **center point** of the tapped hole. Note: When the mouse pointer is positioned over the center point, the center point is displayed in red.

542) Click the **right vertical edge**.

543) Click a **position** above the profile.

544) Enter **25**mm.

545) Click the **Green Check mark** ✅ .

Add a vertical dimension.

546) Click the **center point** of the tapped hole.

547) Click the horizontal **Temporary Axis**.

548) Click a **position** to the right of the profile.

549) Enter **4**mm. Click the **Green Check mark** ✅ .

Display the tapped hole.

550) Click **OK** ✓ from the Hole Position
PropertyManager.

Hide the Temporary Axes.

551) Click **View**, **Hide/Show**, and uncheck **Temporary Axes** from the Menu bar.

Fit GUIDE to the Graphics window.

552) Press the **f** key.

Save the GUIDE.

553) Click **Save** 💾.

GUIDE Part-Linear Pattern Feature

The Linear Pattern 🔲 feature creates multiple
copies of a feature in an array. A copy of a feature
is called an instance. A Linear Pattern feature
requires:

- One or two directions.

- A Feature, Face or Body to pattern.

- Distance between instances.

- Number of instances.

Insert a Linear Pattern feature of tapped holes on the right-angled face of the GUIDE.

Activity: GUIDE Part-Linear Pattern Feature with Geometry Pattern option

Display a Right view.

554) Click **Right view** 🗗 .

Insert a Linear Pattern feature with the Geometry Pattern option.

555) Click the **Linear Pattern** 🔲 feature tool. The Linear Pattern PropertyManager is displayed.

556) Expand GUIDE from the fly-out FeatureManager.

557) Click inside the **Features and Faces** box.

558) Click **M3X0.5 Tapped Hole1** from the fly-out FeatureManager.

559) Click inside the **Pattern Direction** box for Direction 1.

560) Click the **top horizontal edge** for Direction 1. Edge<1> is displayed in the Pattern Direction
box. The direction arrow points to the right.

561) Enter **10**mm in the Direction 1 Spacing box. The direction arrow points to the right. If required, click the Reverse Direction button.

562) Enter **3** in the Number of Instances box for Direction 1.

563) Click inside the **Pattern Direction** box for Direction 2.

564) Click the **left vertical edge** for Direction 2. The direction arrow points upward. If required, click the Reverse Direction button.

565) Enter **12**mm in the Direction 2 Spacing box.

566) Enter **2** in the Number of Instances box.

567) Check **Geometry Pattern** in the Options box.

Display the Linear Pattern feature.

568) Click **OK** from the Linear Pattern PropertyManager. LPattern1 is displayed in the FeatureManager.

Display an Isometric view - Shaded With Edges.

569) Press the **space bar** to display the Orientation dialog box.

570) Click **Isometric view** .

571) Click **Shaded With Edges** from the Heads-up View toolbar.

Save the GUIDE.

572) Click **Save** .

Reuse geometry. The Linear Pattern feature with the Geometry Pattern option saves rebuild time. Rename the seed feature in the part pattern. The seed feature is the first feature for the pattern, Example: M3x0.5 Tapped Hole1. Assemble additional parts to the seed feature in the assembly.

Save time when sketching by utilizing the Grid. Click **Options**, **Document Properties**, **Grid/Snap** to control display and spacing settings. The Display grid check box turns on/off the Grid in the sketch.

Hide geometric relations to clearly display profile lines in the sketch. Check **View**, **Hide/Show**, **Sketch Relations** to hide/display geometric relations in the sketch.

GUIDE Part - Material and Mass Properties

Apply material to the GUIDE part. The Material dialog box helps you manage physical materials. You can work with pre-defined materials, create custom materials, apply materials to parts, and manage favorites.

Calculate the Mass Properties of the GUIDE part using the Mass Properties tool.

Activity: GUIDE Part - Materials Editor and Mass Properties

Apply AISI 304 material to the model.

573) Right-click the **Materials** folder in the FeatureManager.

574) Click **Edit Material**. The Materials dialog box is displayed.

Expand the Steel folder.

575) **Expand** the Steel folder from the SOLIDWORKS Materials box.

576) Select **AISI 304**. View the available materials and information provided in the Materials dialog box. The Materials dialog box is updated from previous years.

577) Click **Apply**.

578) Click **Close** from the Materials dialog box. AISI 304 is displayed in the GUIDE FeatureManager.

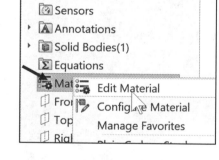

Use the Search feature in the Material dialog box to quickly locate material.

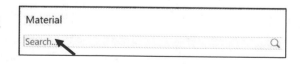

Calculate the Mass Properties
of the GUIDE.

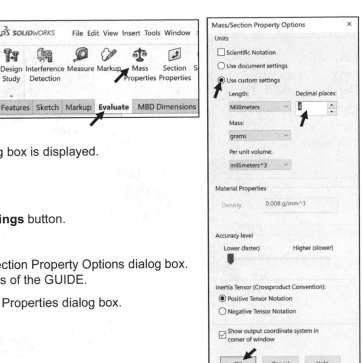

579) Click the **Evaluate** tab
from the
CommandManager.

580) Click the **Mass**
Properties tool.
The Mass Properties dialog box is displayed.

Set the units to display.

581) Click the **Options** button.

582) Click the **Use custom settings** button.

583) Enter **4** in the Decimal box.

584) Click **OK** from the Mass/Section Property Options dialog box.
Review the Mass properties of the GUIDE.

585) Click **Close** from the Mass Properties dialog box.

Save the GUIDE.

586) Click **Save** .

Close the GUIDE.

587) Click **File**, **Close** from the Menu bar.

Add or create a Center of Mass (COM)
point directly in the Mass
Properties box or from the Main menu
for a part or assembly. The position of
the COM point updates when the
model's center of mass changes. For
example, the position of the COM point
updates as you add, modify, delete, or
suppress features or components. The
COM point can be useful when you are
designing assemblies requiring
balanced mass to avoid excessive
vibration.

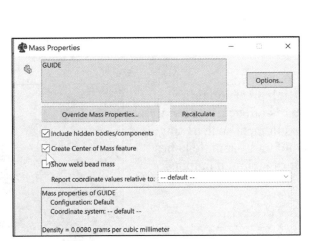

Manufacturing Considerations

CSI Mfg.
(www.compsources.com)
Southboro, MA, USA
provides solutions to their
customers by working with
engineers to obtain their
design goals and to reduce
cost.

Customer Focus
Courtesy of CSI Mfg (www.compsources.com)
Southboro, MA

High precision turned
parts utilize a variety of
equipment such as single
and multi-spindle lathes
for maximum output.

Obtain knowledge about the manufacturing process from your colleagues before you design the part.

Ask questions on material selection, part tolerances, machinability and cost. A good engineer learns from past mistakes of others.

Review the following design guidelines for Turned Parts.

Guideline 1: Select the correct material.

Examples of cost-effective materials
- AISI 303
- BRASS
- AL 2011

Consider part functionality, operating environment, cost, "material and machine time," machinability and surface finish. Utilize standard stock sizes and material that is readily available. Materials are given a machinability rating. Brass (100), AISI 303(65) and AISI 304(30).

Example: Using AISI 304 vs. AISI 303. AISI 303 is easier to machine due to its higher sulfur content.

Guideline 2: Minimize tool changes. Maximize feature manufacturability.

The Extruded Cut, Hole, Chamfer and Fillet features require different tooling operations. Will changes in form or design reduce the number of operations and the cost of machining?

Example:

- Avoid Fillets on inside holes to remove sharp edges. Utilize the Chamfer feature to remove sharp edges.

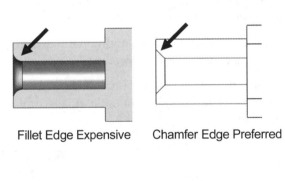

Fillet Edge Expensive   Chamfer Edge Preferred

- Utilize common size holes for fewer tool changes.

- Leave space between the end of an external thread and the shoulder of the rod. Rule of thumb: 1-2 pitch of thread, minimum.

- Leave space at the end of an internal thread for the tool tip endpoint.

💡 To display a cosmetic thread, right-click the Annotations folder in the FeatureManager and click Details. Check the Cosmetic threads box and the Shaded cosmetic thread box. Click OK. Select Shaded With Edges for display modes. View the results.

Cosmetic threads

Guideline 3: Optimize the design for production. Provide dimension, tolerance and precision values that can be manufactured and inspected.

- Check precision on linear, diameter and angular dimensions. Additional non-needed decimal places increase cost to a part.

Example:

The dimension 45 at +/- .5° indicates the dimension is machined between 44.5° and 45.5.

The dimension 45.00 at +/- .05° indicates the dimension is machined between 44.95 and 45.05.

This range of values adds additional machining time and cost to the final product.

Modify the dimension decimal display in the Tolerance/Precision box in the part, assembly or drawing.

Set Document Properties, Dimensions, Tolerance and Precision for the entire document.

In Project 3, create drawings for the GUIDE part. Modify the dimension values and tolerance on the ROD and GUIDE to create a clearance fit.

Review of the GUIDE Part.

The GUIDE part supports the ROD. The ROD moves linearly through the GUIDE. Two slot cuts provide flexibility in locating the GUIDE in the final assembly.

The GUIDE supports a small sensor mounted on the angled right side. You did not have the information on the exact location of the sensor. During the field installation, you receive additional instructions.

The GUIDE part utilized an Extruded Boss/Base feature. The 2D sketched profile utilized the Dynamic Mirror Sketch tool to build symmetry into the part.

The linear and diameter dimensions corresponded to the ROD and GUIDE-CYLINDER assembly.

The Extruded Cut feature created the Slot-Cut. The Mirror feature created a second Slot-Cut symmetric about the Right Plane. The Guide Hole utilized the Extruded Cut feature with the Through All End Condition.

The Hole Wizard created the M3 x0.5 Tapped Hole, the seed feature in the Linear Pattern. The Linear Pattern feature created multiple instances of the M3 x0.5 Tapped Hole.

You reused geometry to save design time and rebuild time. The Mirror feature, Linear Pattern feature, Geometry Pattern option and Sketch Mirror are time saving tools. You defined materials in the part used by other software applications.

Project Summary

You created three parts: ROD, GUIDE and PLATE.

You create the PART-ANSI-MM Part Template for the three parts.

The PLATE part consisted of an Extruded Boss/Base, Extruded Cut, Fillet and Hole Wizard features. You rename the features.

The ROD consisted of an Extruded Boss/Base, Extruded Cut, Chamfer, and Hole Wizard features. The ROD illustrated the process to manipulate design changes with the Rollback bar, Edit Sketch and Edit Feature/Edit Sketch commands. You rename the features.

The GUIDE part utilized an Extruded Boss/Base feature sketch with the Dynamic Mirror Entities tool. The GUIDE also utilized the Extruded Cut, Mirror, Hole Wizard and Linear Pattern features. You rename the features.

All sketches for the Extruded Boss/Base and Extruded Cut features utilized geometric relationships. Incorporate geometric relations such as Horizontal, Vertical, Symmetric, Equal, Collinear and Coradial into your sketches.

The FeatureManager consists of various tabs:

- *FeatureManager design tree* tab.

- *PropertyManager* tab.

- *ConfigurationManager* tab.

- *DimXpertManager* tab.

- *DisplayManager* tab.

- *CAM FeatureManager tree* tab.

- *CAM Operation tree* tab.

- *CAM Tools tree* tab.

Click the direction arrows to expand or collapse the FeatureManager design tree.

CommandManager and FeatureManager tabs and tree folders will vary depending on system set-up and Add-ins.

In Project 3, insert the ROD, GUIDE and PLATE parts into a new assembly. In Project 4, create an assembly drawing and a detailed drawing of the GUIDE part.

Are you ready to start Project 3? Stop! Examine and create additional parts. Perform design changes. Take chances, make mistakes and have fun with the various features and commands in the project exercises.

When you create a new part or assembly, the three standard default Planes (Front, Right and Top) are aligned with specific views. The Plane you select for the Base sketch determines the orientation of the part.

- Utilize the Save As/Save as command to save the file in another file format.

- Utilize the Save as copy and continue command to save the document to a new file name without replacing the active document.

- Utilize the Save as copy and open command to save the document to a new file name that becomes the active document. The original document remains open. References to the original document are not automatically assigned to the copy.

Questions

1. Identify at least four key design areas that you should investigate before starting a project.

2. Why is file management important?

3. Identify the steps in starting a SOLIDWORKS session.

4. Identify the three default Reference planes in SOLIDWORKS.

5. Describe a Base sketch. Provide an example.

6. True or False: The plane or face you select for the Base sketch determines the orientation of the part.

7. True or False: In a fully defined sketch, all sketch entities are displayed in black.

8. True or False: In an over defined sketch, all sketch entities are displayed in blue.

9. True or False: It is considered very poor design practice not to fully define a sketch.

10. Describe a Base feature. Provide an example.

11. Identify the Drafting Standard which is used in the United States.

12. SOLIDWORKS provides system feedback by attaching a symbol to the mouse pointer cursor. The system feedback symbol indicates what you are selecting or what the system is expecting you to select. What is the feedback symbol for a face, edge and vertex?

13. Describe the difference between First Angle Projection and Third Angle Projection.

14. How do you recover from a Rebuild error?

15. When should you apply the Hole Wizard feature vs. the Extruded Cut feature?

16. Describe the difference between an Extruded Boss/Base feature and an Extruded Cut feature.

17. Describe a Fillet feature. Provide an example.

18. Name the command keys used to Copy sketched geometry.

19. Describe the difference between the Edit Feature and Edit Sketch tool.

20. Describe a Chamfer feature. Provide an example.

21. Describe the Rollback bar function. Provide an example.

22. Identify three types of Geometric Relations that you can apply to a sketch.

23. Describe the procedure in creating a Part Document template.

24. Describe the procedure to modify material in an existing part.

25. Describe the procedure to set the Overall drafting standard, units and precision for a Part document.

26. Describe the procedure to sketch an Elliptical profile.

27. Identify the name of the following feature tool icons.

A　　B　　C　　D　　E　　F

A		B		C	D
E		F			

28. Identify the name of the following Sketch tool icons.

A　B　C　D　E　F　G　H　I　J

A	B	C	D
E	F	G	H
I	J		

29. Identify the surfaces with the appropriate letter that would appear in the FRONT view, TOP view and RIGHT view.

FRONT view surfaces: _____

TOP view surfaces: _____

RIGHT view surfaces: _____

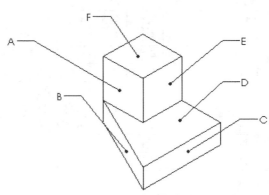

30. Identify the Angle of Projection for the provided illustration.

Angle of Projection_____.

Exercises

Exercise 2.1 and Exercise 2.2

A fast and economical way to manufacture parts is to utilize Steel Stock, Aluminum Stock, Cast Sections, Plastic Rod and Bars and other stock materials.

Stock materials are available in different profiles and various dimensions from tool supply companies.

L-BRACKET   T-SECTION

- Create a new L-BRACKET part and a T-SECTION part. Each part contains an Extruded Base feature. Apply units from the below tables.

- Utilize the Front Plane for the Sketch plane.

Exercise 2.1: L-BRACKET

Ex.:	A:	B:	C:	LENGTH:	Units:
1.1a	3	3	5/8	10	IN
1.1b	4	4	3/4	10	
1.1c	6	4	1/2	25	
1.1d	8	6	3/4	25	
1.1e	12	12	3/2	25	
1.1f	7	4	3/4	30	
1.1g	75	75	10	250	MM
1.1h	75	100	10	250	
1.1i	200	100	25	500	

Exercise 2.2: T-SECTION

Ex.:	A:	B:	C:	LENGTH:	Units:
1.2a	3	3	5/8	30	IN
1.2b	4	4	3/4	30	
1.2c	6	6	1	30	
1.2d	75	75	10	250	MM
1.2e	75	100	10	250	
1.2f	200	100	25	500	

Exercise 2.2a: PART DOCUMENT TEMPLATE

Create a Metric (MMGS) Part document template.

- Use the ISO drafting standard and the MMGS unit system.
- Set the Drawing/Units and precision to the appropriate values.
- Name the Template PART-MM-ISO.

Exercise 2.2b: PART DOCUMENT TEMPLATE

Create an English (IPS) Part document template.

- Use the ANSI drafting standard and the IPS unit system.
- Set Drawing/Units and precision to the appropriate values.
- Name the Template PART-IN-ANSI.

Exercise 2.3: Identify the Sketch plane for the Boss-Extrude1 feature. View the Origin location. Simplify the number of features.

A: Top Plane

B: Front Plane

C: Right Plane

D: Left Plane

Correct answer _____.

Exercise 2.4: Identify the Sketch plane for the Boss-Extrude1 feature. View the Origin location. Simplify the number of features.

A: Top Plane

B: Front Plane

C: Right Plane

D: Left Plane

Correct answer _____.

Exercise 2.5: Identify the Sketch plane for the Boss-Extrude1 feature. View the Origin location. Simplify the number of features.

A: Top Plane

B: Front Plane

C: Right Plane

D: Left Plane

Correct answer _____.

Exercise 2.6: Identify the Sketch plane for the Boss-Extrude1 feature. View the Origin location. Simplify the number of features.

A: Top Plane

B: Front Plane

C: Right Plane

D: Left Plane

Correct answer _____.

Exercise 2.7: Identify the material category for 6061 Alloy.

A: Steel

B: Iron

C: Aluminum Alloys

D: Other Alloys

E: None of the provided

Correct answer _____.

Exercise 2.8: AXLE

Create an AXLE part as illustrated with dual units. Overall drafting standard - ANSI. IPS is the primary unit system.

- Utilize the Front Plane for the Sketch plane.

- Utilize the Mid Plane End Condition. The AXLE is symmetric about the Front Plane. Note the location of the Origin.

Exercise 2.9: SHAFT COLLAR

Create a SHAFT COLLAR part as illustrated with dual system units IPS (inch, pound, second) and MMGS (millimeter, gram, second). Overall drafting standard - ANSI.

- Utilize the Front Plane for the Sketch plane.

Exercise 2.10: FLAT BAR - 3 HOLE

Create the FLAT BAR - 3 HOLE part as illustrated with dual system units IPS (inch, pound, second) and MMGS (millimeter, gram, second). Overall drafting standard - ANSI.

- Utilize the Front Plane for the Sketch plane.

- Utilize the Center point Straight Slot Sketch tool.

- Utilize a Linear Pattern feature for the three holes. The FLAT BAR - 3 HOLE part thickness is 0.09in [2.29mm]. Material 2024 Alloy. Holes are .500in. on center, Radius is .250in., and hole diameter is .190in.

Exercise 2.11: FLAT BAR - 5 HOLE

Create the the FLAT BAR - 5 HOLE part.
Overall drafting standard - ANSI.

- Utilize the Front Plane for the Sketch plane.

- Utilize the Centerpoint Straight Slot Sketch tool.

- Utilize a Linear Pattern feature for the three holes. The FLAT BAR - 5 HOLE part thickness is 0.09in [2.29mm].

- Material 2024 Alloy.

- Holes are .500in. on center, Radius is .250in., and hole diameter is .190in.

Exercise 2.12: SIMPLE BLOCK

Create the illustrated part. Note the location of the Origin. Overall drafting standard - ANSI.

- Calculate the overall mass of the illustrated model. Apply the Mass Properties tool.

- Think about the steps that you would take to build the model.

- Review the provided information carefully.

- Units are represented in the IPS (inch, pound, second) system.

- A = 3.50in, B = .70in

Given:
A = 3.50
B = .70
Material: 1060 Alloy
Density = 0.0975 lb/in^3
Units: IPS
Decimal place: 2

Origin

Exercise 2.13: SIMPLE T

Create the illustrated part. Note the location of the Origin. Overall drafting standard - ANSI.

- Calculate the overall mass of the illustrated model. Apply the Mass Properties tool.

- Think about the steps that you would take to build the model.

- Review the provided information carefully. Units are represented in the IPS (inch, pound, second) system.

- A = 3.00in, B = .75in

Given:
A = 3.00
B = .75
Material: Copper
Density = 0.321 lb/in^3
Units: IPS
Decimal place: 2

Origin

Exercise 2.14: SIMPLE BLOCK STEP

Create the illustrated part. Note the location of the Origin. Overall drafting standard - ANSI.

- Calculate the volume of the part and locate the Center of mass relative to the origin with the provided information.

- Apply the Mass Properties tool.

- Think about the steps that you would take to build the model.

- Review the provided information carefully.

Given:
A = 3.30
B = 2.00
Material: 2014 Alloy
Density = .101 lb/in^3
Units: IPS
Decimal place: 2

Origin

Exercise 2.15:

Create the part from the illustrated ANSI - MMGS Third Angle Projection drawing: Front, Top, Right and Isometric views.

Note the location of the Origin (shown in an Isometric view).

- Apply 1060 Alloy for material.

- Calculate the Volume of the part and locate the Center of mass relative to the origin. Use the Mass Properties tool.

- Think about the steps that you would take to build the model. The part is symmetric about the Front Plane.

Origin

Exercise 2.16:

Create the part from the illustrated ANSI - MMGS Third Angle Projection drawing: Front, Top, Right and Isometric views.

Note the location of the Origin (shown in an Isometric view). Assume Symmetry.

- Apply the Hole Wizard feature.

- Apply 1060 Alloy for material.

- The part is symmetric about the Front Plane.

- Calculate the Volume of the part and locate the Center of mass relative to the origin. Use the Mass Properties tool.

- Think about the steps that you would take to build the model.

Origin

Exercise 2.17: Slider Part

Create the part from the illustrated ANSI - MMGS Third Angle Projection drawing: Front, Top, Right and Isometric view.

Note: The location of the Origin displayed in the Isometric view.

- Apply Cast Alloy steel for material.

- The part is symmetric about the Front Plane.

- Apply Mid Plane for End Condition in Boss-Extrude1.

- Apply Through All for End Condition in Cut-Extrude1.

- Apply Through All for End Condition in Cut-Extrude2.

- Apply Up to Surface for End Condition in Boss Extrude2

- Calculate the Volume of the part and locate the Center of mass.

- Calculate the Volume of the part and locate the Center of mass relative to the origin. Use the Mass Properties tool.

Think about the steps that you would take to build the model. Do you need the Right view for manufacturing? Does it add any valuable information?

Exercise 2.18: COSMETIC THREAD

Apply a Cosmetic thread ¼-28-1 UNC 2A. A Cosmetic thread represents the inner diameter of a thread on a boss or the outer diameter of a thread.

Copy and open the Cosmetic Thread part from the Chapter 2 Homework folder.

Click the bottom edge of the part as illustrated.

Click Insert, Annotations, Cosmetic Thread from the Menu bar menu. View the Cosmetic Thread PropertyManager. Edge<1> is displayed.

Enter ANSI Inch for Standard.

Enter Machine Threads for Type.

Enter ¼-28 for Size.

Select End Condition: Blind.

Enter 1.00in for Depth.

Check the Thread class box.

Select 2A for Thread Class.

Click OK ✔ from the Cosmetic Thread FeatureManager.

Expand the FeatureManager.

View the Cosmetic Thread feature. Right-click the Annotations folder, and click Details.

Check the Cosmetic threads and Shaded cosmetic threads box.

Click OK from the Annotation Properties dialog box.

🔆 ¼-28-1 UNC 2A: ¼ inch major diameter - 28 threads / inch, 1 inch long, Unified National Coarse thread series, Class 2 (General Thread), A -External threads.

Exercise 2.18A: Stud Wizard

Create a stud on a Cylindrical Body. Create an ANSI Inch, Machine thread, ¼-20-2 UNC 2A: ¼ inch major diameter - 20 threads / inch, 2 inch long, Unified National Coarse thread series, Class 2 (General Thread), A -External threads.

Note: Class fit measures the looseness or tightness of external mating threads. You can retrieve and use this information for hole callouts in drawings.

Copy and open the Exercise 2.18A part from the Chapter 2 Homework folder. Use the default setting for UnderCut. The UnderCut option creates an undercut at the end of the threaded portion of the stud to provide clearance. The default values are based on the selections under Standard, but you can override and customize these values.

Click the Creates Stud on a Cylindrical Body icon.

Click the top edge of the part as illustrated. Edge<1> is displayed.

Enter ANSI Inch for Standard.

Enter Machine Threads for Type.

Enter ¼-20 for Size.

Select End Condition: Blind.

Enter 2.00in for Depth.

Check the Thread class box.

Select 2A for Thread Class.

Click OK ✔ from the FeatureManager.

To view the threads, right-click the Annotations folder, and click Details.

Check the Cosmetic threads and Shaded cosmetic threads box. Click OK from the Annotation Properties dialog box.

Exercise 2.18B: Stud Wizard

Create a stud on a Surface. Create an ANSI Inch, Machine thread, ¼-20-2 UNC 2A: ¼ inch major diameter - 20 threads/inch, 2 inch long, Unified National Coarse thread series, Class 2 (General Thread), A -External threads.

Copy and open Exercise 2.18B from the Chapter 2 Homework folder.

Use the default setting for UnderCut. The UnderCut option creates an undercut at the end of the threaded portion of the stud to provide clearance. The default values are based on the selections under Standard, but you can override and customize these values.

Click the Creates Stud on a Surface icon.

Enter 4.00in for Shaft Details.

Enter ANSI Inch for Standard. Enter Machine Threads for Type.

Enter ¼-20 for Size. Select End Condition: Blind.

Enter 2.00in for Depth. Check the Thread class box.

Select 2A for Thread Class.

Click the Position tab.

Click Top view.

Click a position to the right on the top surface.

Click the origin.

Click OK ✔ from the Stud Position PropertyManager. Display an Isometric view.

To view the threads, right-click the Annotations folder, and click Details.

Check the Cosmetic threads and Shaded cosmetic threads box.

Click OK from the Annotation Properties dialog box.

Exercise 2.19: HOLE BLOCK

Create the Hole-Block using the Hole Wizard feature as illustrated. Create the Hole-Block part on the Front Plane. The Hole Wizard feature creates either a 2D or 3D sketch for the placement of the hole in the FeatureManager. You can consecutively place multiple holes of the same type. The Hole Wizard creates 2D sketches for holes unless you select a non-planar face or click the 3D Sketch button in the Hole Position PropertyManager.

Hole Wizard creates two sketches, a sketch of the revolved cut profile of the selected hole type and a sketch of the center placement for the profile. Fully define the sketch for the center placement. Both sketches should be fully defined.

Create a rectangular prism 2 inches wide by 5 inches long by 2 inches high. On the top surface of the prism, place four holes, 1 inch apart.

- Hole #1: Hole Type: Fractional Drill Size, 7/16 diameter, End Condition: Blind, 0.75 inch deep.
- Hole #2: Counterbore hole Type: for 3/8 inch diameter Hex bolt, End Condition: Through All.
- Hole #3: Countersink hole Type: for 3/8 inch diameter Flat head screw (100), 1.5 inch deep.
- Hole #4: Straight Tapped hole Type, Size ¼-20, 1 inch Blind hole deep.

Exercise 2.20: 3D SKETCH/HOLE WIZARD

Create the part using the Hole Wizard feature. Apply the 3D sketch placement method as illustrated in the FeatureManager. Insert and dimension a hole on a cylindrical face.

Copy and open Hole Wizard 2.20 from the Chapter 2 Homework folder.

Note: With a 3D sketch, press the Tab key to move between planes.

Click the Hole Wizard Features tool. The Hole Specification PropertyManager is displayed.

Select the Counterbore Hole Type.

Select ANSI Inch for Standard.

Select Socket Head Cap Screw for fastener Type.

Select 1/4 for Size. Select Normal for Fit.

Select Through All for End Condition.

Enter .100 for Head clearance in the Options box.

Click the Positions Tab. The Hole Position PropertyManager is displayed.

Click the 3D Sketch button. SOLIDWORKS displays a 3D interface with the Point XY tool active.

When the Point tool is active, wherever you click you will create a point.

Dimension the sketch.

Click the cylindrical face of the model as illustrated. The selected face is displayed in blue. This indicates that an OnSurface sketch relation will be created between the sketch point and the cylindrical face. The hole is displayed in the model.

Insert a .25in dimension between the top face and the sketch point.

Locate the point angularly around the cylinder. Apply construction geometry.

Display the Temporary Axes.

Click the Line ✎ Sketch tool. Note: 3D sketch is still activated.

Ctrl+click the top flat face of the model. This moves the red space handle origin to the selected face. This also constrains any new sketch entities to the top flat face. Note the mouse pointer icon.

Move the mouse pointer near the center of the activated top flat face as illustrated. View the small black circle. The circle indicates that the end point of the line will pick up a Coincident relation.

Click the center point of the circle.

Sketch a line so it picks up the **AlongZ sketch relation**. **The cursor displays the relation to be applied**. **This is a very important step!** If needed insert an AlongZ relation.

Create an AlongY sketch relation between the center point of the hole on the cylindrical face and the endpoint of the sketched line as illustrated. The sketch is fully defined.

Add Relations	
↳ˣ	**Along X**
↳ʸ	Along Y
↳ᶻ	Along Z
⋏	Coincident
☒	Fix

Click OK ✔ from the Properties PropertyManager. Click OK ✔ from the Hole Position PropertyManager.

Expand the FeatureManager and view the results. The two sketches are fully defined. One sketch is the hole profile, the other sketch is to define the position of the feature.

Close the model.

💡 You can create a second sketched line and insert an angle dimension between the two lines. This process is used to control the position of the centerpoint of the hole on the cylindrical face as illustrated. Insert an AlongY sketch relation between the centerpoint of the hole on the cylindrical face and the end point of the second control line. Control the hole position with the angular dimension.

Exercise 2.21: Fill Pattern

Create a Polygon Layout Fill Pattern feature. Apply the seed cut option.

1. Open **Fill Pattern 2.21** from the Chapter 2 Homework folder.

2. Click the **Fill Pattern** 🕮 Features tool. The Fill Pattern PropertyManager is displayed.

3. Click the **Front face** of Extrude1. Face<1> is displayed in the Fill Boundary box. The direction arrow points to the right.

4. Click **Polygon** for Pattern Layout.

5. Click **Target spacing**.

6. Enter **15**mm for Loop Spacing.

7. Enter **6** for Polygon sides.

8. Enter **10**mm for Instances Spacing.

9. Enter **10**mm for Margins. View the direction arrow.

10. Click **Create seed cut**.

11. Click **Circle** for Features to Pattern.

12. Enter **4**mm for Diameter.

13. Click **OK** ✔ from the Fill Pattern PropertyManager. Fill Pattern1 is created and is displayed in the FeatureManager.

14. **View** the results.

Exercise 2.22: Fill Pattern

Create a Circular Layout Fill Pattern feature.

1. **Copy** the Chapter 2 Homework folder to your hard drive.

2. Open **Fill Pattern 2.22** from the Chapter 2 Homework folder.

3. Click the **Fill Pattern** Features tool. The Fill Pattern PropertyManager is displayed.

4. Click the **top face** of Boss-Extrude1. Face<1> is displayed in the Fill Boundary box. The direction arrow points to the back.

5. Click **Circular** for Pattern Layout.

6. Enter **.10**in for Loop Spacing.

7. Click **Target spacing**.

8. Enter .**10**in Instance Spacing.

9. Enter **.05**in for Margins. Edge<1> is selected for Pattern Direction.

10. Click **inside** the Features to Pattern box.

11. Click **Boss-Extrude2** from the fly-out FeatureManager. Boss-Extrude2 is displayed in the Features to Pattern box.

12. Click **OK** ✔ from the Fill Pattern PropertyManager. Fill Pattern1 is created and is displayed in the FeatureManager.

13. **View** the results.

Project 3

Fundamentals of Assembly Modeling

Below are the desired outcomes and usage competencies based on the completion of Project 3.

Project Desired Outcomes:	Usage Competencies:
<ul><li>Two assemblies:<ul><li>GUIDE-ROD assembly</li><li>CUSTOMER assembly</li></ul></li></ul>	<ul><li>Insert and edit components. Apply Standard mates and Mechanical mates using the Bottom-up assembly design approach. Utilize the standard and Quick mate procedure.</li></ul>
<ul><li>Use an assembly from SMC USA.<ul><li>GUIDE-CYLINDER</li></ul></li></ul>	<ul><li>Use and assemble components from the internet.</li></ul>
<ul><li>Flange bolt part from the SOLIDWORKS Design Library.</li></ul>	<ul><li>Assemble components from the SOLIDWORKS Design Library.</li></ul>
<ul><li>Two parts:<ul><li>4MMCAPSCREW</li><li>3MMCAPSCREW</li></ul></li></ul>	<ul><li>Apply the Revolved Base, Extruded Cut, Chamfer, Component Pattern features along with the Copy, Edi, and suppress tools.</li><li>Apply and edit material.</li></ul>

Notes:

Project 3 - Fundamentals of Assembly Modeling

Project Objective

Provide an understanding of the Bottom-up assembly design approach. Insert existing parts into an assembly. Orient and position the components in the assembly using Standard and Mechanical mates.

Create the GUIDE-ROD assembly. Utilize the ROD, GUIDE and PLATE parts. The ROD, GUIDE and PLATE parts were created in Project 2.

Create the CUSTOMER assembly. The CUSTOMER assembly consists of two sub-assemblies: GUIDE-ROD and GUIDE-CYLINDER.

Insert the Flange bolt from the SOLIDWORKS Design Library (Add-in) into the GUIDE-ROD assembly.

Utilize a Revolved Base feature to create the 4MMCAPSCREW.

Copy the 4MMCAPSCREW part and modify dimensions to create the 3MMCAPSCREW part.

On the completion of this project, you will be able to:

- Understand the Assembly FeatureManager Syntax.

- Insert parts/sub-components into an assembly.

- Insert and edit Standard and Mechanical mates.

- Rename parts and copy assemblies with internal references.

- Save an assembly with internal references.

- Incorporate design changes into an assembly.

- Insert a sub-component from the SOLIDWORKS Design Library.

- Modify, Edit and Suppress features in an assembly.

- Recover from a Mate error in an assembly.

- Suppress/Un-suppress component features.
- Create an Isometric Exploded view of an assembly.
- Create a Section view of an assembly.

Project Situation

The PLATE, ROD and GUIDE parts were created in Project 2. Perform the following steps:

Step 1: Insert the ROD, GUIDE and PLATE into a GUIDE-ROD assembly.

Step 2: Obtain the customer's GUIDE-CYLINDER assembly from the downloaded model folder. The assembly is obtained to guarantee proper fit between the GUIDE-ROD assembly and the customer's GUIDE-CYLINDER assembly.

Step 3: Create the CUSTOMER assembly. The CUSTOMER assembly combines the GUIDE-ROD assembly with the GUIDE-CYLINDER assembly.

Review the GUIDE-ROD assembly design constraints:

- The ROD requires the ability to travel through the GUIDE.

- The ROD keyway is parallel to the right surface of the GUIDE. The top surface of the GUIDE is parallel to the work area.

- The ROD mounts to the PLATE. The GUIDE mounts to a flat work surface.

The PISTON PLATE is the front plate of the GUIDE-CYLINDER assembly. The PLATE from the GUIDE-ROD assembly mounts to the PISTON PLATE. Create a rough sketch of the conceptual assembly.

Rough Sketch of Design Situation: CUSTOMER assembly

An assembly combines two or more parts. In an assembly, parts are referred to as components. Design constraints directly influence the assembly design process. Other considerations indirectly affect the assembly design, namely cost, manufacturability and serviceability.

Project Overview

Translate the rough conceptual sketch into a SOLIDWORKS assembly.

The GUIDE-ROD is the first assembly.

Determine the first component of the assembly.

The first component is the GUIDE. The GUIDE remains stationary. The GUIDE is a fixed component.

The action of assembling components in SOLIDWORKS is defined as mates.

Mates are relationships between components that simulate the construction of the assembly in a manufacturing environment. Mates affect/restrict the six degrees of freedom in an assembly.

The CUSTOMER assembly combines the GUIDE-CYLINDER assembly with the GUIDE-ROD assembly.

Assembly Modeling Approach

In SOLIDWORKS, components and their assemblies are directly related through a common file structure. Changes in the components directly affect the assembly and vice versa.

Create assemblies using the Bottom-up assembly approach, Top-down assembly approach or a combination of both methods. This project focuses on the Bottom-up assembly approach.

The Bottom-up approach is the traditional method that combines individual components. Based on design criteria, the components are developed independently. The three major steps in a Bottom-up assembly approach are Create each component independent of any other component in the assembly, insert the components into the assembly, and mate the components in the assembly as they relate to the physical constraints of your design.

In the Top-down assembly approach, major design requirements are translated into assemblies, sub-assemblies and components.

In the Top-down approach, you do not need all of the required component design details. Individual relationships are required.

Example: A computer. The inside of a computer can be divided into individual key sub-assemblies such as a motherboard, disk drive, power supply, etc. Relationships between these sub-assemblies must be maintained for proper fit.

Use the Bottom-up design approach for the GUIDE-ROD assembly and the CUSTOMER assembly.

Linear Motion and Rotational Motion

In dynamics, motion of an object is described in linear and rotational terms. Components possess linear motion along the x, y and z-axes and rotational motion around the x, y and z-axes.

In an assembly, each component has six degrees of freedom: three translational (linear) and three rotational.

Mates remove degrees of freedom. All components are rigid bodies. The components do not flex or deform.

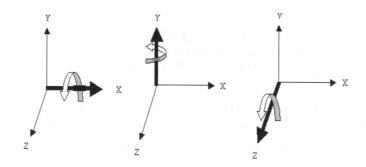

GUIDE-ROD assembly

The GUIDE-ROD assembly consists of six components.

- GUIDE
- ROD
- PLATE
- Flange bolt
- 4MMCAPSCREW
- 3MMCAPSCREW

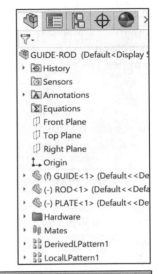

The first component is the GUIDE. The GUIDE is the fixed component in the GUIDE-ROD assembly.

The second component is the ROD. The ROD translates linearly through the GUIDE.

The third component is the PLATE. The PLATE is assembled to the ROD and the GUIDE-CYLINDER assembly.

The fourth component is the Flange bolt. Obtain the Flange bolt from the SOLIDWORKS Design Library.

The fifth component is the 4MM CAPSCREW. Create the 4MM CAPSCREW component with a Revolved Base feature.

A Revolved Base feature requires a centerline for an axis and a sketched profile. The profile is rotated about the axis to create the feature.

The sixth component is the 3MMCAPSCREW. Create the 3MMCAPSCREW from the 4MMCAPSCREW.

Reuse geometry to save time.

Insert a Component Pattern feature of the
3MMCAPSCREW part. A Component Pattern is created in
the assembly. Reference the GUIDE Linear Pattern of
Tapped Holes to locate the 3MMCAPSCREWs.

The book is designed to expose the SOLIDWORKS
user to numerous tools and procedures. It may not always
use the simplest and most direct process.

Activity: Create the GUIDE flange bolt Assembly

Close all SOLIDWORKS documents.

1) Click **Windows**, **Close All** from the Menu bar before you
begin this project.

Open the GUIDE part.

2) Click **Open** from the Menu bar.

3) Select the **ENGDESIGN-W-
SOLIDWORKS\PROJECTS** folder.

4) Double-click **GUIDE**. The GUIDE
FeatureManager is displayed.

Create the GUIDE-ROD assembly.

5) Click **New** ⬜ from the Menu bar. The New
SOLIDWORKS Documents dialog box is
displayed.

6) Double-click **Assembly** from the default
Templates tab. The Begin Assembly
PropertyManager is displayed. **Note**: The
MY-TEMPLATES tab was created in
Project 2.

The Begin Assembly PropertyManager is displayed
if the Start command when creating new assembly option
is selected.

When a part is inserted into an assembly, it is called a component. The Begin Assembly PropertyManager is displayed to the left of the Graphics window. GUIDE is listed in the Part/Assembly to Insert box.

Insert the GUIDE and fix it to the assembly Origin.

7) Click **GUIDE** in the Open documents box.

8) Click **OK** ✔ from the Begin Assembly PropertyManager. The GUIDE part icon is displayed in the assembly FeatureManager. It is fixed to the Assembly origin.

The GUIDE name is added to the assembly FeatureManager with the symbol (f).

The symbol (f) represents a fixed component. A fixed component cannot move and is locked to the assembly Origin. The first component in an assembly should be fixed, fully defined, or mated to an axis.

To fix the first component to the Origin in an assembly, click OK ✔ from the Begin Assembly PropertyManager or click the Origin in the Graphics window.

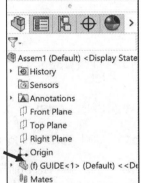

To remove the fixed state (f), right-click the **fixed component name** in the FeatureManager. Click **Float**. The component is free to move.

Your default system document templates may be different if you are a new user of SOLIDWORKS vs. an existing user who has upgraded from a previous version.

CommandManager and FeatureManager tabs and tree folders will vary depending on system set-up and Add-ins.

Save and name the assembly document.

9) Click **Save As** 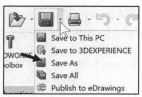.

10) Select the **ENGDESIGN-W-SOLIDWORKS\PROJECTS** folder.

11) Enter **GUIDE-ROD** for File name.

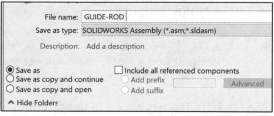

12) Click **Save**.

Close the GUIDE part.

13) Click **Window**, **GUIDE** from the Menu bar.

14) Click **File**, **Close** from the Menu bar. The GUIDE-ROD assembly is the open document.

Select the Ctrl-Tab keys to quickly alternate between open SOLIDWORKS documents. Select inside Close to close the current SOLIDWORKS document. The outside Close exits SOLIDWORKS.

Set Document Properties. Set drafting standard and units.

15) Click **Options** ⚙, **Document Properties** tab.

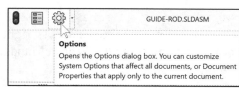

16) Select **ANSI** for Drafting Standard.

17) Click the **Units** folder.

18) Click **MMGS** (millimeter, gram, second) for Unit system. Accept the default settings.

19) Click **OK** from the Document Properties - Units dialog box.

Save the GUIDE-ROD assembly.

20) Click **Save**.

To customize the CommandManager, right-click on an existing tab. Click Customize. Check the option to display in the CommandManager. The options you select are based on the tool types you require for the design.

GUIDE-ROD Assembly-Insert Component

The first component is the foundation of the assembly. The GUIDE is the first component in the GUIDE-ROD assembly. The ROD is the second component in the GUIDE-ROD assembly. Add components to assemblies utilizing the following techniques:

- Utilize the Insert Components Assembly tool.

- Utilize Insert, Component from the Menu bar.

- Drag a component from the SOLIDWORKS Design Library into the Assembly.

- Drag a component from an Open part file into the Assembly.

Activity: GUIDE-ROD Assembly - Insert Component.

Insert the ROD component into the assembly. Use the Insert Components tool.

21) Click the **Assembly** tab from the CommandManager.

22) Click the **Insert Components** Assembly tool. **Note**: Open SOLIDWORKS models are displayed in the Open documents box. If there are no Open models, the Window Open dialog is displayed.

23) Browse to the **ENGDESIGN-W-SOLIDWORKS\PROJECTS** folder.

24) Double-click the **ROD** part. The ROD is displayed in the Graphics window.

The ROD part icon is displayed in the Open documents text box. The mouse pointer displays the ROD component when positioned inside the GUIDE-ROD Graphics window.

25) Click a **position** to the left of the GUIDE as illustrated.

Deactivate the Origin display in the assembly if needed.
26) Click **View**, **Hide/Show**, and uncheck **Origins** from the Main menu.

Re-position the ROD component in the Graphics window.
27) Click and drag the **ROD** in the Graphics window.

28) Move the **ROD** on the right side of the GUIDE as illustrated.

Fit the model to the Graphics window.
29) Press the **f** key.

The component movement in the assembly is determined by its degrees of freedom.

🔆 Degree of freedom is geometry that is not defined by dimensions or relations and is free to move. In a 2D sketch, there are three degrees of freedom: movement along the X axis, Y axis and rotation about the Z axis (the axis normal to the Sketch plane).

Save the GUIDE-ROD assembly.

30) Click **Save** 💾. The ROD component is displayed in the GUIDE-ROD FeatureManager. The ROD part is free to move and rotate.

Review the FeatureManager Syntax.

31) Expand (f) GUIDE<1> in the FeatureManager. The features are displayed in the FeatureManager. **Note**: Base Extrude, Slot Cut, Guide Hole and M3x0.5 Tapped Hole1 contain additional sketches. Features were renamed in Project 2. Example: Boss-Extrude1 was renamed to Base Extrude.

32) Expand (-) ROD<1> in the FeatureManager. The features are displayed in the FeatureManager.

33) Click the **Drop-down arrow** ‣ icon to the left of the GUIDE entry and ROD entry to collapse the list.

An Arrow ‣ icon indicates that additional feature information is available. The Drop-down arrow ‣ icon indicates that the feature list is fully expanded. Manipulating the FeatureManager is an integral part of the assembly. In the step-by-step instructions, expand and collapse are used as follows:

 Expand - Click the Arrow ‣ icon.

 Collapse - Click the Drop-down arrow ‣ icon.

FeatureManager Syntax

How do you distinguish the difference between an assembly and a part in the FeatureManager? Answer: The assembly icon 🔲 contains a blue square block and an upside down yellow "T" extrusion. The part icon 🔲 contains an upside down yellow "T" extrusion. Entries in the FeatureManager design tree have specific definitions. Understanding syntax and states saves time when creating and modifying assemblies. Review the columns below.

First Column: ‣ 🔲 (f) GUIDE<1> (Default<<Default>. A Resolved component (not in lightweight 🔲 state) displays an arrow ‣ icon. The arrow icon indicates that additional feature information is available. A drop-down arrow ‣ icon displays the fully expanded feature list.

Second Column: ▸ 🔧 (f) GUIDE<1> (Default<<Default>. Identifies a component's (part or assembly) relationship with other components in the assembly.

Component or Part States:	
	State:
🔧	**Resolved part**. A part icon indicates a resolved state. A blue part icon indicates a selected, resolved part. The component is fully loaded into memory and all of its features and mates are editable.
🔧	**Lightweight part**. A blue feather on the part icon indicates a lightweight state. When a component is lightweight, only a subset of its model data is loaded in memory.
🔧	**Out-of-Date part**. An eye is displayed on the part icon. The part needs a rebuild in the assembly.
🔧	**Flexible state**. A blue box on the part icon indicates a flexible part in an assembly.
🔧	**Hidden**. A transparent icon indicates that the component is active but invisible.
🔧	**Hidden Lightweight**. A clear feather over a clear part icon indicates the part is hidden and lightweight.
🔧	**Smart Component**. A lightning bolt on the part icon indicates that the component is a Smart Component.
🔧	**Smart Fastener**. A lightning bolt is displayed on the Fastener folder icon.
🔧	**Resolved assembly**. Resolved (or unsuppressed) is the normal state for assembly components. A resolved assembly is fully loaded in memory, fully functional, and fully accessible.
🔧	**Inserted New Component**. A new inserted component into an active assembly.

Third Column: ▸ 🧊 (f) GUIDE<1> (Default<<Default>). The part is fixed (f). You can fix the position of a component so that it cannot move with respect to the assembly Origin. By default, the first part in an assembly is fixed; however, you can float it at any time.

It is recommended that at least one assembly component is either fixed or mated to the assembly planes or Origin. This provides a frame of reference for all other mates and helps prevent unexpected movement of components when mates are added. The Component Properties are:

Component Properties in an assembly:	
Symbol:	**Relationship:**
(-)	A minus sign (–) indicates that the part or assembly is under-defined and requires additional information.
(+)	A plus sign (+) indicates that the part or assembly is over-defined.
None	The Base component is mated to three assembly reference planes. It is fully defined.
(f)	A fixed symbol (f) indicates that the part or assembly does not move.
(?)	A question mark (?) indicates that additional information is required on the part or assembly.

Fourth Column: ▸ 🧊 (f) GUIDE<1> (Default<<Default>). Name of the part.

Fifth Column: ▸ 🧊 (f) GUIDE<1> (Default<<Default>). The symbol <#> indicates the particular inserted instance of a component. The symbol <1> indicates the first inserted instance of a component in the assembly. If you delete a component and reinsert the same component again, the <#> symbol increments by one.

Fifth Column: ▸ 🧊 MOTHERBOARD<1> -> (Default The Resolved state displays the icon with an External reference symbol, "- >". The state of External references is displayed as follows:

- If a part or feature has an external reference, its name is followed by - >. The name of any feature with external references is also followed by - >.

- If an external reference is currently out of context, the feature name and the part name are followed by - >?

- The suffix - >* means that the reference is locked.

- The suffix - >x means that the reference is broken.

There are modeling situations in which unresolved components create rebuild errors. In these situations, issue the forced rebuild, Ctrl+B. The Ctrl+B option rebuilds the model and its features. If the mates still contain rebuild errors, resolve all the components below the entry in the FeatureManager that contains the first error.

Mate Types

Mates provide the ability to create geometric relationships between assembly components. Mates define the allowable directions of rotational or linear motion of the components in the assembly. Move a component within its degrees of freedom in the Graphics window to view the behavior of an assembly.

Mates are solved together as a system. The order in which you add mates does not matter. All mates are solved at the same time. You can suppress mates just as you can suppress features.

The Standard mates tab is the default tab. The mate tab provides the ability to insert a *Standard*, *Advanced*, *Mechanical* mate or to address *Analysis*. The *Analysis* tab provides the ability to assign mate properties for use in SOLIDWORKS Motion and SOLIDWORKS Simulation.

Standard Mates:

Components are assembled with various mate types. Standard Mate types:

- **Coincident**: Locates the selected faces, edges, or planes so they use the same infinite line. A Coincident mate positions two vertices for contact.

- **Parallel**: Locates the selected items to lie in the same direction and to remain a constant distance apart.

- **Perpendicular**: Locates the selected items at a 90° angle to each other.

- **Tangent**: Locates the selected items in a tangent mate. At least one selected item must be a conical, cylindrical or spherical face.

- **Concentric**: Locate the selected items so they can share the same center point. You can prevent the rotation of components that are mated with concentric mates by selecting the Lock Rotation option. Locked concentric mates are indicated by an icon in the FeatureManager design tree.

- **Lock**: Maintains the position and orientation between two components.

- **Distance**: Locates the selected items with a specified distance between them. Use the drop-down arrow box or enter the distance value directly.

- **Angle**: Locates the selected items at the specified angle to each other. Use the drop-down arrow box or enter the angle value directly.

- **Mate alignment**: Toggle the mate alignment as necessary. **Aligned**. Locates the components so the normal or axis vectors for the selected faces point in the same direction. **Anti-Aligned**. Locates the components so the normal or axis vectors for the selected faces point in the opposite direction.

Advanced Mates:

Advanced Mate types:

- **Profile Center**: Mate to Center automatically center-aligns common component types such as rectangular and circular profiles to each other and fully defines the components.

- **Symmetric**: Positions two selected entities to be symmetric about a plane or planar face. A Symmetric Mate does not create a Mirrored Component.

- **Width**: Centers a tab within the width of a groove. You can select geometry to drive the limits of width mates, eliminating the need for numerical input.

- **Path**: Constrains a selected point on a component to a path.

- **Linear/LinearCoupler**: Establishes a relationship between the translation of one component and the translation of another component.

- **Distance (Limit)**: Locates the selected items with a specified distance between them. Use the drop-down arrow box or enter the distance value directly.

- **Angle**: Locates the selected items at the specified angle to each other. Use the drop-down arrow box or enter the angle value directly.

Mechanical Mates:

Mechanical Mate types:

- **Cam**: Forces a plane, cylinder, or point to be tangent or coincident to a series of tangent extruded faces.

- **Slot**: Mate bolts to straight or arced slots and you can mate slots to slots. Select an axis, cylindrical face, or a slot to create slot mates.

- **Hinge**: Limits the movement between two components to one rotational degree of freedom. It has the same effect as adding a Concentric mate plus a Coincident mate.

- **Gear**: Forces two components to rotate relative to one another around selected axes.

- **Rack Pinion**: Provides the ability to have Linear translation of a part; rack causes circular rotation in another part, pinion, and vice versa.

- **Screw**: Constrains two components to be concentric and adds a pitch relationship between the rotation of one component and the translation of the other.

- **Universal Joint**: The rotation of one component (the output shaft) about its axis is driven by the rotation of another component (the input shaft) about its axis.

Mates reflect the physical behavior of a component in an assembly.

Quick Mate:

Quick Mate is a procedure to mate components together. No command (click Mate from the Assembly CommandManager) is required. Hold the Ctrl key down, make the selections. Release the Ctrl key, and a Quick Mate pop-up menu is displayed below the context toolbar. Select the mate and you are finished.

GUIDE-ROD Assembly - Mate the ROD Component

Recall the initial assembly design constraints:

- The ROD requires the ability to travel through the GUIDE.

- The face of the Keyway in the ROD is parallel to the right face of the GUIDE.

Utilize Concentric and Parallel mates between the ROD and GUIDE. The Concentric mate utilizes the cylindrical face of the shaft with the cylindrical face of the GUIDE Hole. The Parallel mate utilizes two planar faces from the ROD Keyway Cut and the right face of the GUIDE.

Concentric Mate - 2 Cylindrical   Parallel - 2 Planar faces

The Concentric and Parallel mate provides the ability for the ROD to translate linearly through the GUIDE Hole. The ROD does not rotate.

You can use the following steps to create a Mate using the Mate tool:

- Click the Mate ✑ icon from the Assembly toolbar.

- Select the Mate tab.

- Select geometry from the first component (usually the part or plane).

- Select the geometry from the second component (usually the assembly or plane).

- Select the Mate type.

- Click OK ✔ to create the Mate PropertyManager. View the created mate.

Activity: GUIDE-ROD Assembly - Mate the ROD Component.

Insert a Concentric mate between two faces. Use the Quick mate procedure.

34) Click the **inside cylindrical face** of the Guide Hole as illustrated.

35) Hold the **Ctrl** key down.

36) Click the **cylindrical face** of the ROD.

37) Release the **Ctrl** key. The Mate Pop-up menu is displayed.

38) Click **Concentric** ◎ from The Mate Pop-up menu. A Concentric mate locates the selected items so they can share the same center point. Note your options (Lock rotation, Flip Mate Alignment). Do not select at this time.

The ROD is Concentric with the GUIDE. The ROD has the ability to move and rotate while remaining concentric to the GUIDE hole.

Move and rotate the ROD in the Graphics window.

39) Click and drag the **ROD** in a horizontal direction. The ROD travels linearly in the GUIDE.

40) Click and drag the **ROD** in a vertical direction. The ROD rotates in the GUIDE.

41) Rotate the **Rod** until the Keyway cut is approximately parallel to the right face of the GUIDE. Display origins in the Assembly.

Recall the second assembly design constraint. The flat end of the ROD must remain parallel to the right surface of the GUIDE.

Insert a Parallel mate between two faces.

42) Click the **Keyway face** of the ROD.

43) Hold the **Ctrl** key down.

44) Click the **flat right face** of the GUIDE.

45) Release the **Ctrl** key. The Mate Pop-up menu is displayed.

46) Click **Parallel** ⟍ from the Mate Pop-up menu.

Move the ROD in the Graphics window.

47) Click and drag the **ROD** in a horizontal direction and position it approximately in the center of the GUIDE.

Hide the GUIDE component in the Graphics window.

48) Right-click on the front face of the **GUIDE** in the Graphics window.

49) Click **Hide Components** from the Context toolbar. The GUIDE component is not displayed in the Graphics window. Note the change in color in the FeatureManager.

Display the Mate types.

50) **Expand** the Mates folder from the FeatureManager.

Display the full Mate names.

51) Drag the **vertical FeatureManager border** to the right. View the two created mates: Concentric1 and Parallel1.

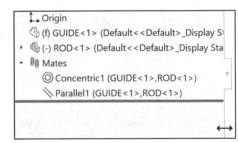

Save the GUIDE-ROD assembly.

52) Click **Save** 🖫.

🔆 Use the Alt key to temporarily hide a face when you need to select an obscured face for mates.

The ROD mates reflect the physical constraints in the GUIDE-ROD assembly. The ROD is under defined; indicated by a minus sign (-).

The Concentric mate allows the ROD to translate freely in the Z direction through the Guide Hole. The Parallel Mate prevents the ROD from rotating in the Guide Hole.

The GUIDE part icon is displayed with no color in the FeatureManager to reflect the Hide state.

GUIDE-ROD Assembly - Mate the PLATE Component

Recall the initial design constraints.

- The ROD is fastened to the PLATE.

- The PLATE part mounts to the GUIDE-CYLINDER PISTON PLATE part.

Apply the Rotate Component ![tool icon] tool to position the PLATE before applying the Mates.

Activity: GUIDE-ROD Assembly - Mate the PLATE Component

Insert the PLATE component into the assembly from the PROJECTS folder.

53) Click the **Insert Components** ![icon] Assembly tool. **Note**: Open SOLIDWORKS models are displayed in the Open documents box. If there are no open models, the Window Open dialog is also displayed.

54) Double-click the **PLATE** part from the PROJECTS folder. **Note**: PLATE was created in Project 2.

55) Click a **position** behind the ROD in the Graphics window as illustrated. The PLATE component is added to the GUIDE-ROD FeatureManager.

Display an Isometric view.
56) Click **Isometric view** ![icon].

Fit the model to the Graphics window.
57) Press the **f** key.

58) **View** the results in the Graphics window.

Rotate the PLATE.

59) Click the **Rotate Component** Assembly tool. The Rotate Component PropertyManager is displayed. View your options.

60) Click and drag the **front face** of the PLATE downward until the PLATE rotates approximately 90°.

61) Click **OK** ✔ from the PropertyManager to deactivate the tool.

62) Click a **position** in the Graphics window, to the right of the PLATE, to deselect any faces or edges.

If needed, deactivate the Origins.

63) Click **View**, **Hide/Show**, uncheck **Origins** from the Menu bar.

Use Selection filters to select difficult individual features such as *faces*, *edges* and *points*. Utilize the Filter Faces tool to select the hidden ROD Back Hole face from the Selection Filter toolbar.

Display the Selection Filter toolbar. Activate the Filter Faces tool.

64) **Right-click** in the Graphics window.

65) Click the **Selection Filters** icon. The Selection Filter toolbar is displayed.

66) Click **Filter Faces** from the Selection Filter drop-down menu as illustrated. The Selection Filter icon is displayed in your mouse pointer.

To deactivate the Filter Faces tool, click Clear All Filters from the Selection Filter toolbar.

💡 The book is designed to expose the SOLIDWORKS user to numerous tools and procedures. It may not always use the simplest and most direct process.

Insert a Concentric mate between two faces.

67) Click **WireFrame** ⬡ from the Heads-up View toolbar.

68) Click the **center inside cylindrical face** from the PLATE Countersink hole.

69) Hold the **Ctrl** key down.

70) Click the **cylindrical face** of the ROD.

71) Release the **Ctrl** key. The Mate Pop-up menu is displayed.

72) Click **Concentric** from the Mate Pop-up menu. Concentric2 is created.

73) Click and drag the **PLATE** behind the ROD.

Insert a Coincident mate between two faces.

74) Use the **middle mouse** button to rotate the model to view the **back face** of the ROD.

75) Click the **back circular face** of the ROD.

76) **Rotate** the model to view the front rectangular face of the PLATE.

77) Hold the **Ctrl** key down.

78) Click the **front rectangular face** of the PLATE.

79) Release the **Ctrl** key. The Mate Pop-up menu is displayed.

80) Click **Coincident** from the Mate Pop-up menu.

Display an Isometric view - Shaded with Edges.

81) Click **Isometric view** 🔲.

82) Click **Shaded With Edges** 🔲 from the Heads-up View toolbar. View the results.

Insert a Parallel mate between two faces.

83) Press the **Shift + z** keys to zoom in on the ROD.

84) Click the ROD **Keyway Cut** flat face.

85) Hold the **Ctrl** key down.

86) Click the PLATE right **rectangular face** as illustrated.

87) Release the **Ctrl** key.

88) Click **Parallel** from the Mate Pop-up menu. Parallel2 is created.

Clear all filters in SOLIDWORKS.
89) **Right-click** in the Graphics window.

90) **Expand** the Selection Filters drop-down menu.

91) Click the **Clear All Filters** icon.

Save the GUIDE-ROD assembly.
92) Click **Save**. View the results.

Mates reflect the physical relations between the PLATE and the ROD. The Concentric mate aligns the PLATE Countersink Hole and the ROD cylindrical face. The Coincident mate eliminates translation between the PLATE front face and the ROD back face.

The Parallel mate removes PLATE rotation about the ROD axis. Create the Parallel mate.

A Distance mate of 0 provides additional flexibility over a Coincident mate. A Distance mate value can be modified. Utilize a Coincident mate when mating faces remain coplanar.

Use the Pack and Go tool to save and gather all related files for a model design (parts, assemblies, drawings, references, design tables, design binder content, decals, appearances, scenes, and SOLIDWORKS Simulation results) into a single folder or zip file. It is one of the best tools to utilize when you are trying to save a large assembly or drawing with references and SOLIDWORKS Toolbox components.

The mouse pointer displays the Filter icon when the Selection Filter is activated. Deactivate Selection Filters when not required.

Activate/Deactivate Filters using the following keys:

Filter for edges	Press e
Filter for faces	Press x
Filter for vertices	Press v
Hide/Show all Filters	F5
Off/On all Selected Filters	F6

Accidentally pressing the e, x or v keys can activate a Filter. If the mouse pointer displays the Filter icon, you cannot select geometry, dimensions or text. Press the F5 key to display the Selection Filter toolbar. Select Clear All Filters.

GUIDE-ROD Assembly - Mate Errors

Mate errors occur when component geometry is over defined. Example: You added a new Concentric Mate between the PLATE bottom Mounting Hole and the ROD cylindrical face.

The ROD back hole cannot physically exist with a Concentric Mate to both the PLATE middle CSK Hole and bottom Mounting Hole.

- Review the design intent. Know the behavior of the components in the assembly.

- Review the messages and symbols in the FeatureManager.

- Utilize Delete, Edit Feature and Undo commands to recover from Mate errors.

To view the mates for more than one component, hold the Ctrl key down, select the components, then right-click, and click the View Mates icon.

A Mate problem displays the following icons:

⚠ **Warning**. The mate is satisfied but is involved in over defining the assembly.

⊗ **Error**. The mate is not satisfied.

Insert the second Concentric mate which will create a Mate error in the following steps.

Activity: GUIDE-ROD Assembly - Mate Errors.

Insert a Concentric mate between two faces.

93) Click the **outside cylindrical face** of the ROD.

94) Hold the **Ctrl** key down.

95) Click the **bottom Mounting Hole inside cylindrical face** of the PLATE.

96) Release the **Ctrl** key.

97) Click **Concentric** from the Mate pop-up menu.

98) Click **OK** from the Mate pop-up menu. A Mate error message is displayed. The components cannot be moved to a position that satisfies this mate.

99) Click **Cancel** from the dialog box.

Review the created Mates.
100) Expand the Mates folder in the FeatureManager.

101) View the created mates and their icons.

If you delete a Mate and then recreate it, the Mate number will be different. View the mate symbols in the Mates folder.

Always save the SOLIDWORKS document (assembly/drawing) and all of its reference files in the same folder location.

Review the mates for a component.
102) Right-click **ROD** from the Graphics window.

103) Click **View Mates** from the Context toolbar. The Mates dialog box is displayed. **Note**: The GUIDE is hidden in the FeatureManager.

Components involved in the mate system for the selected components are vaguely transparent in the Graphics window. Components not involved are hidden.

104) **Close** the pop-up mate dialog box to return to the GUIDE-ROD FeatureManager.

Organize the Mates names. Rename Mate names with descriptive names for clarity.

You can set an option in the Mate PropertyManager so that the first component you select from becomes transparent. Then selecting from the second component is easier, especially if the second component is behind the first. The option is supported for all mate types except those that might have more than one selection from the first component (width, symmetry, linear coupler, cam, and hinge).

GUIDE-ROD Assembly - Collision Detection

The Collision Detection assembly function detects collisions between components as they move or rotate. A collision occurs when geometry on one component coincides with geometry on another component. Place components in a non-colliding position, then test for collisions.

Activity: GUIDE-ROD Assembly - Collision Detection.

Display the GUIDE and apply the Collision Detection tool.

105) Right-click **GUIDE** from the FeatureManager.

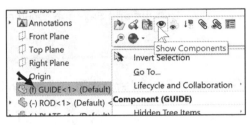

106) Click **Show Components** 👁 from the Context toolbar. The GUIDE is displayed in the Graphics window.

107) Click **Shaded with Edges** 🔲 from the Heads-up View toolbar.

Move the PLATE behind the GUIDE.

108) Click the **Move Component** 🔲 Assembly tool. The Move Component PropertyManager is displayed.

109) Drag the **PLATE** backward until the PLATE clears the GUIDE. The PLATE is free to translate along the Z-axis.

Apply the Collision Detection tool.
110) Click the **Collision Detection** box.

111) Click the **These components** box.

112) Click the **front face** of the PLATE. PLATE-1 is displayed in the Components for collision box.

113) Click the **back face** of the GUIDE. GUIDE-1 is displayed in the Components for collision box.

114) Click the **Stop at collision** box.

115) Check **Highlight faces**, **Sound** and **Ignore complex surfaces** from the Advanced Options box.

116) Click the **Resume Drag** box.

117) Drag the **PLATE** forward. View the results.

Return the PLATE to the original position.
118) Drag the **PLATE** backward until the ROD is approximately halfway through the GUIDE.

119) Click **OK** ✔ from the Move Component PropertyManager.

Save the GUIDE-ROD assembly.
120) Click **Save** 🖫.

121) Click **Rebuild and Save** the document.

GUIDE-ROD Assembly - Modify Component Dimension

Modify part dimensions in the assembly. Utilize Rebuild to update the part and the assembly. You realize from additional documentation that the Slot in the GUIDE is 4mm. Modify the right Slot Cut feature dimensions in the GUIDE-ROD assembly. Rebuild the assembly. The left Mirror Slot Cut and right Slot Cut update with the new value.

Activity: GUIDE-ROD Assembly - Modify Component Dimension

Modify the Slot of the Guide.
122) Double-click on the right **Slot Cut** of the GUIDE in the Graphics window. The Slot Cut dimensions are displayed in the Graphics window.

Modify the radial dimension.
123) Double-click **R3** in the Graphics window.

124) Enter **4**mm.

125) Click **Rebuild** 🔋 from the Modify dialog box.

126) Click the **Green Check mark** ✔ from the Modify dialog box.

Save the GUIDE-ROD assembly.
127) Click **Save** 🖫.

128) Click **Save All**.

SOLIDWORKS Design Library

The SOLIDWORKS Design Library 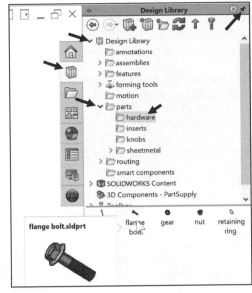 tab in the Task Pane provides a centralized location for reusable elements such as parts, assemblies and sketches. It does not recognize non-reusable elements such as SOLIDWORKS drawings, text files or other non-SOLIDWORKS files.

The Toolbox in the Design Library is an Add-in. 3D Components - PartSupply requires web access and an account.

Your company issued a design policy. The policy states that you are required to only use parts that are presently in the company's parts library. The policy is designed to lower inventory cost, purchasing cost and design time.

In this project, the SOLIDWORKS Design Library parts simulate your company's part library. Utilize a hex flange bolt located in the Hardware folder. Specify a new folder location in the Design Library to locate the components utilized in this project.

The Design Library saves time locating and utilizing components in an assembly. **Note**: In some network installations, depending on your access rights, additions to the Design Library are only valid in new folders.

Activity: GUIDE-ROD Assembly - SOLIDWORKS Design Library.

Open and obtain components from the SOLIDWORKS Design Library.

129) Click the **Design Library** tab from the Task Pane as illustrated. The Design Library menu is displayed in the Graphics window.

Pin the Design Library (Auto Show) to remain open.
130) Click **Pin**.

131) Expand Design Library.

Select the Flange bolt from the parts folder.
132) Expand the parts folder.

133) Click the **hardware** folder. The hardware components are displayed. The flange bolt icon represents a family of similar shaped components in various configurations.

Add the first flange bolt to the assembly.
134) Click and drag the **flange bolt** icon to the right of the GUIDE-ROD assembly in the Graphics window.

135) Release the **mouse button**.

136) Select the 8mm flange bolt, **M8-1.25 x 30**, from the drop-down menu.

137) Click **OK** from the Select a Configuration dialog box.

Insert the second flange bolt.
138) Click a **position** to the left of the GUIDE-ROD assembly.

139) Click **Cancel** ✖ from the Insert Component PropertyManager. Two flange bolts are displayed in the Graphics window.

Rotate the flange bolts.
140) Click the **shaft** of the first flange bolt.

141) Click the **Rotate Component** 🔄 Assembly tool. The Rotate Component PropertyManager is displayed.

142) Click and drag the **shaft** of the first flange bolt in a vertical direction as illustrated. The flange bolt rotates.

143) Click and drag the **shaft** of the second flange bolt in a vertical direction.

144) Click **OK** ✔ from the Rotate Component PropertyManager.

Un-pin the Design Library.

145) Click **Pin** .

Display an Isometric view. Save the GUIDE-ROD assembly.

146) Click **Isometric view** from the Heads-up View toolbar.

147) Click **Save** . **Note** if needed, use the Save As command to break all references and save the toolbox flange bolt to your active folder.

148) Click **Yes** to rebuild. The two flange bolts are displayed in the FeatureManager.

✎	flange bolt
◀	GUIDE
◀	GUIDE-ROD
◥	PLATE
✎	ROD

GUIDE-ROD Assembly - Insert Mates for Flange Bolts.

Insert a Center Slot mate for the first flange bolt. Select an axis, cylindrical face or slot to create a slot mate.

149) Click the **cylindrical face** of the first flange bolt as illustrated.

150) Hold the Ctrl key down.

151) Click the **right inside flat face** of the slot as illustrated.

152) Release the Ctrl key. The Mate pop-up menu is displayed.

Thread Cut of flange bolt<1>

Slot Cut of GUIDE<1>

153) Click **Slot** 🔗 mate from the Mate pop-up menu.

154) Click the **Slot position** icon.

155) Select **Center in Slot** from the drop-down menu. This option centers the component in the slot.

156) Click **OK** ✔ from the Mate pop-up menu. Slot1 is created.

Create a Coincident mate between the first flange bolt head and the right slot cut top face.

157) Click the **right slot cut top face** as illustrated.

158) **Rotate** the model to view the bottom face of the first flange bolt as illustrated.

159) **Hold** the Ctrl key down.

160) Click the **bottom face** of the first flange bolt as illustrated.

161) **Release** the Ctrl key. The mate pop-up menu is displayed.

162) Click **Coincident** from the Mate pop-up menu.

163) **Rotate** the model and view the results in the Graphics window.

View the created Mates.

164) **Expand** the Mates folder in the Assembly PropertyManager. View the created mates.

Display the Mates in the FeatureManager to check that the components and the Mate types correspond to your design intent.

Display an Isometric view.

165) Press the **space bar** to display the Orientation dialog box.

166) Click **Isometric view** . The first flange bolt is free to rotate about its centerline in the Slot Cut.

Insert a Parallel mate between two faces.

167) Click the **front face** of the hex head.

168) Hold the **Ctrl** key down.

169) Click the **front face** of the GUIDE.

170) Release the **Ctrl** key. The Mate pop-up menu is displayed.

171) Click **Parallel** from the Mate pop-up menu.

If you delete a Mate and then recreate it, the Mate numbers will be different (increase).

Insert a Center Slot mate for the second flange bolt. Use the Quick Mate procedure.

172) Click the **cylindrical face** of the second Flange bolt as illustrated.

173) Hold the **Ctrl** key down.

174) Click the **inside face** of the second slot.

175) Release the **Ctrl** key. The Mate pop-up menu is displayed.

176) Click **Slot** 🔗 from the Mate pop-up menu.

177) Click the **Slot position** 🔗 icon.

178) Select **Center in Slot** from the Slot Position drop-down menu. This option centers the component in the slot.

179) Click **OK** ✔ from the Mate pop-up menu. Slot2 is created.

Create a Coincident mate between the second Flange bolt head and the right slot cut top face.

180) Click the **left slot cut top face** as illustrated.

181) **Rotate** the model to view the bottom face of the second flange bolt as illustrated.

182) Hold the **Ctrl** key down.

183) Click the **bottom face** of the second flange bolt as illustrated. The Mate pop-up menu is displayed.

184) Click **Coincident** for mate type.

185) Rotate the model and view the results in the Graphics window.

Display an Isometric view.
186) Press the **space bar** to display the Orientation dialog box.

187) Click **Dimetric view**. The second flange bolt is free to rotate about its centerline in the Slot Cut.

Insert a Parallel mate between two faces.
188) Click the **front face** of the hex head as illustrated.

189) Hold the **Ctrl** key down.

190) Click the **front face** of the GUIDE as illustrated.

191) Release the **Ctrl** key. The Mate pop-up menu is displayed.

192) Click **Parallel** ⟍ from the Mate Pop-up menu.

View the created Mates.
193) Expand the Mates folder in the Assembly PropertyManager. View the created mates.

Later, delete the second flange bolt and apply a Linear Component Pattern tool to create the second flange bolt from the first flange bolt.

Save the GUIDE-ROD assembly.

194) Click **Save** .

▸ 🎨 (-) ROD<1> (Default) <<Default>_Display
▸ 🎨 (-) PLATE<1> (Default) <<Default>_Displ
▸ 🎨 flange bolt<1> (M8-1.25 x 30) <<M8-1.2
▸ 🎨 flange bolt<2> (M8-1.25 x 30) <<M8-1.2
▾ 🕕 Mates
   ◎ Concentric1 (GUIDE<1>,ROD<1>)
   ⟍ Parallel1 (GUIDE<1>,ROD<1>)
   ◎ Concentric2 (ROD<1>,PLATE<1>)
   ⋏ Coincident1 (ROD<1>,PLATE<1>)
   ⟍ Parallel2 (ROD<1>,PLATE<1>)
   ⬯ Slot1 (flange bolt<1>,GUIDE<1>)
   ⋏ Coincident2 (GUIDE<1>,flange bolt<1
   ⟍ Parallel3 (GUIDE<1>,flange bolt<1>)
   ⬯ Slot2 (flange bolt<2>,GUIDE<1>)
   ⋏ Coincident3 (GUIDE<1>,flange bolt<2
   ⟍ Parallel4 (GUIDE<1>,flange bolt<2>)

Copy components directly in the Graphics window. Hold the Ctrl key down. Click and drag the component from the FeatureManager directly into the Graphics window to create a new instance (copy). Release the Ctrl key. Release the mouse button.

Socket Head Cap Screw Part

The PLATE mounts to the PISTON PLATE of the GUIDE CYLINDER assembly with two M4x0.7 Socket Head Cap Screws. Create a simplified version of the 4MMCAPSCREW based on the ANSI B 18.3.1M-1986 standard.

How do you determine the overall length of the 4MMCAPSCREW? Answer: The depth of the PLATE plus the required blind depth of the PISTON PLATE provided by the manufacturer.

When using fasteners to connect two plates, a design rule of thumb is to use a minimum of 75% to 85% of the second plate's blind depth. Select a common overall length available from your supplier.

The 4MMCAPSSCREW utilizes the Revolved Base, Chamfer and Extruded Cut feature. The Base feature is a Revolved feature. The Revolved Base 🦴 feature creates the head and shaft of the 4MMCAPSCREW.

The Chamfer 🔶 feature inserts two end cuts. The Extruded Cut 🔲 feature utilizes a hex profile. Utilize the Polygon Sketch ⬡ tool to create the hexagon.

Activity: Create Socket Head Cap Screw - 4MMCAPSCREW Part.

Create the 4MMCAPSCREW.

195) Click **New** ⬜ from the Menu bar.

196) Click the **MY-TEMPLATES** tab from the New SOLIDWORKS Document dialog box.

197) Double-click **PART-MM-ANSI**. The Part FeatureManager is displayed.

Save the part. Enter part name. Enter part description.

198) Click **Save As** 📑.

199) Select the **ENGDESIGN-W-SOLIDWORKS\PROJECTS** folder.

200) Enter **4MMCAPSCREW** for File name.

201) Enter **CAP SCREW, 4MM** for Description.

202) Click **Save**. The 4MMCAPSCREW FeatureManager is displayed.

Create the Base sketch on the Front Plane.
203) Right-click **Front Plane** from the FeatureManager.

204) Click **Sketch** ⌐ from the Context toolbar.

205) Click the **Centerline** ⤢ Sketch tool. The Insert Line PropertyManager is displayed.

Sketch a vertical centerline.

206) Click the **Origin** ⌐ in the Graphics window. The Origin is coincident with the first point.

207) Click a **position** directly above the Origin as illustrated (approximately 25mm).

Sketch the illustrated profile of the 4MMCAPSCREW using the Line Sketch tool.

208) Click the **Line** ✎ Sketch tool. The Insert Line PropertyManager is displayed. **Note:** You can right-click the Line sketch tool from the Sketch Entities box.

209) Click the **Origin** ⌊.

210) Click a **position** to the right of the Origin to create a horizontal line.

211) Sketch the first **vertical line** to the right of the centerline.

212) Sketch the second **horizontal line**.

213) Sketch the second **vertical line**.

214) Sketch the third **horizontal line**. The endpoint of the line is Coincident with the centerline. The centerline extends above the third horizontal line.

Deselect the Line Sketch tool.

215) Right-click **Select** in the Graphics window to deselect the Line Sketch tool.

A diameter dimension for the revolved sketch requires a centerline, profile line and a dimension position to the left of the centerline. A dimension position directly below the bottom horizontal line creates a radial dimension.

💡 Insert smaller dimensions first, then larger dimensions to maintain the shape of the sketch profile.

Fit the sketch to the Graphics window.
216) Press the **f** key.

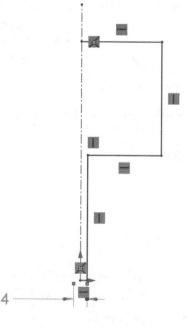

Insert dimensions on the Base Sketch.

217) Click the **Smart Dimension** ✸ from the Sketch tool.

Add a bottom diameter dimension.
218) Click the **centerline**.

219) Click the **first vertical line**.

220) Click a **position** below and to the left of the Origin to create a diameter dimension.

221) Enter **4**mm. Click the **Green Check mark** ✔ from the Modify dialog box.

Add a vertical dimension.
222) Click the **second vertical line**.

223) Click a **position** to the right of the profile.

224) Enter **4**mm.

225) Click the **Green Check mark** ✔ from the Modify dialog box.

Add a top diameter dimension.
226) Click the **centerline**.

227) Click the **second vertical line**.

228) Click a **position** to the left of the Origin and above the second horizontal line to create a diameter dimension.

229) Enter **7**mm.

230) Click the **Green Check mark** ✔ from the Modify dialog box.

Create an overall vertical dimension.
231) Click the **top horizontal line**.

232) Click the **Origin**.

233) Click a **position** to the right of the profile.

234) Enter **14**mm.

235) Click the **Green Check mark** ✔.

Deselect the Smart Dimension Sketch tool.
236) Right-click **Select**.

237) Click the **centerline** in the Graphics window for axis of revolution. The Line Properties PropertyManager is displayed.

Create the first feature. Insert a Revolved Base feature.

238) Click **Revolved Boss/Base** 🌶 from the Features toolbar.

239) Click **Yes** to the question, "The sketch is currently open. A non-thin revolution feature requires a closed sketch. Would you like the sketch to be automatically closed?" The Revolve PropertyManager is displayed.

Note: The "Yes" button causes a vertical line to be automatically sketched from the top left point to the Origin. The Graphics window displays a preview of the Revolved Base feature.

240) Accept the default options. Click **OK** ✔ from the Revolve PropertyManager. Revolve1 is displayed in the FeatureManager.

Fit the model to the Graphics window.
241) Press the **f** key.

Insert a Chamfer feature.

242) Click the **Chamfer** ⬦ feature tool.

243) Select **Angle Distance** for Chamfer type.

244) Click the **top circular edge** of the 4MMCAPSCREW head as illustrated.

245) Click the **bottom circular edge** of the 4MMCAPSCREW shaft as illustrated. The selected entities are displayed in the Chamfer Parameters box.

246) Enter **0.40**mm in the Distance box.

247) Enter **45.00** for angle. Accept the default settings.

248) Click **OK** ✔ from the Chamfer PropertyManager. Chamfer1 is displayed in the FeatureManager.

Save the 4MMCAPSCREW part.

249) Click **Save** 💾 .

💡 Angle distance values for the Chamfer feature can be entered directly in the Pop-up box from the Graphics window.

Rotate the view. Create the Hex Extruded Cut feature.

250) Rotate the **middle mouse button** to display the top circular face of 4MMCAPSCREW.

Insert a sketch.

251) Right-click the **top circular face** of Revolve1 for the Sketch plane as illustrated.

252) Click **Sketch** 📐 from the Context toolbar.

Display a Top view.

253) Click **Top view** ⬜.

Sketch a hexagon.

254) Click the **Polygon** ⬡ Sketch tool. The Polygon PropertyManager is displayed.

255) Click the **Origin** ↳. Click a **position** to the right as illustrated.

Insert a Horizontal relation.

256) Right-click **Select** to deselect the Polygon Sketch tool.

257) Click the **Origin** ↳.

258) Hold the **Ctrl** key down.

259) Click the **right point** of the hexagon.

260) Release the **Ctrl** key. The Properties PropertyManager is displayed.

261) Right-click **Make Horizontal** from the Pop-up Context toolbar.

262) Click **OK** ✔ from the Properties PropertyManager.

Add a dimension using the Smart Dimension tool.

263) Click the **Smart Dimension** 🖋 Sketch tool.

264) Click the **inscribed circle**.

265) Click a position **diagonally** to the right of the profile. Enter **2**mm.

266) Click the **Green Check mark** ✔.

Insert an Extruded Cut feature.

267) Click the **Extruded Cut** feature tool. The Cut-Extrude PropertyManager is displayed.

268) Enter **4**mm for the Depth. Accept the default settings.

269) Click **OK** ✔ from the Cut-Extrude PropertyManager. Cut-Extrude1 is displayed in the FeatureManager.

Display an Isometric view.
270) Press the **space bar** to display the Orientation dialog box.

271) Click **Isometric view** 🔲.

Save the 4MMCAPSCREW.

272) Click **Save** 💾. View the results.

SmartMates

A SmartMate is a Mate that automatically occurs when a component is placed into an assembly. The mouse pointer displays a SmartMate feedback symbol when common geometry and relationships exist between the component and the assembly.

SmartMates are Concentric or Coincident. A Concentric SmartMate assumes that the geometry on the component has the same center as the geometry on an assembled reference. A Coincident Plane SmartMate assumes that a plane on the component lies along a plane on the assembly. As the component is dragged into place, the mouse pointer provides feedback such as:

Mating Entities:	*Type of Mate*:	*Icon Feedback*:
• Two linear edges	Coincident	.
• Two planner faces	Coincident	.
• Two vertices	Coincident	.
• Two conical faces	Concentric	.
• Two circular edges	Coincident/Concentric	.

Coincident/Concentric SmartMate

The most common SmartMate between a screw/bolt and a hole is the Coincident/Concentric SmartMate. The following technique utilizes two windows. The first window contains the 4MMCAPSCREW part and the second window contains the GUIDE-ROD assembly. Zoom in on both windows to view the Mate reference geometry. Drag the part by the shoulder edge into the assembly window. View the mouse pointer for Coincident/Concentric feedback ⌖. Release the mouse pointer on the circular edge of the PLATE Mounting Hole.

Activity: Insert - Coincident/Concentric SmartMate

Display the 4MMCAPSCREW part and the GUIDE-ROD assembly.
273) Click **Window**, **Tile Horizontally** from the Menu bar.

274) Zoom in on the PLATE to view the Mounting Holes.

Insert the first 4MMCAPSCREW.
275) Click and drag the **circular edge** of the 4MMCAPSCREW part into the GUIDE-ROD assembly Graphics window.

276) Release the mouse pointer on the **top circular edge** of the PLATE. The mouse pointer displays the Coincident/Concentric circular edges feedback icon.

Insert the second 4MMCAPSCREW.
277) Click and drag the **circular edge** of the 4MMCAPSCREW part into the GUIDE-ROD assembly Graphics window.

278) Release the mouse pointer on the **bottom circular edge** of the PLATE. The mouse pointer displays the Coincident/Concentric circular edges feedback ⌖ icon.

279) Maximize the GUIDE-ROD assembly.

Fit the GUIDE-ROD assembly to the Graphics window.
280) Press the **f** key.

💡 **Zoom in** before dragging a component into the assembly to select the correct circular edge for a Coincident/Concentric Mate.

The circular edge of the 4MMCAPSCREW produces both the Coincident/Concentric Mate. The cylindrical face of the 4MMCAPSCREW produces only a Concentric Mate. View the mouse pointer for the correct feedback.

Insert a Parallel mate for the first 4MMCAPSCREW.

281) Click the **Mate** 📎 Assembly tool. The Mate PropertyManager is displayed.

282) Click the **Standard** mate tab.

283) **Expand** the GUIDE-ROD icon from the fly-out FeatureManager.

284) Click **Right Plane** of the 4MMCAPSCREW<1> in the fly-out FeatureManager as illustrated.

285) Click **Right Plane** of the 4MMCAPSCREW<2> in the fly-out FeatureManager. The selected planes are displayed in the Mate Selections box.

286) Click **Parallel** ⟍ .

287) Click **OK** ✔ from the Parallel PropertyManager.

Fit the assembly to the Graphics window. Display an Isometric view.
288) Press the **f** key.

289) Click **Isometric view** 🧊 .

Click the GUIDE-ROD assembly.
290) Click **Save** 💾 .

The ROD, PLATE and 4MMCAPSCREWS are free to translate along the z-axis. Their component status remains under defined, (-) in the GUIDE-ROD FeatureManager.

The GUIDE is fixed (f) to the assembly Origin. The flange bolts are fully defined since they are mated only to the GUIDE.

Understand how Mates reflect the physical behavior in an assembly. Move and rotate components to test the mate behavior. The correct mate selection minimizes rebuild time and errors in the assembly.

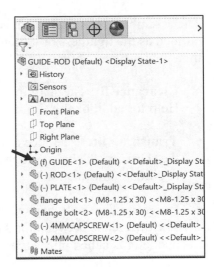

Tolerance and Fit

The ROD travels through the GUIDE in the GUIDE-ROD assembly. The shaft diameter of the ROD is 10mm. The hole diameter in the GUIDE is 10mm. A 10mm ROD cannot be inserted into a 10mm GUIDE hole without great difficulty. (Press Fit.)

Note: The 10mm dimension is the nominal dimension. The nominal dimension is approximately the size of a feature that corresponds to a common fraction or whole number.

Tolerance is the difference between the maximum and minimum variation (Limits) of a nominal dimension and the actual manufactured dimension.

Example: A ROD has a nominal dimension of 100mm with a tolerance of ± 2mm (100mm ± 2mm). This translates to a part with a possible manufactured dimension range between 98mm (lower limit) to 102mm (upper limit). The total ROD tolerance is 4mm.

Note: Design rule of thumb: Design with the maximum permissible tolerance. Tolerance flexibility saves in manufacturing time and cost.

The assembled relationship between the ROD and the GUIDE is called the fit. The fit is defined as the tightness or looseness between two components. This project discusses three major types of fits:

- **Clearance fit**: The shaft diameter is less than the hole diameter. See SOLIDWORKS Help for additional information.

- **Transition fit**: Clearance or interference can exist between the shaft and the hole. See SOLIDWORKS Help for additional information.

- **Press fit**: The shaft diameter is larger than the hole diameter. The difference between the shaft diameter and the hole diameter is called interference. See SOLIDWORKS Help for additional information.

You require a Clearance fit between the shaft of the ROD and the Guide Hole of the GUIDE. There are multiple categories for Clearance fits. Dimension the GUIDE hole and ROD shaft for a Sliding Clearance fit. All below dimensions are in millimeters.

Use the following values:

Hole	Maximum	10.015mm.
	Minimum	10.000mm.
Shaft	Maximum	9.995mm.
	Minimum	9.986mm.
Fit	Maximum	10.015 - 9.986 = .029 Max. Hole - Min. Shaft.
	Minimum	10.000 - 9.995 = .005 Min. Hole - Max. Shaft.

Calculate the maximum variation:

Hole Max: 10.015 - Hole Min: 10.000 = .015 Hole Max. Variation.

Select features from the FeatureManager and the Graphics window. In the next activity, locate feature dimensions with the FeatureManager for the GUIDE.

Locate feature dimensions in the Graphics window for the ROD. Select the dimension text, and then apply the Tolerance/Precision through options in the Dimension PropertyManager.

Activity: Address - Tolerance and Fit.

Locate the dimension in the FeatureManager.
291) Expand GUIDE from the GUIDE-ROD FeatureManager.

292) Double-click **Guide Hole** from the FeatureManager to display the dimensions.

Add the maximum and minimum Guide Hole dimensions.
293) Click the ∅**10.00** dimension. The Dimension PropertyManager is displayed.

294) Select **Limit** from the Tolerance/Precision box.

295) Select **.123** for three place Precision.

296) Enter **0.015**mm for Maximum Variation.

297) Enter **0.000**mm for Minimum Variation.

298) Click **OK** from the Dimension PropertyManager.

Add maximum and minimum Shaft dimensions.
299) Double-click the **Base-Extrude (ROD)** component from the Graphics window.

300) Click the ∅**10.00** diameter dimension.

301) Select **Limit** from the Tolerance/Precision box.

302) Select **.123** for three place Precision.

303) Enter -**0.005**mm for Maximum Variation.

304) Enter -**0.014**mm for Minimum Variation.

305) Click **OK** from the Dimension PropertyManager.

Save the GUIDE-ROD assembly.
306) Click **Save**. Click **Save All**.

ISO symbol Hole/Shaft Classification is applied to an individual dimension for Fit, Fit with tolerance, or Fit (tolerance only) types. Classification can be User Defined, Clearance, Transitional or Press.

For a hole or shaft dimension, select a classification from the list. The Hole/Shaft designation for a Sliding Fit is H7/g6.

The values for Maximum and Minimum tolerances are calculated automatically based on the diameter of the Hole/Shaft and the Fit Classification.

 Utilize Hole/Shaft Classification early in the design process. If the dimension changes, then the tolerance updates. The Hole/Shaft Classification propagates to the details in the drawing. Create the drawing in Project 4. See SOLIDWORKS Help for additional information.

Review of the GUIDE-ROD Assembly.

The GUIDE-ROD assembly combined the GUIDE, ROD and PLATE components. The GUIDE was the first component inserted into the GUIDE-ROD assembly.

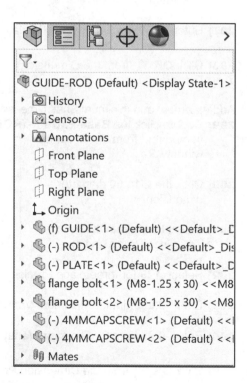

Mates removed degrees of freedom. Concentric, Coincident and Parallel Mates were utilized to position the ROD and PLATE with respect to the GUIDE.

The flange bolts were obtained from the Design Library. You utilized a Revolved Base feature to create the 4MMCAPSCREW.

The Revolved Base feature contained an axis, sketched profile and an angle of revolution. The Polygon Sketch tool was utilized to create the hexagon Extruded Cut feature.

The 4MMCAPSCREW utilized the Concentric/Coincident SmartMate option.

Explode View

The Explode View illustrates how to assemble the components in an assembly. Create an Explode View with a few simple steps in the ROD-GUIDE assembly. Click and drag components in the Graphics window.

The Manipulator icon indicates the direction to explode. Select an alternate component edge for the Explode direction. Drag the component in the Graphics window or enter an exact value in the Explode distance box. In this activity, manipulate the top-level components in the assembly.

In the project exercises, create exploded views for each sub-assembly and utilize the Re-use sub-assembly Explode option in the top-level assembly.

Access the Explode view option as follows:

- Right-click the configuration name in the ConfigurationManager.

- Select the Exploded View tool in the Assembly toolbar.

- Select Insert, Exploded View from the Menu bar.

The Exploded View feature uses the Explode PropertyManager as illustrated. There are two Explode Step types: Regular step (translate and rotates) and Radial step.

The Radial step provides the ability to explode components aligned radially/cylindrically about an axis. You can also explode radially by diverging along an axis.

You can create exploded view steps that rotate a component with or without linear translation. Use the rotation and translation handles of the triad. You can include rotation and translation in the same explode step. You can also edit the explode step translation distance and rotation angle values in the PropertyManager.

Activity: GUIDE-ROD Assembly-Explode View.

Insert an Explode view in the Assembly.

307) Click the **Exploded View** Assembly tool. The Explode PropertyManager is displayed.

308) Click **Regular step** in the Add a Step box.

Fit the model to the Graphics window.
309) Press the **f** key.

Create the first explode step.
310) Click the **PLATE** component in the Graphics window. The selected entity is displayed in the Explode Step Components box.

311) Enter **100**mm in the Explode Distance box. The direction of the explode view is towards the back. If required, click the **Reverse direction** button.

312) Click **Add Step**. View the results in the Graphics window. Explode Step1 is created.

If the exact Explode distance value is not required, click and drag the Manipulator handle as illustrated in the next step.

Create Explode Step2.
313) Click the **ROD** component from the Graphics window.

Fit the model to the Graphics window.
314) Press the **f** key.

315) Drag the **manipulator handle** backward to position between the PLATE and the GUIDE as illustrated.

316) Click **Done**. Explode Step2 is created.

Create Explode Step3.

317) Click the **left flange bolt** from the Graphics window.

318) Drag the **vertical manipulator handle** upward above the GUIDE.

319) Click **Done**. Explode Step3 is created.

Create Explode Step4.

320) Click the **right flange bolt** from the Graphics window.

321) Drag the **vertical manipulator handle** upward above the GUIDE.

322) Click **Done**. Explode Step4 is created.

323) Click **OK** ✔ from the Explode PropertyManager.

Display an Isometric view. Save the model.

324) Click **Isometric view** 🔲.

325) Click **Save** 💾.

Split the FeatureManager to view the Exploded Steps.
326) Position the **mouse pointer** at the top of the
FeatureManager. The mouse pointer displays the Split bar
⫢.

327) Drag the **Split bar** halfway down to display two
FeatureManager windows.

328) Expand Default.

329) Expand Exploded View1 to display the four exploded steps.

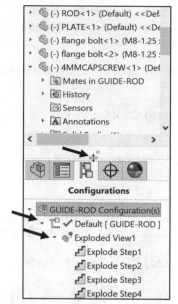

Fit the Exploded view to the Graphics window.
330) Press the **f** key.

Remove the Exploded state.
331) Right-click in the **Graphics window**.

332) Click **Collapse** from the Pop-up menu.

Display an Isometric view.
333) Click **Isometric view** .

Animate the Exploded view.
334) Right-click **Exploded View1** from the
ConfigurationManager.

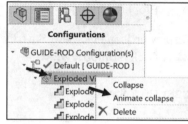

335) Click **Animate explode**. The Animation
Controller dialog box is displayed. Play is
selected by default. View the animation.

336) Click **Pause** .

337) Click **End** .

Close the Animation Controller.
338) Close ☒ the Animation Controller. The
GUIDE-ROD is in the Exploded state.

339) Right-click **Collapse** as illustrated.

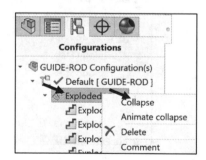

Display the FeatureManager. Return to a single FeatureManager.
340) Drag the **Split bar** upward to display one FeatureManager
window.

341) Click the **Assembly FeatureManager** tab.

Save the GUIDE-ROD assembly.
342) Click **Save** .

Click the Motion Study # tab at the bottom of the
Graphics window. Select Basic Motion from the
MotionManager. Click the Animation Wizard to create a
simple animation. Click Play to view the Animation Wizard.
Click the Model tab at the bottom of the Graphics window to
return to the SOLIDWORKS screen.

Note: The SOLIDWORKS Animator application is required to record a motion file
through the Animation Controller. Play the animation files through the Windows Media
Player. The time required to create the animation file depends on the number of
components in the assembly and the options selected.

Stop the animation before closing the Animation Controller toolbar to avoid issues.
Utilize SOLIDWORKS Animator for additional control over the Explode/Collapse
motion in the assembly.

Create the Exploded steps in the order that you would disassemble the assembly.
Collapse the assembly to return to the original assembled position. Your animations will
appear more realistic based on the order of the Explode steps.

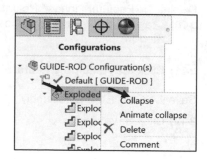

Review of the GUIDE-ROD assembly Exploded View.

You created an Explode View in the GUIDE-ROD assembly. The Explode View displayed the assembly with its components separated from one another.

The Explode View animation illustrated how to assemble and disassemble the GUIDE-ROD assembly through the collapse and explode states.

Section View

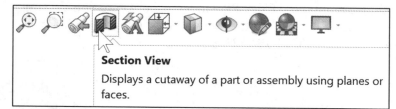

A Section view displays the internal cross section of a component or assembly.

Section View
Displays a cutaway of a part or assembly using planes or faces.

The Section view dissects a model like a knife slicing through a stick of butter. Section views can be performed anywhere in a model. The location of the cut corresponds to the Section plane. A Section plane is a planar face or reference plane.

Use a Section view to determine the interference between the 10mm Guide Hole and the Linear Pattern of 3mm holes in the GUIDE.

Activity: GUIDE-Section View.

Open the GUIDE part.
343) Right-click the **GUIDE** component in the Graphics window.

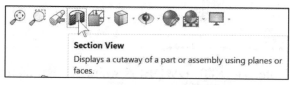

Section View
Displays a cutaway of a part or assembly using planes or faces.

344) Click **Open Part** from the shortcut toolbar. The GUIDE part is displayed in the Graphics window.

Insert a Section View on the Front Plane of the GUIDE.
345) Click **Front Plane** from the GUIDE FeatureManager.

346) Click **Section View** from the Heads-up View toolbar in the Graphics window. The Section View PropertyManager is displayed. View your options.

Save the Section View.

347) Click **Save** from the Section View PropertyManager. The Save As dialog box is displayed. SectionView is the default view name.

348) Click **Save**.

Click the drop-down arrow from the View Orientation tool as illustrated. SectionView1 is saved as a custom view.

Display and Pin the View Orientation dialog box.

349) Press the **space** bar. View the Orientation box with the new view.

350) Close the Orientation dialog box.

Return to a full standard view.

351) Click **Front view** from the Heads-up View toolbar.

352) Click **Section view** from the Heads-up View toolbar to return to a full view.

353) Click **Hidden Lines Visible** from the Heads-up View toolbar.

Save the GUIDE part.

354) Click **Save**.

The Section view tool detects potential problems before manufacturing. The Section view tool determines the interference between the 10mm Guide hole and the Linear Pattern of 3mm holes in the GUIDE.

The GUIDE, ROD and PLATE components all share a common file structure with the GUIDE-ROD assembly. What component do you modify? How will changes affect other components? Let's analyze the interference problem and determine a solution.

Analyze an Interference Problem

An interference problem exists between the 10mm Guide Hole and the 3mm tapped holes. Review your design options: 1.) Reposition the Guide Hole, 2.) Modify the size of the Guide Hole, 3.) Adjust the length of the 3mm holes or 4.) Reposition the Linear Pattern feature.

The first three options affect other components in the assembly. The GUIDE-CYLINDER assembly and PLATE determine the Guide Hole location. The ROD diameter determines the size of the Guide Hole. The position sensor requires the current depth of the 3mm Holes. Reposition the Linear Pattern feature by modifying the first 3mm Thru Hole.

Activity: GUIDE-Analyze an Interference Problem

Modify the Thru Hole dimensions.
355) Expand the M3x0.5 Tapped Hole1 feature from the FeatureManager.

Display an Isometric view - Shaded With Edges.
356) Click **Isometric view** 🔲.

357) Click **Shaded With Edges** 🔲 from the Heads-up View toolbar.

358) Double-click **Sketch4** from the FeatureManager to display the position dimensions.

359) Double-click the **4.00**mm dimension created from the temporary axis of the Guide Hole.

360) Enter **6.00**mm.

Rebuild the model.
361) Click **Rebuild** 🔘 from the Modify dialog box.

362) Click the **Green Check mark** ✔ from the Modify dialog box.

363) Click **Exit Sketch**.

Close the GUIDE part.
364) Click **File**, **Close** from the Menu bar. Click **Yes** to Save changes to the GUIDE.

365) Click **Wireframe** 🔲 from the Heads-up View toolbar.

366) Click **Front view** . The GUIDE-ROD assembly updates to display the changes to the GUIDE part.

Return to an Isometric view - Shaded With Edges.
367) Click **Isometric view** .

368) Click **Shaded With Edges** .

Save the GUIDE-ROD assembly.
369) Click **Save** .

SOLIDWORKS provides an Interference Detection tool and a Clearance Verification tool to analyze assemblies. See SOLIDWORKS help for additional information.

Analyze issues at the part level first. Working at the part level reduces rebuild time and complexity. Return to the assembly and review the modifications.

The last component to insert to the GUIDE-CYLINDER is the 3MMCAPSCREW. Utilize the Save As/Save as copy and open option to copy the 4MMCAPSCREW to create the 3MMCAPSCREW part.

Save As/Save as Copy Options

Conserve design time and cost. Modify existing parts and assemblies to create new parts and assemblies. Utilize the Save As/Save as command to save the file in another file format.

Utilize the Save as copy and continue command to save the document to a new file name without replacing the active document. Utilize the Save as copy and open command to save the document to a new file name that becomes the active document. The original document remains open. References to the original document are not automatically assigned to the copy.

The 4MMCAPSCREW was created earlier. The GUIDE requires 3MMCAPSCREWs to fasten the sensor to the 3mm Tapped Hole Linear Pattern.

Start with the 4MMCAPSCREW. Utilize the Save As/Save as copy and open command option. Enter the 3MMCAPSCREW for the new file name.

The Save as copy and open command prevents the 3MMCAPSCREWS from replacing the 4MMCAPSCREWS in the GUIDE-CYLINDER assembly.

Important: Check the Save as copy and open command to save the document to a new file name that becomes the active document. The original document remains open. References to the original document are not automatically assigned to the copy.

The 3MMCAPSCREW is the new part name. Modify the dimensions of the Revolved Base feature to create the 3MMCAPSCREW.

Activity: 4MMCAPSCREW-Save As/Save as copy and open Option.

Open the 4MMCAPSCREW part.

370) Right-click the **4MMCAPSCREW front face** component in the GUIDE-ROD assembly Graphics window.

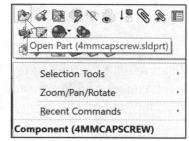

371) Click **Open Part** from the Context toolbar. The 4MMCAPSCREW part is displayed.

Apply the Save as copy and open option.

372) Click **File**, **Save As** from the Menu bar. A SOLIDWORKS dialog box is displayed.

373) Click **Save as copy and open** from the SOLIDWORKS dialog box.

374) Select the **ENGDESIGN-W-SOLIDWORKS\PROJECTS** folder.

375) Enter **3MMCAPSCREW** for File name.

376) Enter **CAP SCREW, 3MM** for Description.

377) Click **Save**. The 3MMCAPSCREW part is the active document. **Note**: The 4MMCAPSCREW document is still open.

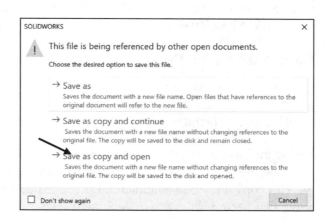

Modify the 3MMCAPSCREW part.
378) Double-click **Revolve1** in the FeatureManager. Dimensions are displayed.

Fit the model to the Graphics window.
379) Press the **f** key.

Modify the Revolve dimensions.
380) Double-click the vertical dimension **4.00**mm.

381) Enter **3.00**mm.

382) Double-click the depth dimension **14.00**mm.

383) Enter **9.00**mm.

384) Double-click the diameter dimension **4.00**mm.

385) Enter **3.00**mm.

386) Double-click the diameter dimension **7.00**mm.

387) Enter **5.50**mm.

Save the 3MMCAPSCREW part.
388) Click **SAVE**.

FeatureManager tabs and tree folders will vary depending on your SOLIDWORKS system setup and Add-ins.

The GUIDE Linear Pattern of 3mm tapped holes requires six 3MMCAPSCREWs. Do you remember the seed feature in the Linear Pattern?

The first 3mm tapped hole is the seed feature. The seed feature is required for a Component Pattern in the GUIDE-ROD assembly.

GUIDE-ROD Assembly - Pattern Driven Component Pattern

There are various tools to define a pattern in an assembly:

- Linear Component Pattern

- Circular Component Pattern

- Pattern Driven Component Pattern

- Sketch Driven Component Pattern

- Curve Driven Component Pattern

- Chain Component Pattern

- Mirror Components

Utilize the Pattern Driven Component Pattern tool to create instances of the 3MMCAPSCREW. Insert the 3MMCAPSCREW part into the GUIDE-ROD assembly.

Activity: GUIDE-ROD Assembly-Pattern Driven Component Pattern.

Insert and mate the 3MMCAPSCREW.
389) Click **Window**, **Tile Horizontally** from the Menu bar. The 3MMCAPSCREW and the GUIDE-ROD assembly are displayed.

The 3MMCAPSCREW and the GUIDE-ROD assembly are the open documents. Close all other documents. Select the Close ☒ icon. Click Window, Tile Horizontally again to display the two open documents.

390) **Zoom in** on the bottom left circular edge of the GUIDE left Tapped Hole.

391) Click and drag the **bottom circular edge** of the 3MMCAPSCREW into the GUIDE-ROD assembly.

392) Release the mouse button on the **bottom left circular edge** of the GUIDE left Tapped Hole. The mouse pointer displays the Coincident/Concentric Circular edges feedback symbol. The 3MMCAPSCREW part is positioned in the bottom left Tapped Hole.

Insert a Pattern Driven Component Pattern.
393) Maximize the GUIDE-ROD assembly.

Fit the GUIDE-ROD assembly to the Graphics window.
394) Press the **f** key.

395) Click the **Pattern Driven Component Pattern** tool from the Consolidated Assembly toolbar. The Pattern Driven PropertyManager is displayed. 3MMCAPSCREW<1> is displayed in the Components to Pattern box.

396) Click inside the **Driving Feature or Component** box.

397) Expand GUIDE in the GUIDE-ROD fly-out FeatureManager.

398) Click **LPattern1** under GUIDE<1> from the fly-out FeatureManager. LPattern1@GUIDE-1 is displayed in the Driving Feature or Component to Pattern box.

399) Click **OK** from the Pattern Driven PropertyManager. The DerivedLPattern1 is displayed in the FeatureManager.

Display an Isometric view.
400) Click **Isometric view** .

Save the GUIDE-ROD assembly.
401) Click **Save** .

Reuse geometry. Utilize patterns early in the design process. A Linear Pattern feature in the part is utilized as a Pattern Driven Component Pattern in the assembly.

Review the Pattern Driven Component Pattern feature for the 3MMCAPSCREWs.

You utilized the Save As/Save as copy and open option to copy the 4MMCAPSCREW to the 3MMCAPSCREW. The 3MMCAPSCREW was mated to the GUIDE 3MM Tapped Hole. You utilized the Pattern Driven Component Pattern tool to create an array of 3MMCAPSCREWs. The Pattern Driven Component Pattern was derived from the GUIDE Linear Pattern feature of Tapped Holes.

Folders and Suppressed Components

The Assembly FeatureManager entries increase as components are inserted into the Graphics window of an assembly. Folders reduce the length of the Assembly FeatureManager in the part and assembly. Folders also organize groups of similar components. Organize hardware in the assembly into folders.

Suppress features, parts and assemblies that are not displayed. During model rebuilding, suppressed features and components are not calculated. This saves rebuilding time for complex models. The names of the suppressed features and components are displayed in light gray.

Create a folder in the FeatureManager named Hardware. Drag all individual bolts and screws into the Hardware Folder. The DerivedLPattern1 feature and the LocalLPattern1 feature cannot be dragged into a folder. Suppress the Hardware folder. Suppress the DerivedLPattern1 and LocalLPattern1 feature.

Activity: Folders and Suppressed Components.

Create a new folder in the FeatureManager.

402) Right-click ▸ 🗎 flange bolt<1> (M8-1.25 x 30<<M8-1.25 x in the GUIDE-ROD FeatureManager.

403) Click **Create New Folder**. Folder1 is displayed in the FeatureManager.

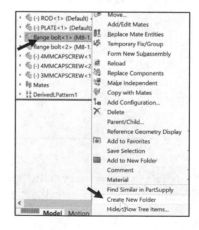

404) Enter **Hardware** for folder name.

405) Click and drag

> flange bolt<1> (M8-1.25 x 30<<M8-1.25 x

from the FeatureManager into the Hardware folder. The mouse pointer displays the Move Component icon.

406) Repeat the above process for the **flange bolt <2>**, **4MMCAPSCREW<1>**, **4MMCAPSCREW<2>** and **3MMCAPSCREW<1>** components.

407) **Expand** the Hardware folder. View the components.

Suppress the Folder and Pattern.
408) Right-click the **Hardware** folder in the FeatureManager.

409) Click **Suppress**.

410) Right-click **DerivedLPattern1** in the FeatureManager.

411) Click **Suppress**.

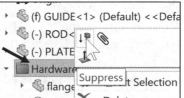

Display an Isometric view.
412) Click **Isometric view** .

Save the GUIDE-ROD assembly.
413) Click **Save** . View the FeatureManager.

Close all parts and assemblies.
414) Click **Window**, **Close All** from the Menu bar.

Standardized folder names, such as Hardware or Fillets, help colleagues recognize folder names. Place a set of continuous features or components into an individual folder.

In the next section, create the CUSTOMER Assembly from the MGPM12-1010 assembly. The manufacturer of the MGPM12-1010 assembly is SMC of USA. Utilize the assembly model that is located in the **ENGDESIGN-W-SOLIDWORKS\VENDOR COMPONENTS** folder for the next section.

CUSTOMER Assembly

Three documents for this project are contained in the **ENGDESIGN-W-SOLIDWORKS\VENDOR COMPONENTS** folder. Copy the folder directly to your hard drive. Work directly from your hard drive.

The MGPM12-1010_MGPTube(12M) part and MGPM12-1010_MGPRod(12M) part contain references to the MGPM12-1010 assembly.

Create the CUSTOMER assembly. The CUSTOMER assembly combines the GUIDE-ROD assembly and the GUIDE-CYLINDER (MGPM12-1010) assembly. The GUIDE-ROD assembly is fixed to the CUSTOMER assembly Origin.

Activity: Create the CUSTOMER Assembly.

Copy the ENGDESIGN-W-SOLIDWORKS\VENDOR COMPONENTS folder.

415) Copy the **ENGDESIGN-W-SOLIDWORKS\VENDOR COMPONENTS** folder to your hard drive (PROJECTS folder). Open the MGPM12-1010 assembly from your hard drive.

416) Double-click the **MGPM12-1010** assembly from the ENGDESIGN-W-SOLIDWORKS\VENDOR-COMPONENTS folder. The MGPM12-1010 FeatureManager is displayed.

417) If needed, click **Yes** to rebuild.

Fit the model to the Graphics window.
418) Press the **f** key.

FH-M4-25ZI.SLDPRT
FH-M5-30ZI.SLDPRT
MGPM12-1010.SLDASM
MGPM12-1010_MGPRod(12M).sldprt
MGPM12-1010_MGPTube(12M).sldprt

Open the GUIDE-ROD assembly.

419) Double-click the **GUIDE-ROD** assembly from the PROJECTS folder.

Create the CUSTOMER assembly.

420) Click **New** ⬜ from the Menu bar.

421) Double-click **Assembly** from the Templates tab. The Begin Assembly PropertyManager is displayed.

422) Double-click **GUIDE-ROD** from the Open documents box.

423) Click **OK** ✔ from the Begin Assembly PropertyManager to fix the GUIDE-ROD to the Origin.

Save the assembly. Enter assembly name. Enter assembly description.

424) Click **Save As** 📄.

425) Select **ENGDESIGN-W-SOLIDWORKS\PROJECTS** for folder. Enter **CUSTOMER** for file name.

426) Enter **GUIDE-ROD AND GUIDE-CYLINDER ASSEMBLY** for description.

427) Click **Save** 💾.

GUIDE-ROD is the first component in the CUSTOMER assembly. Insert the second component.

Activity: CUSTOMER Assembly-Insert Component

Insert the MGPM12-1010 assembly.

428) Click the **Insert Components** 📦 Assembly tool. The Insert Component PropertyManager is displayed.

429) Click **MGPM12-1010** from the Part/Assembly box.

430) Click a **position** in the Graphics window behind the GUIDE-ROD assembly as illustrated. The MGPM12-1010<1> component is added to the CUSTOMER FeatureManager.

Fit the model to the Graphics window.
431) Press the **f** key.

Reposition the MGPM12-1010<1> component.
432) Click the **MGPM12-1010<1>** component in the FeatureManager. It's highlighted in the Graphics window.

433) Click the **Rotate Component** Assembly tool. The Rotate Component PropertyManager is displayed.

434) Select **By Delta XYZ** from the Rotate drop-down menu.

435) Enter **90** in the delta Z box.

436) Click **Apply**. The MGPM12-1010<1> component rotated 90 degrees in the Graphics window.

437) Click **OK** ✔ from the Rotate Component PropertyManager.

Fit the model to the Graphics window.
438) Press the **f** key.

439) Click and drag the **MGPM12-1010 assembly** behind the GUIDE-ROD assembly. The PISTON PLATE is behind the GUIDE-ROD assembly.

Zoom in on the PISTON PLATE component.

440) Click **Zoom to Area** from the Heads-up View toolbar.

441) **Zoom in** on the PISTON PLATE and the top MountHoles face of the PLATE as illustrated.

442) Click **Zoom to Area** to deactivate.

Insert a Concentric mate between two faces.
443) Click the **inside MountHoles** face of the PLATE.

444) Hold the **Ctrl** key down.

445) Click the **inside Mounting Hole** face of the PISTON PLATE. Both faces are selected.

446) Release the **Ctrl** key. The Mate pop-up menu is displayed.

447) Click **Concentric** from the Mate pop-up menu.

The bottom hole feature contains the Slot Cut and a Mounting Hole. The Mounting Hole is hidden. Utilize the Select Other option to view your options.

Insert a Concentric mate between two faces. Utilize the Select Other tool to view your options.

448) Click the **Mate** Assembly tool. The Mate PropertyManager is displayed.

449) Click the **Standard** mate tab.

450) Click **View**, **Hide/Show**, and check **Temporary Axes** from the Menu bar.

Experience the Select Other tool. Insert a Concentric mate.

451) Click **Zoom to Area** . **Zoom in** on the bottom Slot and the MountHoles. Do not select the Slot.

452) Click **Zoom to Area** to deactivate.

453) Right-click a **position** behind the Slot Cut on the Piston Plate.

454) Right-click **Select Other**. The Select Other dialog box is displayed. The Select Other box lists geometry in the selected region of the Graphics window. There are edges, faces and axes. The list order and geometry entries depend on the selection location of the mouse pointer. Your geometry will vary.

455) **Close** the Select Other dialog box.

456) Clear **all Selections** in the Mate entities box.

457) Position the **mouse pointer** over the Mounting Hole in the Piston Plate.

458) Click the inside **Mounting Hole** face as illustrated.

459) Click the **bottom inside Mounting Hole face** of the PLATE.

460) Click **Concentric** mate.

461) Click **OK** from the Concentric Mate PropertyManager.

Deactivate the Temporary Axes.
462) Click **View**, **Hide/Show**, un-check **Temporary Axes** from the Menu bar.

Fit the model to the Graphics window.
463) Press the **f** key.

Insert a Coincident mate between two faces.
464) Click and drag the **PISTON PLATE** backward to create a gap.

465) Click the **front face** of the PISTON PLATE. **Rotate** the view until you can see the back face of the PLATE.

466) Hold the **Ctrl** key down. Click the **back face** of the PLATE.

467) Release the **Ctrl** key. The Mate pop-up menu is displayed.

468) Click **Coincident** from the Mate pop-up menu.

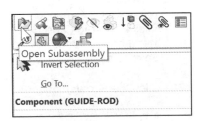

Un-suppress all components and Sub-assemblies.
469) Right-click the **GUIDE-ROD** component in the CUSTOMER FeatureManager.

470) Click **Open Sub-assembly** from the Context toolbar. The GUIDE-ROD Assembly FeatureManager is displayed.

471) Right-click the **Hardware** folder in the FeatureManager.

472) Click **UnSuppress**. Right-click **DerivedLPattern1** in the FeatureManager. Click **UnSuppress**.

Return to the CUSTOMER assembly.
473) Press **Ctrl+Tab**.

474) Select **CUSTOMER assembly**.

Display an Isometric view.

475) Click **Isometric view** .

Save the CUSTOMER assembly in the ENGDESIGN-W-SOLIDWORKS\PROJECTS file folder.

476) Click **Save** .

Copy the CUSTOMER Assembly - Apply Pack and Go

Copying an assembly in SOLIDWORKS is not the same as copying a document in Microsoft Word. The CUSTOMER assembly contains references to its parts and other sub-assemblies. Your task is to provide a copy of the CUSTOMER assembly to a colleague for review. Copy the CUSTOMER assembly and all of the components (references) into a different file folder named **ENGDESIGN-W-SOLIDWORKS\CopiedModels**. Reference all component file locations to the new file folder. Apply the SOLIDWORKS Pack and Go tool.

Activity: CUSTOMER Assembly - SOLIDWORKS Pack and Go tool.

Apply the SOLIDWORKS Pack and Go tool to save the CUSTOMER assembly to the ENGDESIGN-W-SOLIDWORKS\CopiedModels folder.

477) Click **File**, **Pack and Go** from the Menu bar. Note the save options: Save to folder or Save to Zip file.

478) Click the **Browse** button.

479) Select the **ENGDESIGN-W-SOLIDWORKS\ CopiedModels** folder.

Save the model.
480) Click **Save**. The CUSTOMER assembly remains open.

Close all files.
481) Click **Windows**, **Close All** from the Menu bar.

482) Click **Yes**. View the saved assembly in the CopiedModels folder.

Review of the CUSTOMER Assembly.

The CUSTOMER assembly combined the GUIDE-ROD assembly and the MGPM12-1010 assembly. In the design process, you decided to obtain the MGPM12-1010 assembly in SOLIDWORKS format from an outside source.

The GUIDE-ROD assembly is the first component inserted into the CUSTOMER assembly. The GUIDE-ROD assembly is fixed (f) to the CUSTOMER assembly Origin.

You inserted and mated the MGPM12-1010 assembly to the GUIDE-ROD assembly. The flange bolts and cap screws utilized SmartMates to create Concentric and Coincident mates.

In the GUIDE-ROD assembly, you utilized a Pattern Driven Component Pattern for the flange bolt and 3MMCAPSCREW. You modified and redefined the flange bolt mates.

The Save As option with the Reference option copied the CUSTOMER assembly and all references to the ENGDESIGN-W-SOLIDWORKS\CopiedModels folder.

Point at the Center of Mass

Add a center of mass (COM) point to parts, assemblies or drawings.

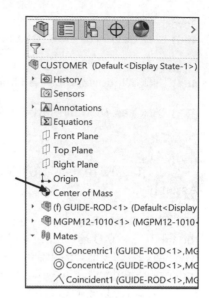

COM points added in component documents also appear in the assembly document. In a drawing document of parts or assemblies that contain a COM point, you can display and reference the COM point.

Add COM to a part or assembly by clicking **Center of Mass** (Reference Geometry toolbar) or **Insert, Reference Geometry, Center of Mass** ⊕ or checking the **Create Center of Mass feature** box in the Mass Properties dialog box.

The center of mass of the model is displayed in the Graphics window and in the FeatureManager design tree just below the origin.

The position of the **COM** point ⊕ updates when the model's center of mass changes. The COM point can be suppressed and unsuppressed for configurations.

You can measure distances and add reference dimensions between the COM point and entities such as vertices, edges and faces.

🔆 If you want to display a reference point where the CG was located at some particular point in the FeatureManager, you can insert a Center of Mass Reference Point. See SOLIDWORKS Help for additional information.

Add a center of mass (COM) point to a drawing view. The center of mass is a selectable entity in drawings, and you can reference it to create dimensions.

In a drawing document, click **Insert**, **Model Items**. The Model Items PropertyManager is displayed. Under Reference Geometry, click the **Center of Mass** icon. Enter any needed additional information. Click **OK** from the Model Items PropertyManager. View the results in the drawing.

The part or assembly **needs to have a COM before** you can view the COM in the drawing. To view the center of mass in a drawing, click **View**, **Hide/Show**, **Center of mass**.

Project Summary

You created the GUIDE-ROD assembly from the following parts: GUIDE, ROD, PLATE, Flange Bolt, 3MMCAPSCREW and 4MMCAPSCREW.

The components were oriented and positioned in the assembly using Concentric, Coincident, Parallel and Concentric/Coincident SmartMates.

Mates were redefined in the GUIDE-ROD assembly to center the Flange Bolts in the Slot Cut of the GUIDE. In the GUIDE-ROD assembly, you utilized the Pattern Driven Component Pattern feature.

The CUSTOMER assembly contained the MGPM12-1010 assembly and the GUIDE-ROD assembly. The assemblies utilized a Bottom-up design approach. In a Bottom-up design approach, you possess all the required design information for the individual components.

Project 3 is completed. In Project 4, create an assembly drawing and a detailed drawing of the GUIDE.

Think design intent. When do you use the various End Conditions and Geometric sketch relations? What are you trying to do with the design? How does the component fit into an assembly?

Questions

1. Describe an assembly or sub-assembly.

2. What are Mates and why are they important in assembling components?

3. Name and describe three types of Fits.

4. Name and describe two assembly modeling techniques in SOLIDWORKS.

5. Describe Dynamic motion.

6. In an assembly, each component has_____# degrees of freedom? Name them.

7. True or False. A fixed component cannot move and is locked to the Origin.

8. Identify the procedure to create a Revolved Base feature.

9. Describe the different types of SmartMates. Utilize online help to view the mouse pointer feedback icons for different SmartMate types.

10. Identify the process to insert a component into an assembly from the Design Library.

11. Describe a Section view.

12. What are Suppressed features and components? Provide an example.

13. True or False. If you receive a Mate Error you should always delete the component from the assembly.

Exercises

Exercise 3.1: Weight-Hook Assembly

Create the Weight-Hook assembly. The Weight-Hook assembly
has two components: WEIGHT and HOOK.

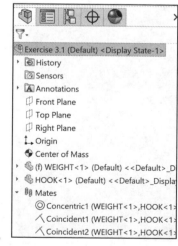

- Create a new assembly document. Copy and insert the
  WEIGHT part from the Chapter 3 Homework folder.

- Fix the WEIGHT to the Origin as illustrated in the Assem1
  FeatureManager.

- Copy and insert the HOOK part from the
  Chapter 3 Homework folder into the
  assembly.

- Insert a Concentric mate between the
  inside top cylindrical face of the
  WEIGHT and the cylindrical face of the
  thread. Concentric is the default mate.

- Insert the first Coincident mate between
  the top edge of the circular hole of the
  WEIGHT and the **top point** of the thread
  as illustrated. The HOOK can rotate in
  the WEIGHT.

Fix the position of the HOOK.

- Insert the second Coincident mate between the Right
  Plane of the WEIGHT and the Right Plane of the
  HOOK. Coincident is the default mate.

- Expand the Mates folder and
  view the created mates. Note:
  Do not display Tangent
  Edges or Origins in the final
  assembly.

- Calculate the Mass and
  Volume of the assembly.

- Identify the Center of Mass
  for the assembly.

Exercise 3.2: Weight-Link Assembly

Create the Weight-Link assembly. The Weight-Link assembly has two components and a sub-assembly: Axle component, FLATBAR component and the Weight-Hook sub-assembly that you created in Exercise 3.1.

- Create a new assembly document. Copy and insert the Axle part from the Chapter 3 Homework folder.

- Fix the Axle component to the Origin of the assembly.

- Copy and insert the FLATBAR part from the Chapter 3 Homework folder.

- Insert a Concentric mate between the Axle cylindrical face and the FLATBAR inside face of the top circle.

- Insert a Coincident mate between the Front Plane of the Axle and the Front Plane of the FLATBAR.

- Insert a Coincident mate between the Right Plane of the Axle and the Top Plane of the FLATBAR. Position the FLATBAR as illustrated.

- Insert the Weight-Hook sub-assembly that you created in exercise 3.1.

- Insert a Tangent mate between the inside bottom cylindrical face of the FLATBAR and the top circular face of the HOOK in the Weight-Hook assembly. Tangent mate is selected by default. ClickFlip Mate Alignment if needed.

- Insert a Coincident mate between the Front Plane of the FLATBAR and the Front Plane of the Weight-Hook sub-assembly. Coincident mate is selected by default. The Weight-Hook sub-assembly is free to move in the bottom circular hole of the FLATBAR.

Exercise 3.3: Counter Weight Assembly

Create the Counter Weight assembly as illustrated using SmartMates and Standard mates. Copy all files from the Chapter 3 Homework\Counter-Weight folder to your hard drive. The Counter Weight consists of the following items:

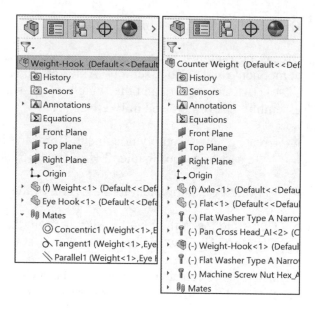

- Weight-Hook sub-assembly

- Weight

- Eye Hook

- Axle component

- Flat component

- Flat Washer Type A from the SOLIDWORKS Toolbox

- Pan Cross Head Screw from the SOLIDWORKS Toolbox

- Flat Washer Type A from the SOLIDWORKS toolbox

- Machine Screw Nut Hex from the SOLIDWORKS Toolbox

Fix the Axle component to the Origin of the assembly.

Insert other components and sub-assemblies. Apply all needed mates. Use SmartMates with the Flat Washer Type A Narrow_AI, Machine Screw Nut Hex_AI and the Pan Cross Head_AI components.

Use a Distance mate to fit the Axle in the middle of the Flat. Note a Symmetric mate could replace the Distance mate. Think about the design of the assembly.

The symbol (f) represents a fixed component. A fixed component cannot move and is fixed to the assembly Origin.

Exercise 3.4: Binder Clip Assembly

- Create a simple Gem binder clip.

- Create an ANSI - IPS model.

- Create two components - BASE and HANDLE.

- Apply material (plan carbon steel) to each component and address all needed mates. Think about where you would start.

- Note: The cuts in the main body are made before folding the metal and strengthening the material during manufacturing. You do not need to model this exactly. The key to this exercise is to apply Advanced distance mates (Limit mates) to limit the movement of the handle.

- What would the Base Sketch for each component be?

- Approximate the dimensions from a small or large Gem binder clip.

- View the sample Assembly FeatureManager. Your Assembly FeatureManager can (should) be different. This is just one way to create the assembly.

🔆 Use the Quick mate procedure to apply Advanced distance angle (LimitAngle) mates.

Exercise 3.5: Limit Mate Assembly (Advanced Mate Type)

- Copy all files from the Chapter 3 Homework\Limit Mate folder to your hard drive. **Note**: This assembly uses Toolbar components.

- Open the Limit Mate assembly. View the created mates.

- Create the final assembly with all needed mates for proper movement.

- Insert a Distance (Advanced Limit Mate) to restrict the movement of the Slide Component - lower and upper movement.

- Use SOLIDWORKS Help for additional information.

A Distance (Limit Mate) is an Advanced Mate type. Limit mates allow components to move within a range of values for distance and angle. You specify a starting distance or angle as well as a maximum and minimum value.

- Save the model and move the slide to view the results in the Graphics window.

- Think about how you would use this mate type in other assemblies.

Exercise 3.6: Screw Mate Assembly (Mechanical Mate Type)

- Copy all files from the Chapter 3 Homework\Screw Mate folder to your hard drive. Create the final assembly with all needed mates for proper movement. **Note**: This assembly uses Toolbar components.

- Insert a Screw mate between the inside Face of the Base and the Face of the vice_Screw. A Screw is a Mechanical Mate type.

- Calculate the mass and volume of the assembly.

A Screw mate constrains two components to be concentric. It adds a pitch relationship between the rotation of one component and the translation of the other. Translation of one component along the axis causes rotation of the other component according to the pitch relationship. Likewise, rotation of one component causes translation of the other component. Use SOLIDWORKS Help if needed.

- Rotate the handle and view the results. Think about how you would use this mate type in other assemblies.

Use the Select Other tool (See SOLIDWORKS Help) to select faces and edges that are hidden in an assembly.

Exercise 3.7: Angle Mate Assembly

- Copy all files from the Chapter 3 Homework\Angle Mate folder to your hard drive.

- Create the final assembly with all needed mates for proper movement.

- Move the Handle in the assembly. The Handle is free to rotate. Set the angle of the Handle.

- Insert an Angle mate (165 degrees) between the Handle and the Side of the valve using Planes. An Angle mate places the selected items at the specified angle to each other.

- The Handle has a 165-degree Angle mate to restrict flow through the valve. Think about how you would use this mate type in other assemblies.

Exercise 3.8: Angle Mate Assembly (Cont:)

Create two end caps (lids) for the ball valve using the Top-down Assembly method. Note: The Reference-In-Content symbols in the FeatureManager.

- Modify the Appearance of the body to observe the change - enhance visualization.

- Apply the Select-other tool to obtain access to hidden faces and edges.

Exercise 3.9: 4 Bar Linkage Assembly

Create the 4-bar linkage assembly as illustrated. The four-bar linkage assembly has five simple components.

Create the five simple components. Assume dimensions.

Insert all needed mates.

In an assembly, fix (f) the first component to the origin or fully define it to the three default sketch planes of the Assembly.

Insert additional components and insert needed mates to simulate the movement of a 4 bar linkage assembly.

Read the section on Coincident, Concentric and Distance mates in SOLIDWORKS Help.

Create a base with text for extra credit.

Exercise 3.9A: Gear Mate Assembly

Create the Gear Mate assembly as illustrated.

View the provided ppt and motion file in the Chapter 3 Homework\Gear Mate folder.

The Gear assembly consists of a **Base Plate** two **Shafts** for the gears, and two different gears from the **SOLIDWORKS toolbox**.

Use the SOLIDWORKS toolbox to insert the needed gear components. You create the **Base Plate** and **Shafts**.

Insert all needed mates for proper movement.

Gear mate: Forces two components to rotate relative to one another around selected axes. The Gear mate provides the ability to establish gear type relations between components without making the components physically mesh.

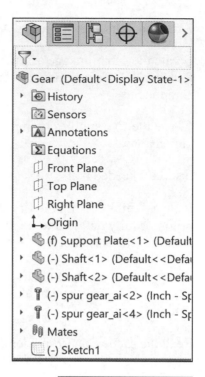

SOLIDWORKS provides the ability to modify the gear ratio without changing the size of the gears. Align the components before adding the Mechanical gear mate.

Exercise 3.9B: Cam Assembly

Copy all files from the Chapter 3 Homework\Cam Mate folder.

Create the final assembly with all needed mates for proper movement.

Open the assembly.

View the model. The model contains two cams.

Insert a tangent Cam follower mate.

1. Click the **Mate** ✎ tool from the Assembly tab.

2. **Pin** (Keep Visible icon) the Mate PropertyManager.

3. Click the **Mechanical** tab.

4. Click **Cam Follower** ⌀ mate.

5. Click inside of the **Cam Path** box.

6. Click the **outside face** of cam-s<1> as illustrated.

7. Click the **face of Link-c<1>** as illustrated. Face<2>@Link-c-1 is displayed.

8. Click **OK** ✔ from the CamMateTangent1 PropertyManager.

Insert a Coincident Cam follower mate.

9. Click **Cam Follower** 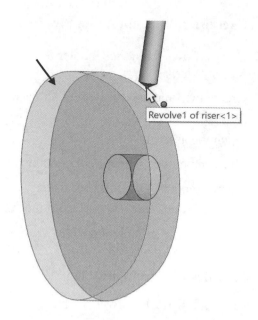 mate from the Mechanical Mates box.

10. Click the **outside face** of cam-s2<1>.

11. Click the vertex of **riser<1>** as illustrated. Vertex<1>@riser-1 is displayed.

12. Click **OK** ✔ from the CamMateCoincident1 PropertyManager.

13. **Un-pin** the Mate PropertyManager.

14. Click **OK** ✔ from the Mate PropertyManager.

15. **Expand** the Mates folder.

16. **View** the new created mates.

17. Display an **Isometric** view.

18. **Rotate** the cams. View the results in the Graphics window.

19. **Close** the model.

When creating a cam follower mate, make sure that your Spline or Extruded Boss/Base feature, which you use to form the cam contact face, does nothing but form the face.

Exercise 3.9C: Slot Assembly

Copy all files from the Chapter 3 Homework\Slot Mate folder. **Note**: This assembly uses Toolbar components.

Open the GUIDE-ROD assembly. Create two Slot mates for the two flange bolts. You can mate bolts to straight or arced slots and you can mate slots to slots. Use the Quick Mate procedure.

Insert a Slot mate (Center in Slot option) between the first flange bolt and the inside face of the right slot.

Insert the second slot mate (Center in Slot option) between the second flange bolt and the inside face of the left slot.

Insert a Coincident mate between the top face of the GUIDE and the bottom face of the second flange bolt cap.

Insert a Coincident mate between the top face of the GUIDE and the bottom face of the first flange bolt cap.

Insert a parallel mate between the front face of the first hex head flange bolt and the front face of the GUIDE.

Insert a parallel mate between the front face of the second hex head flange bolt and the front face of the GUIDE.

Expand the Mates folder. View the created mates.

Notes:

Project 4

Fundamentals of Drawing

Below are the desired outcomes and usage competencies based on the completion of Project 4.

Project Desired Outcomes:	Usage Competencies:
• B-ANSI-MM Drawing Template.	• Generate a Drawing Template with Document Properties and Sheet Properties.
• CUSTOM-B Sheet Format.	• Produce a Sheet Format with Custom Sheet Properties, Title block, Company logo and more.
• GUIDE SUPPORT Drawing.	• Create Standard Orthographic, Auxiliary, Detail and Section views. • Insert, create, and modify dimensions and annotations.
• GUIDE-ROD Drawing with a Bill of Materials.	• Knowledge to develop and incorporate a Bill of Materials with Custom Properties. Insert an Isometric exploded view with a Bill of Materials

Notes:

Project 4 - Fundamentals of Drawing

Project Objective

Provide an understanding of Drawing Templates, Part drawings, Assembly drawings, details and annotations.

Create a B-ANSI-MM Drawing Template. Create a CUSTOM-B Sheet Format. The Drawing Template contains Document Property settings. The Sheet Format contains a Company logo, Title block, Revision table and Sheet information.

Create the GUIDE SUPPORT drawing. Display Standard Orthographic, Section, Auxiliary and Detail drawing views. Insert, create and modify part and component dimensions.

Create an Isometric exploded GUIDE-ROD assembly drawing with a Bill of Materials. Obtain knowledge to develop and incorporate a Bill of Materials with Custom Properties.

On the completion of this project, you will be able to:

- Create a new Drawing Template.

- Generate a customized Sheet Format with Custom Properties.

- Open, Save and Close Drawing documents.

- Produce a Bill of Materials with Custom Properties.

- Insert and position views on a Multi Sheet drawing.

- Set the Dimension Layers.

- Insert, move and modify dimensions from a part into a drawing view.

- Insert Annotations: Center Mark, Centerline, Notes, Hole Callouts and Balloons.

- Use Edit Sheet Format and Edit Sheet mode.

- Insert a Revision table.

- Modify dimensioning scheme.

- Create parametric drawing notes.

- Link notes in the Title block to SOLIDWORKS properties.

- Rename parts and drawings.

- Insert a Center of Mass point.

Project Situation

The individual parts and assembly are completed. What is the next step? You are required to create drawings for various internal departments, namely production, purchasing, engineering, inspection and manufacturing. Each drawing contains unique information and specific footnotes. Example: A manufacturing drawing would require information on assembly, Bill of Materials, fabrication techniques and references to other relative documents.

Project Overview

Generate two drawings in this project:

- GUIDE SUPPORT drawing with a customized Sheet Format.

- GUIDE-ROD assembly drawing with a Bill of Materials.

The GUIDE drawing contains three Standard Orthographic views and an Isometric view. Do you remember what the three Principle Orthographic Standard views are? They are Top, Front and Right side (Third Angle Projection).

Three new views are introduced in this project: Detailed view, Section view and Auxiliary view. Orient the views to fit the drawing sheet. Incorporate the GUIDE dimensions into the drawing with inserted Annotations.

The GUIDE-ROD assembly drawing contains an Isometric Exploded view.

The drawing contains a Bill of Materials with balloon text with bent leaders and magnetic lines.

Both drawings utilize a custom Sheet Format containing a Company logo, Title block and Sheet information.

There are two major design modes used to develop a SOLIDWORKS drawing:

- Edit Sheet Format

- Edit Sheet

The Edit Sheet Format mode provides the ability to:

- Define the Title block size and text headings.

- Incorporate a company logo.

- Add a picture, design, company text and more.

The Edit Sheet mode provides the ability to:

- Add or modify drawing views.

- Add or modify drawing view dimensions.

- Add or modify text and more.

Drawing Template and Sheet Format

The foundation of a SOLIDWORKS drawing document is the Drawing Template. Drawing size, drawing standards, company information, manufacturing, and or assembly requirements, units and other properties are defined in the Drawing Template.

The Sheet Format is incorporated into the Drawing Template. The Sheet Format contains the border, Title block information, revision block information, company name, and or company logo information, Custom Properties, and SOLIDWORKS Properties. Custom Properties and SOLIDWORKS Properties are shared values between documents.

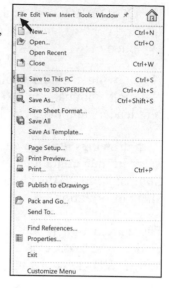

Utilize the standard B (ANSI) Landscape size Drawing Template with no Sheet Format. Set the Units, Font and Layers.

Modify a B (ANSI) Landscape size Sheet Format to create a Custom Sheet Format and Custom Drawing Template.

1. Set Sheet Properties and Document Properties for the Drawing Template.

2. Insert Custom Properties: CompanyName, Revision, Number, DrawnBy, DrawnDate, Company Logo, Third Angle Projection Logo etc. for the Sheet Format.

3. Save the Custom Drawing Template and Custom Sheet Format in the MY-TEMPLATE folder.

Views from the part or assembly are inserted into the SOLIDWORKS Drawing.

A Third Angle Projection scheme is illustrated in this project. For non-ANSI dimension standards, the dimensioning techniques are the same, even if the displayed arrows and text size are different.

For printers supporting millimeter paper sizes, select A3 (ANSI) Landscape (420mm x 297mm).

The default Drawing Templates with Sheet Format displayed contain predefined Title block Notes linked to Custom Properties and SOLIDWORKS Properties.

Activity: Create a Drawing Template.

Close all documents.

1) Click **Window**, **Close All** from the Menu bar.

Create a B (ANSI) Landscape, Third Angle Projection drawing document.

2) Click **New** ⬜ from the Menu bar.

3) Double-click **Drawing** from the Templates tab.

4) If needed **uncheck** the **Only show standard formats** box.

5) Select **B (ANSI) Landscape** from the Standard sheet size box.

6) Click **OK** from the Sheet Format/Size dialog. Accept the default sheet properties.

7) Click **Cancel** ✖ from the Model View PropertyManager. Draw1 is the default drawing name. Sheet1 is the default first sheet name.

You can define drawing sheet zones on a sheet format for the purpose of providing locations where drawing views and annotations reside on the drawing.

The B (ANSI) Landscape Standard Sheet border defines the drawing size, 17″ x 11″ (431.8mm x 279.4mm).

Zone letters and numbers will vary depending on your sheet format size.

Set Drawing Size, Sheet Properties and Document Properties for the Drawing Template. Sheet Properties control Sheet Size, Sheet Scale and Type of Projection. Document Properties control the display of dimensions, annotations and symbols in the drawing.

Set Sheet Properties. View your options in the Sheet Properties dialog box.

8)   **Expand** Sheet1 in the Drawing FeatureManager.

9)   Right-click **Sheet 1. View** your options.

10)   Click **Properties** 📇 . The Sheet Properties dialog box is displayed.

11)   Select Sheet Scale **1:1**.

12)   Select **B (ANSI) Landscape** from the Standard sheet size box.

13)   Select **Third Angle** for Type of projection.

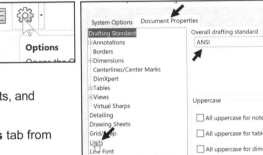

14)   Click **Apply Changes** from the Sheet Properties dialog box.

Set Document Properties. Set drafting standard, units, and precision.

15)   Click **Options** ⚙ `·`, **Document Properties** tab from the Menu bar.

16)   Select **ANSI** for Overall drafting standard.

17)   Click the **Units** folder.

18)   Select **MMGS** for Unit system.

19)   Select **.12** for basic unit length decimal place.

20)   Select **None** for basic unit angle decimal place.

Detailing options provide the ability to address dimensioning standards, text style, center marks, extension lines, arrow styles, tolerance and precision.

There are numerous text styles and sizes available in SOLIDWORKS. Companies develop drawing format standards and use specific text height for Metric and English drawings.

Numerous engineering drawings use the following format:

- Font: Century Gothic - All capital letters.

- Text height: .125in. or 3mm for drawings up to B (ANSI) Size, 17in. x 22in.

- Text height: .156in. or 5mm for drawings larger than B (ANSI) Size, 17in x 22in.

- Arrowheads: Solid filled with a 1:3 ratio of arrow width to arrow height.

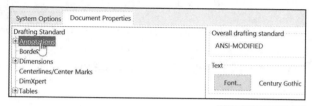

Set the Annotations font height.
21) Click the **Annotations** folder.

22) Click the **Font** button.

23) Click the **Units** button.

24) Enter **3.00**mm for Height.

25) Enter **1.00**mm for Space.

26) Click **OK**.

Change the Arrow Height.
27) Click the **Dimensions** folder from the Document Properties column as illustrated.

28) Enter **1**mm for arrow Height.

29) Enter **3**mm for arrow Width.

30) Enter **6**mm for arrow Length.

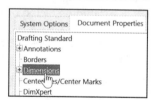

Set Section/Views size.
31) **Expand** the Views folder from the Document Properties column.

32) Click the **Section** folder.

33) Enter **2**mm for arrow Height.

34) Enter **6**mm for arrow Width.

35) Enter **12**mm for arrow Length.

36) Click **OK** from the Document Properties - Section dialog box.

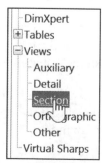

Drawing Layers organize dimensions, annotations and geometry. Create a new drawing layer to contain dimensions and notes. Create a second drawing layer to contain hidden feature dimensions.

Select the Show/Hide eye 👁 to turn On/Off Layers. Dimensions placed on the hidden layers are turned on and off for clarity and can be recalled for parametric annotations.

Display the Layer toolbar.
37) **Right-click** in the gray area in the Menu bar.

38) Click **Toolbars**. Check **Layer**. The Layer toolbar is displayed. Click and drag the Layer toolbar to the center of the Graphics window.

39) Click the **Layer Properties** 📁 file folder from the Layer toolbar. The Layers dialog box is displayed.

Create the Dimension Layer.
40) Click the **New** button. Enter **Dim** in the Name column.

41) **Double-click** under the Description column.

42) Enter **Dimensions** in the Description column.

Create the Notes Layer.
43) Click the **New** button. Enter **Notes** for Name. **Double-click** under the Description column.

44) Enter **General Notes** for Description.

Create the Hidden Dims Layer.
45) Click the **New** button. Enter **Hidden Dims** for Name. **Double-click** under the Description column. Enter **Hidden Insert Dimensions** for Description.

Dimensions placed on the Hidden Dims Layer are not displayed on the drawing until the Hidden Dims Layer status is On. Set the Layer Color to locate dimensions on this layer easily.

Turn the Hidden 👁 Insert Dimension Layer Off.
46) Click **On/Off** 👁. The row is displayed in black.

Layers					
Name	Description	👁	🖨	🔵	Style
FORMAT		👁	🖨	⬛	
Dim	Dimensions	👁	🖨	⬛	
Notes	General Notes	👁	🖨	⬛	
Hidden Dims	Hidden Insert Dimensions	👁			

Set the Layer 🔵 Color.
47) Click the **small black square** in the Hidden Dims row as illustrated.

Select a Color Swatch from the Color dialog box.

48) Select **Dark Blue**.

49) Click **OK**.

50) Click **OK** from the Layers dialog box.

The current Layer is Hidden Dims. Set the current
Layer to None before saving the Drawing
Template.

Set None for Layer.

51) Click the **Layer drop-down arrow**.

52) Click **None**. None is displayed in the Layer toolbar.

53) Drag the **Layer toolbar** out of the Graphics area.

The Drawing Template contains the drawing Size, Document Properties
and Layers. The Overall drafting standard is ANSI and the Units are in
millimeters. The current Layer is set to None. The Drawing Template
requires a Sheet Format. The Sheet Format contains Title block
information. The Title block contains vital part or assembly information.
Each company may have a unique version of a Title block.

Sheet Format and Title block

The Sheet Format contains the Title block, Revision block, Company logo, Custom
Properties, Zones, etc. The Title block contains text fields linked to System Properties
and Custom Properties.

System Properties are determined from the SOLIDWORKS documents. Custom Property
values are assigned to named variables. Save time. Utilize System Properties and define
Custom Properties in your Sheet Formats.

System Properties Linked to fields in default Sheet Formats:	Custom Properties of drawings linked to fields in default Sheet Formats:		Custom Properties of parts and assemblies linked to fields in default Sheet Formats:
SW-File Name (in DWG. NO. field):	CompanyName:	EngineeringApproval:	Description (in TITLE field):
SW-Sheet Scale:	CheckedBy:	EngAppDate:	Weight:
SW-Current Sheet:	CheckedDate:	ManufacturingApproval:	Material:
SW-Total Sheets:	DrawnBy:	MfgAppDate:	Finish:
	DrawnDate:	QAApproval:	Revision:
	EngineeringApproval:	QAAppDate:	

Utilize the standard landscape B (ANSI) Sheet Format (17in. x 11in.) or the standard-A (ANSI) Sheet Format (420mm x 297mm) to create a Custom Sheet Format.

Activity: Sheet Format and Title block.

Display the standard B (ANSI) Landscape Sheet Format.
54) Right-click in the **Graphics window**.

55) **Expand** the drop-down menu.

56) Click **Properties**. The Sheet Properties dialog box is displayed.

57) Click the **Standard sheet size** box.

58) Select **B (ANSI) Landscape**.

59) Check the **Display sheet format** box. The default Sheet Format, b - landscape.slddrt is displayed.

60) Click **Apply Changes** from the Sheet Properties dialog box. Note: If there are not changes, click Cancel.

The default Sheet Format is displayed in the Graphics window. The FeatureManager displays Draw1, Sheet 1.

A SOLIDWORKS drawing contains two edit modes:

1. Edit Sheet

2. Edit Sheet Format

Insert views and dimensions in the Edit Sheet mode.

Modify the Sheet Format text, lines or Title block information in the Edit Sheet Format mode.

The CompanyName Custom Property is located in the Title block above the TITLE box. There is no value defined for CompanyName. A small text box indicates an empty field. Define a value for the Custom Property CompanyName. Example: D&M ENGINEERING.

Activate the Edit Sheet Format mode.

61)   Right-click in the **Graphics window**.

62)   Click **Edit Sheet Format**. The Title block lines turn blue.

View the right side of the Title block.

63)   Click **Zoom to Area** 🔍. **Zoom in** on the Sheet Format Title block.

64)   Click **Zoom to Area** 🔍 to deactivate.

Define CompanyName Custom Property.

65)   Hover over the **center position** above the TITLE box as illustrated. The Note text $PRP:"COMPANYNAME" is displayed.

66)   Click the **position**.

67)   Click **File**, **Properties** from the Menu bar. The Summary Information dialog box is displayed.

68)   Click the **Custom** tab.

69)   Click inside the **Property Name** box.

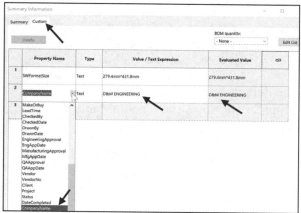

70)   Click the **drop-down arrow** ⌄ in the Property Name box.

71)   Select **CompanyName** from the Property list.

72)   Enter **D&M ENGINEERING** (your company or school name) in the Value/Text Expression box.

73)   Click inside the **Evaluated Value** box. The CompanyName is displayed in the Evaluated Value box.

74)   Click **OK**. The Custom Property "$PRP:COMPANYNAME," Value "D&M ENGINEERING" is displayed in the Title block.

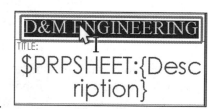

Modify the font size.

75)   Double-click **D&M ENGINEERING**. The Formatting dialog box and the Note PropertyManager is displayed.

76)   Click the **drop-down arrows** to set the Text Font and Height.

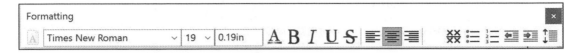

77)   Click the **Style buttons** and **Justification buttons** to modify the selected text.

78)   Click **OK** ✔ from the Note PropertyManager. View the results.

💡 Click a position outside the selected text box to save and exit the text.

The Tolerance block is located in the Title block. The Tolerance block provides information to the manufacturer on the minimum and maximum variation for each dimension on the drawing. If a specific tolerance or note is provided on the drawing, the specific tolerance or note will override the information in the Tolerance block.

General tolerance values are based on the design requirements and the manufacturing process.

💡 Create Sheet Formats for different parts types. Example: sheet metal parts, plastic parts and high precision machined parts. Create Sheet Formats for each category of parts that are manufactured with unique sets of Title block notes.

Modify the Tolerance block in the Sheet Format for ASME Y14.5-2009 machined, millimeter parts. Delete unnecessary text. The FRACTIONAL text refers to inches. The BEND text refers to sheet metal parts. The Three Decimal Place text is not required for this millimeter part.

Modify the Tolerance Note.

79)   Double-click the text **INTERPRET GEOMETRIC TOLERANCING PER:**

80)   Enter **ASME Y14.5** as illustrated.

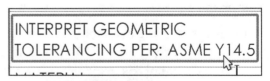

81)   Click **OK** ✔ from the Note PropertyManager.

82)   Double-click inside the **Tolerance block** text. The Formatting dialog box and the Note PropertyManager is displayed.

83)   Delete the text **INCHES**.

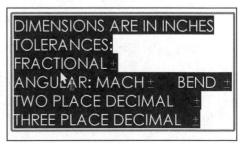

84) Enter **MILLIMETERS**.

85) Delete the line **FRACTIONAL +-**.

86) Delete the text **BEND +-**.

Enter ANGULAR tolerance.
87) Click a **position** at the end of the line.

88) Enter **0**. Click **Add Symbol** from the Text Format box.

89) Select **Degree** from the Modifying Symbols library.

90) Enter **30′** for minutes of a degree.

Modify the TWO and THREE PLACE DECIMAL LINES.
91) Delete the **TWO** and **THREE PLACE DECIMAL lines**.

92) Enter **ONE PLACE DECIMAL +- 0.5**.

93) Enter **TWO PLACE DECIMAL +- 0.15**.

94) Click **OK** from the Note PropertyManager.

95) Right-click **Edit Sheet** in the Graphics window.

Fit the drawing to the Graphics window. Save Draw1.
96) Press the **f** key.

97) Click **Save**. Accept the default name.

98) **View** the results.

Various symbols are available through the Symbol button in the Text dialog box. The $\pm$ symbol is located in the Modify Symbols list. The $\pm$ symbol is displayed as <MOD-PM>. The degree symbol $°$ is displayed as <MOD-DEG>.

Interpretation of tolerances is as follows:

- The angular dimension 110 is machined between 109.5 and 110.5.

- The dimension 2.5 is machined between 2.0 and 3.0.

- The Guide Hole dimension 10.000/10.015 is machined according to the specific tolerance on the drawing.

Company Logo

A Company logo is normally located in the Title block of the drawing. Create your own Company logo or copy and paste an existing picture.

If you have your own logo, skip the process of copying and applying the Logo.jpg file below. Insert the logo in the Edit Sheet Format mode.

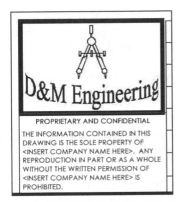

Activity: Insert a Company Logo.

Insert a Company or School Logo.

99) Right-click **Edit Sheet Format** in the Graphics window.

100) Click **Insert**, **Picture** from the Menu bar. The Open dialog box is displayed.

101) Select the **ENGDESIGN-W-SOLIDWORKS\LOGO folder**.

102) Double-click the **Logo.jpg** file. The Sketch Picture PropertyManager is displayed.

103) Un-check the **Enable scale tool** box.

104) Un-check the **Lock aspect ratio** box.

105) Drag the picture handles to size the **picture** to the left side of the Title block.

106) Click **OK** ✔ from the Sketch Picture PropertyManager.

Text can be added to create a custom logo. You can insert a picture or an object.

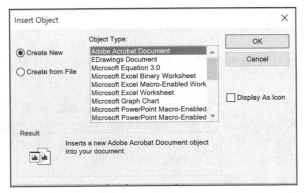

Return to the Edit Sheet mode.
107) Right-click in the
Graphics window.

108) Click **Edit Sheet**. The
Title block is displayed
in black.

Fit the Sheet Format to the Graphics window.
109) Press the **f** key.

Draw1 displays Editing Sheet1 in the Status bar. The Title block is displayed in black
when in Edit Sheet mode.

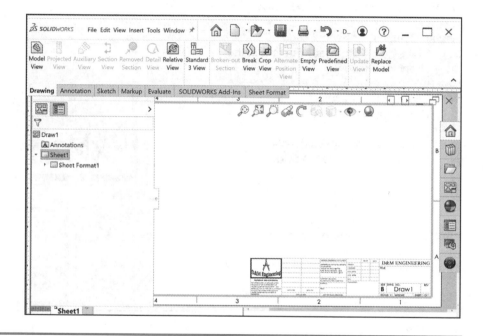

Save Sheet Format and Save As Drawing Template

Save the drawing document in the Graphics window in two forms: Sheet Format and
Drawing Template. Save the Sheet Format as a custom Sheet Format named
CUSTOM-B.

Use the CUSTOM-B (ANSI) Sheet Format for the drawings in this project. The Sheet
format file extension is .slddrt.

The Drawing Template can be displayed with or without the Sheet Format. Combine the
Sheet Format with the Drawing Template to create a custom Drawing Template named
B-ANSI-MM. Utilize the File, Save As option to save a Drawing Template. The Drawing
Template file extension is .drwdot.

☀ Select the Save as type option first, then select the Save in folder to avoid saving in default SOLIDWORKS installation directories.

Talk to your IT department *before you set* any locations on a network system. The procedure below is designed for your personal computer.

Note: If you do not create the templates or the MY-TEMPLATES tab, use a standard SOLIDWORKS drawing template and address the needed Document properties and Custom properties.

Activity: Save Sheet Format. Utilize the Save As Option for the Drawing Template.

Save the Sheet Format.

110) Click **File**, **Save Sheet Format** from the Menu bar. The Save Sheet Format dialog box is displayed. The file extension for Sheet Format is .slddrt.

111) Select **ENGDESIGN-W-SOLIDWORKS\MY-SHEETFORMAT** for Save In File Folder. The MY-SHEETFORMAT folder was download with the model folders.

112) Enter **CUSTOM-B** for File name.

113) Click **Save** from the Save Sheet Format dialog box.

Save the Drawing Template.

114) Click **Save As** 📄.

115) Click **Drawing Templates (*.drwdot)** from the Save as type box.

116) Select **ENGDESIGN-W-SOLIDWORKS\MY-TEMPLATES** for Save In File Folder. The MY-TEMPLATES folder was downloaded with the model folders.

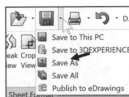

117) Enter **B-ANSI-MM** for File name.

118) Click **Save**.

If you created the MY-TEMPLATES tab in Project 2, skip this section. Set System Options - File Locations.

119) Click **Options** from the Menu bar. The System Options General dialog box is displayed.

120) Click the **File Locations** folder.

Add a new tab.

121) Click the **Add** button.

122) Browse to the **ENGDESIGN-W-SOLIDWORKS\MY-TEMPLATES** folder. **Note**: The folder and files were downloaded from the SDC Publications website (www.SDCpublications.com/downloads/978-1-63057-630-1).

123) Click **Select Folder**. The folder is added.

124) Position the folder at the bottom of the column. Click the **Move Down** button.

125) Click **OK** from the System Options, File Locations dialog box.

126) Click **Yes** to the SOLIDWORKS dialog box. The SOLIDWORKS dialog box displays the new Folder location.

127) Click **Yes**. Click **Yes**.

Close all files.

128) Click **Window**, **Close All** from the Menu bar. SOLIDWORKS remains open, and no documents are displayed.

💡 Talk to your IT department *before you set* any locations on a network system. Your document templates and tutorial location will vary depending on your network system setup.

💡 Combine customized Drawing Templates and Sheet Formats to match your company's drawing standards. Save the empty Drawing Template and Sheet Format separately to reuse information.

Note: If you do not create the templates or the MY-TEMPLATES tab, use a standard SOLIDWORKS drawing template and address the needed document properties and custom properties.

Review Drawing Templates.

A custom Drawing Template was created from the default Drawing Template. You modified Sheet Properties and Document Properties to control the Sheet size, scale, annotations, dimensions and layers.

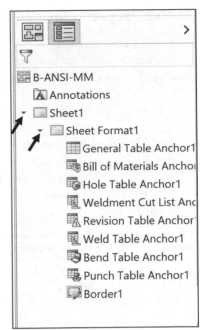

The Sheet Format contained Title block and Custom Property information. You inserted a Company Logo and modified the Title block.

The Save Sheet Format option was utilized to save the CUSTOM-B.SLDDRT Sheet Format. The File, Save As option was utilized to save the B-ANSI-MM.DRWDOT Template. The Sheet Format was saved in the MY-SHEETFORMATS folder and Drawing Template was saved in the MY-TEMPLATES folder.

GUIDE Part - Modify

A drawing contains part views, geometric dimensioning and tolerances, centerlines, center marks, notes, custom properties and other related information. Perform the following tasks before starting the GUIDE drawing:

- Verify the part. The drawing requires the associated part.

- View dimensions in each part. Step through each feature of the part and review all dimensions.

- Review the dimension scheme to determine the required dimensions and notes to manufacture the part.

Activity: GUIDE Part - Modify.

Open the GUIDE part.

129) Click **Open** from the Menu bar.

130) Select the **folder** that the GUIDE document is in.

Modify the dimensions.
131) Double-click **GUIDE**.

132) Click **Hidden Lines Visible** from the Heads-up View toolbar.

133) Double-click **Base-Extrude** from the GUIDE FeatureManager.

134) Click the **80.00** dimension.

135) Enter **100**mm. Click **inside** the Graphics window.

Display a Shaded With Edges view. Save the GUIDE part.
136) Click **Shaded With Edges** from the Heads-up View toolbar.

137) Click **Save**.

Review part history with the Rollback bar to understand how the part was created. Position the Rollback bar at the top of the FeatureManager. Drag the Rollback bar below each feature. When working between features, right-click in the FeatureManager. Select Roll to Previous\Roll to End.

GUIDE Part - Drawing

The GUIDE drawing consists of multiple views, dimensions and annotations. The GUIDE part was designed for symmetry. Add or redefine dimensions in the drawing to adhere to a Drawing Standard. Add dimensions and notes to the drawing in order to correctly manufacture the part.

Address the dimensions for three features:

- The right Slot Cut is not dimensioned to an ASME Y14 standard.

- No dimensions exist for the left Mirror Slot Cut.

- The Guide Hole and Linear Pattern of Tapped Holes require notes.

The GUIDE part remains open. Create a new GUIDE drawing. Utilize the B-ANSI-MM Drawing Template. Utilize the Model View 🖱 tool to insert the Front view into Sheet1. Utilize the Auto-start Projected View option to project the Top, Right and Isometric views from the Front view. **Note**: You can also use the View Palette in the Task Pane.

Activity: GUIDE Part - Drawing.

Create the GUIDE Drawing.

138) Click **New** ⬜ from the Menu bar.

139) Double-click **B-ANSI-MM** from the MY-TEMPLATES tab. The Model View PropertyManager is displayed.

140) Click **Cancel** ✕ from the Model View PropertyManager. The Draw# FeatureManager is displayed.

Review the Draw# FeatureManager.
141) **Expand** Sheet1 from the FeatureManager.

142) **Expand** Sheet Format1 from the FeatureManager. View your options.

🔆 The current drawing name is Draw2 if the second new drawing is created in the same session of SOLIDWORKS. The current sheet name is Sheet1. Sheet1 is the current Sheet.

Insert four Drawing Views.

143) Click **Model View** from the Drawing tab in the CommandManager. The Model View PropertyManager is displayed.

144) Double-click **GUIDE** from the Part/Assembly to Insert box.

Insert four views.

145) Check the **Create multiple views** box.

146) Click **Front, Top** and **Right** from the Orientation box. **Note**: All four views are selected. *Isometric should be selected by default.

147) Click **OK** ✔ from the Model View PropertyManager. Four views are displayed on Sheet1.

148) Click inside the **Isometric view boundary**. Drawing View4 PropertyManager is displayed.

149) Click **Shaded With Edges** 🔲 from the Display Style box.

150) Click **OK** ✔ from the Drawing View4 PropertyManager. If required, hide any annotations, origins or dimensions as illustrated.

💡 By default, the Center marks-holes box is checked under **Options**, **Document Properties**, **Detailing** from the Menu bar.

Save the GUIDE drawing. Enter name.

151) Click **Save As** 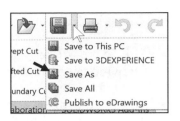.

152) Select **PROJECTS** for Save in folder.

153) Enter **GUIDE** for file name. Drawing is the default Save as type.

154) Click **Save**. Note: The correct display modes need to be selected, dimensions to be added, along with Center Marks, Centerlines, additional views, Custom Properties, etc.

The DWG. NO. box in the Title block displays the part File name, GUIDE. The TITLE: box in the Title block displays GUIDE SUPPORT.

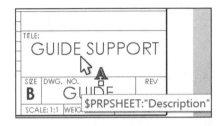

Predefined text in the CUSTOM-B Sheet Format links the Properties: $PRPSHEET:"Description" and $PRP:"SW-FileName." The Properties were defined in the GUIDE part utilizing File, Save As. Properties in the Title block are passed from the part to the drawing.

Always confirm the File name and Save in folder. Projects deal with multiple File names and folders. Select Save as type from the drop down list. Do not enter the extension. The file extension is entered automatically.

Each drawing has a unique file name. Drawing file names end with a .slddrw suffix. Part file names end with a .sldprt suffix. A drawing or part file can have the same prefix. A drawing or part file **can't have the same suffix**.

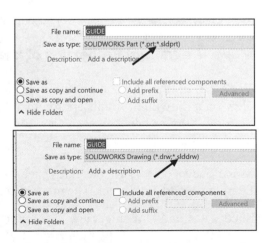

Example: Drawing file name GUIDE.slddrw. Part file name GUIDE.sldprt. The current file name is GUIDE.slddrw.

The GUIDE drawing contains three Principle views (Standard Orthographic views): Front, Top, Right and an Isometric view. You created the views with the Model View option.

Drawing views can be inserted as follows:

- Utilize the **Model View** tool from the Drawing tab in the CommandManager.

- **Drag** and **drop** an active part view from the View Palette located in the Task Pane as illustrated. With an open part, drag and drop the selected view into the active drawing sheet.

- **Drag** and **drop** a part into the drawing to create three Standard Views.

- **Predefine** views in a custom Drawing Template.

- Drag a **hyperlink** through Internet Explorer.

The Top view and Right view are projected off the view you place in the Front view location. Any view can be dragged and dropped into the Front view location of a drawing.

The View Palette from the Task Pane populates when you:

- Click Make Drawing from Part/Assembly.

- Browse to a document from the View Palette.

- Select from a list of open documents in the View Palette.

Move Views and Properties of the Sheet

The GUIDE drawing contains four views. Reposition the view on a drawing. Provide approximately 1in. - 2in. (25mm - 50mm) between each view for dimension placement.

Move Views on Sheet1 to create space for additional Drawing View placement. The mouse pointer provides feedback in both the Drawing Sheet and Drawing View modes. The mouse pointer displays the Drawing Sheet icon when the Sheet properties and commands are executed.

The mouse pointer displays the Drawing View icon when the View properties and commands are executed.

View the mouse pointer for feedback to select Sheet, View, and Component and Edge properties in the Drawing.

Sheet Properties

- Sheet Properties display properties of the selected sheet. Right-click in the sheet boundary to view the available commands.

View Properties

- View Properties display properties of the selected view. Right-click inside the view boundary. Modify the View Properties in the Display Style box or the View Toolbar.

Component Properties

- Component Properties display properties of the selected component. Right-click on the face of the component. View the available options.

Edge Properties

- Edge Properties display properties of the selected geometry. Right-click on an edge inside the view boundary. View the available options.

Base Extrude of GUIDE

Activity: Move Views and Properties of the Sheet.

Modify and move the front view.

155) Click inside the **Drawing View1** (Front) view boundary. The mouse pointer displays the Drawing View ⬚ icon. The view boundary is displayed in blue.

156) Click **Hidden Lines Visible** ⬜ from the Display Style box.

157) Position the **mouse pointer** on the edge of the Front view until the Drawing Move View ✥ icon is displayed.

158) Click and drag **Drawing View1** in an upward vertical direction. The Top and Right view move aligned to Drawing View1 (Front).

Modify the top view.

159) Click inside the **Top view boundary** (Drawing View3). The Drawing View3 PropertyManager is displayed.

160) Click **Hidden Lines Removed** ⬜ from the Display Style box. Various display styles provide the ability to select and view features of a part.

161) Click **OK** ✔ from the Drawing View3 PropertyManager. Later, address Centerlines, Center Marks, Tangent Edges Removed, Display styles, Customer properties, etc. to finish the drawing.

Select the dashed view boundary to move a view in the drawing.

Auxiliary View, Section View and Detail View

The GUIDE drawing requires additional views to document the part. Insert an Auxiliary view, Section view and Detail view from the Drawing tab in the CommandManager. Review the following view terminology before you begin the next activity.

Auxiliary View:

The Auxiliary view drawing tool provides the ability to display a plane parallel to an angled plane with true dimensions.

A primary Auxiliary view is hinged to one of the six Principle Orthographic views. Create a Primary Auxiliary view that references the angled edge in the Front view.

Section View:

The Section view drawing tool provides the ability to display the interior features of a part. Define a cutting plane with a sketched line in a view perpendicular to the Section view.

Use the Section view sketch mode in conjunction with the Section tool user interface to create a Section view in the Top view.

Detail View:

The Detail view ⌀ⒶⒶ drawing tool provides the ability to enlarge an area of an existing view. Specify location, shape and scale.

Create a Detail view from a Section view with a 3:2 scale.

💡 The book is designed to expose the new user to many tools, techniques and procedures. It does not always use the most direct tool or process.

DETAIL C
SCALE 3 : 2

💡 You can add multiple breaks to models using the Model Break View tool. The model breaks are saved as configurations. Model break views are helpful when you need to shorten components especially for technical and marketing purposes. Additionally, the Model Break View tool lets you display breaks on drawing views per ASME Y14.3.

Activity: Auxiliary Drawing View.

Insert an Auxiliary drawing view.
162) Click the **Drawing** tab from the CommandManager.

163) Click the **Auxiliary View** 𝄜 drawing tool. The Auxiliary View PropertyManager is displayed.

164) Click the **right angled edge** of the GUIDE in the Front view as illustrated.

VIEW A-A

165) Click a **position** to the right and up from the Front view as illustrated.

Position the Auxiliary View.
166) Click and drag the **section line A-A midpoint** toward Drawing View1. The default label A is displayed in the Arrow box.

167) Click the **OK** ✔ from the PropertyManager.

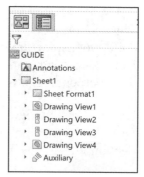

Rename the new view.
168) **Rename** the new view (Drawing View#) to Auxiliary in the FeatureManager as illustrated.

Activity: Section Drawing View.

Insert a Section drawing view.

169) Click the **Section View** ↩ drawing tool. The Section View PropertyManager is displayed. The Section tab is selected by default.

170) Click the **Horizontal** Cutting Line button as illustrated.

171) Locate the **midpoint of the left vertical line** as illustrated.

172) Click the **midpoint**. Note: Your options with the Pop-up menu.

173) Click the **OK** ✓ icon from the pop-up menu.

Position the Section drawing View.

174) Click a **position** above the Top view. The section arrows point downward.

175) If needed, check **Flip direction** from the Section Line box. The section arrows point upward. If required, enter **B** for Section View Name in the Label box.

176) Click **OK** ✓ from the Section View B-B PropertyManager. Section View B-B is displayed in the FeatureManager. If required, hide any annotations.

177) If needed, insert a **CenterMark** in the Section view. Utilize the CenterMark tool under the Annotations tab.

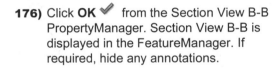

SECTION B-B

Save the drawing.

178) Click **Save** 💾.

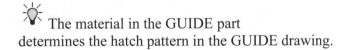

🔆 The material in the GUIDE part determines the hatch pattern in the GUIDE drawing.

Activity: Detail Drawing View.

Insert a Detail drawing view.

179) Click the **Detail View** drawing tool. The Detail View PropertyManager is displayed.

Sketch a Detail circle.

180) Click the **center point** of the Guide Hole in the Section view.

181) Click a **position** to the lower left of the Guide Hole to complete the circle.

Position the Detail View.

182) Click a **position** to the right of the Section View. If required, enter **C** for Detail View Name in the Label box.

183) Select **Per Standard** for Detail Circle Style.

Modify the drawing view scale.

184) Check **Use custom scale** option.

185) Enter **3:2** in the Scale box. Click **OK** ✔ from the Detail View C PropertyManager. Detail View C is displayed in the FeatureManager. **Note**: If needed, insert a CenterMark in the Detail view. Utilize the CenterMark tool under the Annotations tab.

SECTION B-B

DETAIL C
SCALE 3 : 2

Save the GUIDE drawing.

186) Click **Save** 🖫.

Partial Auxiliary Drawing View - Crop Drawing View

Create a Partial Auxiliary view from the Full Auxiliary view. Sketch a closed profile in the active Auxiliary view. Create the Profile with a closed Spline. Create a Partial Auxiliary view.

Crop the view. The 6mm dimension references the centerline from the Guide Hole. For Quality Assurance and Inspection of the GUIDE part, add a dimension that references the Temporary Axis of the Guide Hole. Sketch a centerline collinear with the Temporary Axis.

Activity: Partial Auxiliary Drawing View - Crop Drawing View.

Select the view.

187) Click **Zoom to Area** 🔍 from the Heads-up View toolbar.

188) Zoom in on the Auxiliary view.

189) Click **Zoom to Area** 🔍 to deactivate.

190) Click inside the **Auxiliary view boundary**. The Auxiliary PropertyManager is displayed.

Display Hidden Lines Removed.
191) Click **Hidden Lines Removed** ⬚.

Sketch a closed Spline profile.
192) Click the **Sketch** tab from the CommandManager.

193) Click the **Spline** 〜 Sketch tool.

194) Click **five or more positions** clockwise to create the closed Spline as illustrated. The first point is Coincident with the last point. The Sketch profile is closed.

Insert a Partial Auxiliary Drawing View.

195) Click the **Crop View** 🔲 drawing tool from the Drawing tab in the CommandManager.

196) Click **OK** ✔ from the Spline PropertyManager. The Crop View is displayed on Sheet1.

VIEW A-A

Display the Temporary Axes and insert a sketched centerline.

197) Click inside the **Auxiliary view boundary**. The Auxiliary PropertyManager is displayed.

198) Click **View**, **Hide/Show**, check **Temporary Axes** from the Menu bar. The Temporary Axis for the Guide Hole is displayed.

199) Click **Sketch** tab from the CommandManager.

200) Click the **Centerline** Sketch tool. The Insert Line PropertyManager is displayed.

VIEW A-A

201) Sketch a **centerline** parallel above the Temporary axis. The centerline extends approximately 5mm to the left and right of the profile lines. If needed insert a Parallel relation to the temporary axis.

202) Right-click **Select** to deselect the Centerline Sketch tool.

Add a Collinear relation between the centerline and temporary axis.

203) Click the **centerline**. The Line Properties PropertyManager is displayed.

204) Hold the **Ctrl** key down.

205) Click the **Temporary Axis**. The mouse pointer displays the Axis feedback icon. The Properties PropertyManager is displayed. Axis<1> and Line1 are displayed in the Selected Entities box.

206) Release the **Ctrl** key.

207) Click **Collinear** from the Add Relations box.

208) Click **OK** from the Properties PropertyManager.

Hide the Temporary Axis.

209) Click **View**, **Hide/Show**, uncheck **Temporary Axes** from the Menu bar.

Move the views to allow for ample spacing for dimensions and notes.

VIEW A-A

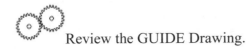 Review the GUIDE Drawing.

You created a new drawing, GUIDE with the B-ANSI-MM Drawing Template. The GUIDE drawing utilized the GUIDE part in the Model View PropertyManager. The Model View PropertyManager allowed new views to be inserted with a View Orientation. You selected Front, Top, Right and Isometric to position the GUIDE views.

Additional views were required to fully detail the GUIDE. You inserted the Auxiliary Section, Detail, Partial Auxiliary and Crop view. You moved the views by dragging the view boundary. The next step is to insert the dimensions and annotations to detail the GUIDE drawing.

Display Modes and Performance

Display modes for a Drawing view are similar to a part. When applying Shaded With Edges, select either Tangent edges removed or Tangent edges As phantom from the System Options section.

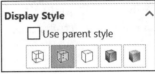

Mechanical details require either the Hidden Lines Visible mode or the Hidden Lines Removed display mode. Select Shaded/Hidden Lines Removed to display Auxiliary views to avoid confusion.

Wireframe   Hidden Line Visible   Hidden Line Removed   Shaded With Edges

Tangent Edges Visible provides clarity for feature edges. To address the ASME Y14.5 standard, use Tangent Edges With Font (Phantom lines) or Tangent Edges Removed. Right-click in the view boundary to access the Tangent Edge options.

Drawing views can be displayed in High quality and Draft quality. In High quality, all model information is loaded into memory. By default, drawing views are displayed in High quality.

In Draft quality, only minimum model information is loaded into memory. Utilize Draft quality for large assemblies to increase performance.

💡 Utilize **Options**, **System Options**, **Drawings**, **Display Style** to control the quality of a view.

By default, SOLIDWORKS will populate Section and Detail views before other views on your drawing.

Use the Pack and Go tool to save and gather all related files for a model design (parts, assemblies, drawings, references, design tables, Design Binder content, decals, appearances, and scenes and SOLIDWORKS Simulation results) into a single folder or zip file. It's one of the best tools to utilize when you are trying to save a large assembly or drawing with references and SOLIDWORKS Toolbox components.

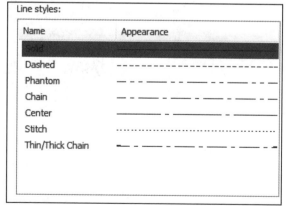

Detail Drawing

The design intent of this project is to work with dimensions inserted from parts and to incorporate them into the drawings. Explore methods to move, hide and recreate dimensions to adhere to a drawing standard.

There are other solutions to the dimensioning schemes illustrated in this project. Detail drawings require dimensions, annotations, tolerance, materials, Engineering Change Orders, authorization, etc. to release the part to manufacturing and other notes prior to production.

Review a hypothetical "worst case" drawing situation. You just inserted dimensions from a part into a drawing. The dimensions, extension lines and arrows are not in the correct locations. How can you address the position of these details? Answer: Dimension to an ASME Y14.5M standard.

No.	Situation:
1	Extension line crosses dimension line. Dimensions not evenly spaced.
2	Largest dimension placed closest to profile.
3	Leader lines overlapping.
4	Extension line crossing arrowhead.
5	Arrow gap too large.
6	Dimension pointing to feature in another view. Missing dimension – inserted into Detail view (not shown).
7	Dimension text over centerline, too close to profile.
8	Dimension from other view – leader line too long.
9	Dimension inside section lines.
10	No visible gap.
11	Arrows overlapping text.
12	Incorrect decimal display with whole number (millimeter), no specified tolerance.

Worst Case Drawing Situation

The ASME Y14.5 2009 standard defines an engineering drawing standard. Review the twelve changes made to the drawing to meet the standard.

Dimensions are displayed in MILLIMETERS.

No.	Preferred Application of the Dimensions:
1	Extension lines do not cross unless situation is unavoidable. Stagger dimension text.
2	Largest dimension placed farthest from profile. Dimensions are evenly spaced and grouped.
3	Arrowheads do not overlap.
4	Break extension lines that cross close to arrowhead.
5	Flip arrows to the inside.
6	Move dimensions to the view that displays the outline of the feature. Ensure that all dimensions are accounted for.
7	Move text off of reference geometry (centerline).
8	Drag dimensions into their correct view boundary. Create reference dimensions if required. Slant extension lines to clearly illustrate feature.
9	Locate dimensions outside off section lines.
10	Create a visible gap between extension lines and profile lines.
11	Arrows do not overlap the text.
12	Whole numbers displayed with no zero and no decimal point (millimeter).

Apply these dimension practices to the GUIDE drawing. Manufacturing utilizes detailed drawings. A mistake on a drawing can cost your company a substantial loss in revenue. The mistake could result in a customer liability lawsuit.

As the designer, dimension and annotate your parts clearly to avoid common problems and mistakes.

Insert Dimensions from the part.
Dimensions you created for each part feature
are inserted into the drawing.

Select the first dimensions to display for the
Front view. Do not select the Import Items
into all Views option for complex drawings.
Dimension text is cluttered and difficult to
locate.

Follow a systematic, "one view at a time"
approach for complex drawings. Insert part
feature dimensions onto the Dim Layer for
this project.

Activity: Detail Drawing - Insert Model Items.

Set the Dimension Layer.
210) Right-click in the drawing sheet.

211) Click **Change Layer**. The Change Layer dialog
box is displayed.

212) Select **Dim** from the drop-down menu.

Insert dimensions into Drawing View1 (Front).
213) Click inside the **Drawing View1** boundary. The Drawing View1
PropertyManager is displayed.

214) Click the **Model Items** tool from the Annotation tab in the
CommandManager. The Model Items PropertyManager is displayed.

215) Select **Entire model** from the Source box. Drawing View1 is displayed
in the Destination box. At this time, do not click the Hole callout button
to import the Hole Wizard information into the drawing. Accept the
default settings as illustrated.

216) Click **OK** from the Model Items PropertyManager. Dimensions are
displayed in the Front view.

In SOLIDWORKS, inserted dimensions in the drawing are
displayed in gray. Imported dimensions from the part are displayed in
black.

Drawing dimension location is dependent on:

- Feature dimension creation.

- Selected drawing views.

Note: The Import items into all views option, first inserts dimensions into Section Views and Detail Views. The remaining dimensions are distributed among the visible views on the drawing.

Move Dimensions in the Same View

Move dimensions within the same view. Use the mouse pointer to drag dimensions and leader lines to a new location.

Leader lines reference the size of the profile. A gap must exist between the profile lines and the leader lines. Shorten the leader lines to maintain a drawing standard. Use the blue Arrow control buttons to flip the dimension arrows.

Insert part dimensions into the Top view. The Top view displays crowded dimensions. Move the overall dimensions. Move the Slot Cut dimensions.

Place dimensions in the view where they display the most detail. Move dimensions to the Auxiliary View. Hide the diameter dimensions and add Hole Callouts. Display the view with Hidden Lines Removed.

 Illustrations may vary depending on your SOLIDWORKS release version.

Activity: Detail Drawing – Modify Linear Precision. Move Dimensions.

Modify linear precision. Move the linear dimensions in Drawing View1.

217) Modify linear **precision** to zero in the Drawing document.

218) **Zoom to area** on Drawing View1.

219) Click and drag the vertical dimension text **10**, **29** and **50** to the right as illustrated.

Create a gap between the extension lines and the profile lines for the 10mm vertical dimension.

220) Click the vertical dimension text **10**. The vertical dimension text, extension lines and the profile lines are displayed in blue.

221) Click and drag the **square blue endpoints** approximately 10mms from the right vertex as illustrated. A gap is created between the extension line and the profile.

A gap exists between the profile line and the leader lines. Drag the blue endpoints to a vertex, to create a gap.

💡 The smallest linear dimension should be placed closest to the profile.

💡 Note the default settings for a drawing document under Dimensions for **Leading zeroes**, **Trailing zeroes**, **Dimensions**, **Tolerances** and **Properties**.

Fit the drawing to the Graphics window.

222) Press the **f** key.

Insert dimensions into Drawing View3.

223) Click inside the **Drawing View3** boundary. The Drawing View3 PropertyManager is displayed.

224) Click **Hidden Lines Removed**.

225) Click the **Model Items** tool from the Annotation tab in the CommandManager. The Model Items PropertyManager is displayed.

226) Select **Entire model** from the Source box. Drawing View3 is displayed in the Destination box. Accept the default settings.

227) Click **OK** ✔ from the Model Items PropertyManager.

Move the vertical dimensions.

228) Click and drag the two vertical Slot Cut dimensions, **10** to the right of the Section arrow as illustrated.

229) Flip the **arrows** to the inside.

230) Click the **dimension text** and drag the text outside the leader lines. Hide all other dimensions and annotations as illustrated.

231) If needed, enter **2X** for the Radius as illustrated.

Fit the drawing to the Graphics window.

232) Press the **f** key.

Save the GUIDE drawing.

233) Click **Save** 💾.

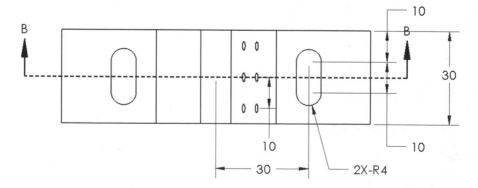

Insert dimensions into the Auxiliary View.
234) Click **inside** the Auxiliary view boundary.

235) Click the **Model Items** Drawing tool. The Model Items PropertyManager is displayed.

236) Select **Entire Model** from the Source box. Auxiliary is displayed in the Destination box. View your options from the Dimensions box.

237) Click the **Hole Wizard Locations** button from the Dimensions box. **Note**: Do not click the Hole Callout button at this time. You will manually create the Hole Wizard Callout in the drawing.

238) Click **OK** ✔ from the Model Items PropertyManager.

239) **Move** the dimensions off the view as illustrated. If needed zoom in on the dimensions to move them.

The dimensions for the Linear Pattern of Holes are determined from the initial Hole Wizard position dimensions and the Linear Pattern dimensions. Your dimensioning standard requires the distance between the holes in a pattern.

Do not over dimension. In the next steps, hide a few existing dimensions and add a new dimension.

In the next section, if needed, click **Options**, **Document Properties, Dimensions** from the Menu bar. Uncheck the **Add parentheses by default** box. Click **OK**.

VIEW A-A

Hide the following dimensions.
240) Right-click **25**.

241) Click **Hide**. Right-click **6**. Click **Hide**.

242) Right-click **M3x0.5**. Click **Hide**. Note: if required, hide any other dimensions or annotations. Do not hide the 12 dimension.

VIEW A-A

☀ To show a hidden dimension, click **View**, **Hide/Show**, **Annotations** from the Menu bar.

Add a dimension.

243) Click **Smart Dimension** ↖ from the Annotation toolbar. The Dimension PropertyManager is displayed.

244) Click **Smart dimensioning** in the Dimension Assist Tools box.

245) Click the **center point** of the bottom left hole. Click the **center point** of the bottom right hole.

246) Click a **position** below the profile as illustrated.

247) Check **OK** ✔ from the Dimension PropertyManager.

Save the drawing.
248) Click **Save** 💾. View the results.

VIEW A-A

Move Dimensions to a Different View

Move the linear dimension 10 that defines the Linear Hole Pattern feature from Drawing View3 (Top) to the Auxiliary view. When moving dimensions from one view to another, utilize the Shift key and only drag the dimension text. Release the dimension text inside the view boundary. The text will not switch views if positioned outside the view boundary.

Activity: Move Dimensions to a Different View.

Move dimensions from the Top view to the Auxiliary view.
249) Press the **z key** approximately 4 times to view the dimensions in the Top view.

250) Hold the **Shift** key down. Click and drag the vertical dimension **10** between the 2 holes from the Top view to the Auxiliary view. Release the **mouse button** and the **Shift** key when the mouse pointer is inside the Auxiliary view boundary.

251) Click and drag the **dimensions** off the Auxiliary view.

252) Click and drag the **VIEW A-A** text off the view boundary. View the results.

Save the drawing.

253) Click **Save** 💾. View the results.

Dimension Holes and the Hole Callout

Simple holes and other circular geometry are dimensioned in various ways: Diameter, Radius and Linear (between two straight lines).

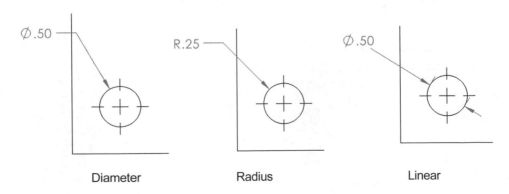

The holes in the Auxiliary view require a diameter dimension and a note to represent the six holes. Use the Hole Callout to dimension the holes. The Hole Callout function creates additional notes required to dimension the holes.

The dimension standard symbols are displayed automatically when you use the Hole Wizard feature.

Activity: Dimension Holes and the Hole Callout.

Dimension the Linear Pattern of Holes.
254) Click the **Annotation** tab from the CommandManager.

255) Click the **Hole Callout** ⌴⌀ tool. The Hole Callout tool inserts information from the Hole Wizard.

256) Click the **circumference** of the lower left circle in the Auxiliary view as illustrated. The tool tip M3x0.5 Tapped Hole1 of GUIDE is displayed.

257) Click a **position** to the bottom left of the Auxiliary view.

💡 The Hole Callout text displayed in the Dimension Text box depends on the options utilized in the Hole Wizard feature and the Linear Pattern feature.

Set Hole Callout precision.
258) Select **.12** from the Primary Unit Tolerance/Precision box.

259) Click **OK** ✔ from the Dimension PropertyManager. The Hole Callout is deactivated.

💡 Display the Hole Wizard Callout information automatically by selecting the Hole callout button in the Model Items PropertyManager.

VIEW A-A

Know inch/mm decimal display. The ASME Y14.5-2009 standard states:

- For millimeter dimensions <1, display the leading zero. Remove trailing zeroes.

- For inch dimensions <1, delete the leading zero. The dimension is displayed with the same number of decimal places as its tolerance.

Symbols are located on the bottom of the Dimension Text dialog box. The current text is displayed in the text box. Example:

- <NUM_INST>: Number of Instances in a Pattern.

- <MOD-DIAM>: Diameter symbol ⌀.

- <HOLE-DEPTH>: Deep symbol ⏚.

- <HOLE-SPOT>: Counterbore symbol ⊔.

- <DIM>: Dimension value 3.

Fit the Drawing to the Graphics window.
260) Press the **f** key.

Insert dimension text.
261) Click inside the **Drawing View1** boundary. The Drawing View1 PropertyManager is displayed.

262) Click the Guide Hole dimension ⌀**10.015/10.000** in Drawing View1.

263) Enter text **THRU** in the Dimension box. THRU is displayed on the drawing in blue.

264) Click **OK** ✔ from the Dimension PropertyManager.

Save the drawing.
265) Click **Save** 🖫.

 Access Notes, Hole Callouts and other Dimension Annotations through the Annotations toolbar menu. Access Annotations with the right mouse button, and click Annotations.

Center Marks and Centerlines

Hole centerlines are composed of alternating long and short dash lines. Centerlines indicate symmetry. Centerlines also identify the center of a circle, axes or cylindrical geometry.

Center Marks represents two perpendicular intersecting centerlines. The default Size and display of the Center Mark is set in Options, Document Properties, Dimensions, Centerlines/Center Marks as illustrated.

Center Marks are inserted on view creation by default. Modify the position and size of the Center Marks. The following three steps illustrate how to create a new Center Mark.

Activity: Center Marks and Centerlines.

Insert a Center mark into the Front view. If a Center mark exists, skip the next few steps.

266) Click inside the **Drawing View1** boundary.

267) Click the **Center Mark** ⊕ tool from the Annotation tab in the CommandManager. The Center Mark PropertyManager is displayed.

268) Click the **Guide Hole circumference**. The Center mark is inserted.

269) Click **OK** ✔ from the Center Mark PropertyManager.

Dimension standards require a gap between the Center Mark and the end points of the leader line. Currently, the leader line overlaps the Center Mark.

Insert Centerlines into the drawing views.

270) Click the **Centerline** tool from the Annotation tab in the CommandManager. The Centerline PropertyManager is displayed.

271) Check the **Select View** box in the Auto Insert dialog box.

272) Click inside the Front view, **Drawing View1** boundary.

273) Click inside the **Detail view** boundary. Centerlines are displayed.

274) Click inside the **Section view** boundary. Centerlines are displayed.

275) Click **OK** ✔ from the Centerline PropertyManager.

Save the drawing.
276) Click **Save** 💾.

Activate the Top view and insert Center marks.
277) Click inside the **Drawing View3** boundary.

278) Click the **Center Mark** ⊕ tool from the Annotation tab in the CommandManager. The Center Mark PropertyManager is displayed.

279) Uncheck the **Slot center mark** box.

280) Click the **four arcs**. The center marks are displayed.

281) Uncheck the **Use document defaults** box.

282) Enter **1**mm for Mark size.

283) Un-check the **Extended lines** box. Accept the default settings.

284) Click **OK** ✔ from the Center Mark PropertyManager. View the results.

285) Click **Save** 💾. View the results.

Insert an Annotation Centerline.

286) Click the **Centerline** tool from the Annotation tab. The Centerline PropertyManager is displayed.

287) Click the **left edge** of the GUIDE hole.

288) Click the **right edge** of the GUIDE hole. The centerline is displayed.

289) Click **OK** ✅ from the Centerline PropertyManager.

Save the drawing.

290) Click **Save** 💾. View the results.

🔅 Click the dimension Palette

rollover button to display the dimension palette. Use the dimension palette in the Graphics window to save mouse travel to the Dimension PropertyManager. Click on a dimension in a drawing view, and modify it directly from the dimension palette.

Modify the Dimension Scheme

The current feature dimension scheme represents the design intent of the GUIDE part. The Mirror Entities Sketch tool built symmetry into the Extruded Base sketch. The Mirror feature built symmetry into the Slot Cuts.

The current dimension scheme for the Slot Cut differs from the ASME 14.5M Dimension Standard for a slot. Redefine the dimensions for the Slot Cut according to the ASME 14.5 Standard.

The ASME 14.5 Standard requires an outside dimension of a slot. The radius value is not dimensioned. The left Slot Cut was created with the Mirror feature.

Create a centerline and dimension to complete the detailing of the Slot Cut. Sketch the vertical dimension, 10. The default arc conditions are measured from arc center point to arc center point. The dimension extension lines are tangent to the top arc and bottom arc.

Activity: Modify the Dimension Scheme.

Modify the Slot Cut dimension scheme.

291) Right-click **Change Layer** in the Graphics window. The Change document Layers dialog box is displayed.

292) Click the **Layer Properties** icon from the Change document layer dialog box. The Layers dialog box is displayed.

Set the Show/Hide icon to modify the status of the Layers.

293) Click the **Dims** Layer Show. Dims is the current layer.

294) Click the **Notes** Layer Show.

295) Click the **Hidden Dims** Layer Hide.

296) Click the **Format** Layer Show.

297) Click **OK** from the Layers dialog box.

298) Click **Zoom to Area** from the Heads-up View toolbar.

299) Zoom in on the right slot of Drawing View3 as illustrated.

Hide the dimension between the two arc center points.

300) Click the **10** dimension on the right bottom side of Drawing View3.

301) Click the **Others** tab in the Dimension PropertyManager.

302) Select **Hidden Dims** from the Layer box in the Dimension PropertyManager.

303) Click **OK** from the Dimension PropertyManager. The 10 dimension is not displayed.

Create a vertical dimension.

304) Click the **Smart Dimension** Sketch tool. The Dimension PropertyManager is displayed.

305) Click **Smart dimensioning** from the Dimension Assist Tools box.

306) Click the top of the **top right arc**. Do not select the Center Mark of the Slot Cut.

307) Click the bottom of the **bottom right arc**. Do not select the Center Mark of the Slot Cut.

308) Click a **position** to the right of Drawing View3.

309) Click the **Leaders** tab in the Dimension PropertyManager.

310) Click **Max** for First arc condition.

311) Click **Max** for Second arc condition.

312) Click **OK** ✔ from the Dimension PropertyManager.

Modify the Radius text to 2X R.
313) Click **2X R4** in Drawing View2.

314) Delete **R<DIM>** in the Dimension text box.

315) Click **Yes** to the Confirm dimension value text override message.

316) Enter **R** for Dimension text.

317) Click **OK** ✔ from the Dimension PropertyManager.

Add dimensions.

318) Click the **Smart Dimension** 🖉 tool.

319) Click **Smart dimensioning** from the Dimension Assist Tools box.

320) Click the **left vertical line** of the right Slot Cut.

321) Click the **right vertical line** of the right Slot Cut.

322) Click a **position** below the horizontal profile line.

Drawing dimensions are added in the drawing document. Model dimensions are inserted from the part. Utilize Smart Dimension in the GUIDE drawing to create drawing dimensions.

323) Click the **centerline**.

324) Click the left **Slot Cut arc top center point**.

325) Click a **position** above the top horizontal line as illustrated. 30 is the dimension.

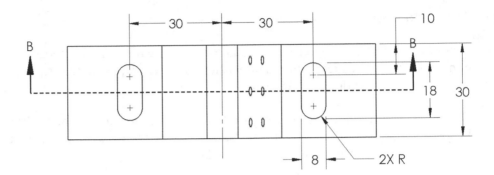

Insert a Reference dimension in the Section view.

326) Click the **left centerline** in the Section view.

327) Click the **right centerline** in the Section view.

328) Click a **position** below the bottom horizontal line.

329) Click the **Add Parenthesis** box in the Dimension Text.

330) Click **OK** ✔ from the Dimension PropertyManager.

Save the drawing.

331) Click **Save** 💾.

Design parts to maximize symmetry and geometric relations to fully define sketches. Minimize dimensions in the part. Insert dimensions in the drawing to adhere to machining requirements and Dimension Standards.

Review Dimensions and Annotations.

You inserted part dimensions and annotations into the drawing. Dimensions were moved to new positions. Leader lines and dimension text were repositioned. Annotations were edited to reflect the drawing standard.

Centerlines and Center Marks were inserted into each view and were modified in the PropertyManager. You modified Hole Callouts, dimensions, annotations and referenced dimensions to conform to the drawing standard.

GUIDE Part - Insert an Additional Feature

The design process is dynamic. We do not live in a static world. Create and add an edge Fillet feature to the GUIDE part. Insert dimensions into Drawing View1; (Front).

Activity: GUIDE Part - Insert an Additional Feature.

Open the Guide part.
332) Right-click inside the **Drawing View1** boundary.

333) Click **Open Part** from the Content toolbar.

If the View menu is not displayed, right-click Select. Tools in the Annotation toolbar remain active until they are deactivated.

Display Hidden Lines to select edges to fillet.
334) Click **Wire Frame** from the Heads-up View toolbar.

Display an Isometric view.
335) Click **Trimetric view** from the Heads-up View toolbar or from the Orientation dialog box.

Create a Fillet feature.

336) Click the **Fillet** feature tool. The Fillet
PropertyManager is displayed.

337) Click the **Manual** tab.

338) Click the **Constant Size** Fillet type.

339) Enter **1**mm in the Radius box. Symmetric is
selected by default.

340) Click the **hidden edge** as illustrated.

341) Click the other **3 edges**. Each edge is added to
the Items to Fillet list. Accept the default settings.

342) Click **OK** from the Fillet PropertyManager.
Fillet1 is displayed in the FeatureManager.

Save the GUIDE part.

343) Click **Save**.

Open the Drawing.

344) Right-click the **GUIDE Part** GUIDE (Default) icon
in the FeatureManager.

345) Click **Open Drawing**. The GUIDE drawing is
displayed.

Fit the drawing to the Graphics window.

346) Press the **f** key. Note: Hide any unwanted
annotations.

Display the GUIDE features.

347) Expand Drawing View1 from the FeatureManager.

348) Expand the GUIDE part from the FeatureManager.

349) Click **Fillet1** from the Drawing FeatureManager. View the Fillet
feature in the drawing.

The Model Items, Import into drawing view, Import items into
all views option inserts dimensions into the Section view and
Detail view before other types of views on the drawing. The
Fillet feature dimension in the part is displayed in the Section
view.

Insert the Fillet dimension in the Section view.
350) Click **inside** the Section view.

351) Click the **Smart Dimension** ↖ tool.

352) Click Fillet **feature** as illustrated.

353) Position the **dimension** up to the left.

354) Click **OK** ✔ from the Dimension PropertyManager. View the results.

Save the GUIDE drawing.
355) Click **Save** 💾.

R1

General Notes and Parametric Notes

Plan ahead for general drawing notes. Notes provide relative part or assembly information. Example: Material type, material finish, special manufacturing procedure or considerations, preferred supplier, etc.

Below are a few helpful guidelines to create general drawing notes:

- Use capital letters.

- Use left text justification.

- Font size should be the same size as the dimension text.

Create Parametric notes by selecting dimensions in the drawing. Example: Specify the Fillet radius of the GUIDE as a note in the drawing. If the radius is modified, the corresponding note is also modified.

☀ Hide superfluous feature dimensions. Do not delete feature dimensions. Recall a hidden dimension by using the View, Hide/Show, Hide/Show Annotations from the Menu bar. Utilize a Layer to Hide/Show superfluous feature dimensions with the Layer on/off icon.

Activity: General Notes and Parametric Notes.

Insert a Parametric drawing note.

356) Click the **Notes** layer from the Layer toolbar.

357) Click the **Note** A tool from the Annotation tab in the CommandManager. The Note PropertyManager is displayed.

358) Click a **start point** in the lower left corner of the Graphics window, to the left of the Title block, below Drawing View1.

Type two lines of notes in the Note text box.

359) Line 1: Enter **1. ALL ROUNDS**.

360) Press the **Space** key.

361) Click **R1**. The radius value R1 is added to the text box.

362) Press the **Space** key.

363) Enter **MM**.

364) Press the **Enter** key.

365) Line 2: Enter **2. REMOVE ALL BURRS**.

366) Click **OK** ✔ from the Note PropertyManager.

Do not double dimension a drawing with a note and the corresponding dimension. Hide the radial dimension.

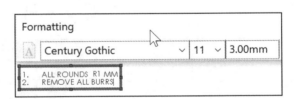

1.   ALL ROUNDS R1 MM
2.   REMOVE ALL BURRS

Hide the R1 dimension in the drawing.
367) Click the **R1** text in the Section view.

368) Click **Hidden Dims** from the Layer drop-down menu.

369) Click **OK** ✔ from the Dimension PropertyManager. R1 is hidden.

370) View the updated drawing.

An Engineering Change Order (ECO) or Engineering Change Notice (ECN) is written for each modification in the drawing. Modifications to the drawing are listed in the Revision Table.

 Obtain the ECO number before you institute the change to manufacturing.

Revision Table

The Revision Table provides a drawing history. Changes to the document are recorded systematically in the Revision Table. Insert a Revision Table into the GUIDE drawing. The default columns are as follows: Zone, Rev, Description, Date and Approved.

Zone utilizes the row letter and column number contained in the drawing border. Position the Rev letter in the Zone area. Enter the Zone letter/number.

Enter a Description that corresponds to the ECO number. Modify the date if required. Enter the initials/name of the engineering manager who approved the revision.

Activity: Revision Table.

Insert a Revision table.

371) Click **None** from the Layer drop-down menu.

372) Click the **Revision Table** tool from the Consolidated toolbar in the CommandManager. The Revision Table PropertyManager is displayed. Accept the defaults.

373) Click **OK** ✔ from the Revision Table PropertyManager. The Revision table is displayed in the top right corner of the Sheet.

374) Zoom in on the first row in the table.

Insert a row. Create the first revision.

375) Right-click the **Revision Table**.

376) Click **Revisions**, **Add Revision**. The Revision letter A and the current date are displayed.

The Revision Symbol is displayed on the mouse pointer.

377) Click the **top face** of the GUIDE in the Isometric view to place the

icon as illustrated.

378) Click **OK** ✔ from the Revision Symbol PropertyManager.

379) Double-click the **text box** under the Description column.

380) Enter **ECO 32510 RELEASE TO MANUFACTURING** for DESCRIPTION.

381) Click a **position** outside of the text box.

382) Double-click the **text box** under APPROVED.

383) Enter Documentation Control Manager's Initials, **DCP**.

384) Click a **position** outside of the text box.

385) **View** the results. The ZONE box is updated by the software.

 Zone letters and numbers will vary depending on your sheet size.

Part Number and Document Properties

Engineers manage the parts they create and modify. Each part requires a Part Number and Part Name. A part number is a numeric representation of the part. Each part has a unique number. Each drawing has a unique number. Drawings incorporate numerous part numbers or assembly numbers.

There are software applications that incorporate unique part numbers to create and perform:

- Bill of Materials.

- Manufacturing procedures.

- Cost analysis.

- Inventory control/Just in Time (JIT).

You are required to procure the part and drawing numbers from the documentation control manager.

Utilize the following prefix codes to categories, created parts and drawings. The part name, part number and drawing numbers are as follows:

Category:	Prefix:	Part Name:	Part Number:	Drawing Number:
Machined Parts	56-	GUIDE	56-A26	56-22222
		ROD	56-A27	56-22223
		PLATE	56-A28	56-22224
Purchased Parts	99-	FLANGE BOLT	99-FBM8x1.25	999-551-8
Assemblies	10-	GUIDE-ROD	10-A123	10-50123

Link notes in the Title block to SOLIDWORKS Properties. The title of the drawing is linked to the GUIDE Part Description, GUIDE-SUPPORT.

Activity: Part Number and Document Properties.

Enter Summary information for the GUIDE drawing.

386) Click **File**, **Properties** from the Menu bar. The Summary Information dialog box is displayed.

387) Click the **Summary** tab.

388) Enter your **initials** for Author. Example: DCP.

389) Enter **GUIDE, ROD** for Keywords.

390) Enter **GUIDE for customer ABC** for Comments.

Select the Custom tab. Add additional drawing information.

391) Click the **Custom** tab. View the information in the rows.

392) Click inside the **Property Name** cell as illustrated.

393) Select **DrawnBy** from the drop-down menu.

394) Click inside the **Value / Text Expression** cell.

395) Enter your **initials**. Example: DCP.

396) Click inside the **Property Name** cell.

397) Select **DrawnDate** from the drop-down menu.

398) Click inside the **Value / Text Expression** cell.

399) Enter the **date**.

400) Click inside the **Property Name** cell.

401) Select **Number** from the drop-down menu

402) Click inside the **Value / Text Expression** cell.

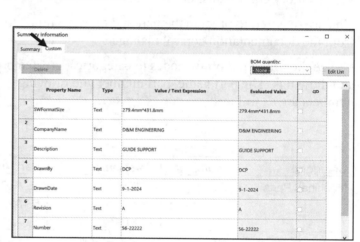

403) Enter **56-22222**.

404) Click **OK** from the Summary Information dialog box.

The DrawnBy, DrawnDate and Revision Property are updated in the Title block.

Link the Number Property to the DWG. NO.
405) Click inside **Sheet1**.

406) Right-click **Edit Sheet Format**.

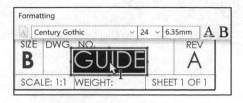

Edit the DWG NO.
407) Double-click the **GUIDE** text in the DWG NO. box. The Formatting and Note PropertyManager is displayed.

408) Delete **GUIDE**. **Note**: The Formatting dialog is still active.

409) Click the **Link to Property** 📇 icon in the Properties dialog box.

410) Select **Number** from the Link to Property drop-down menu.

411) Click **OK** from the Link to Property box. The number is displayed in the DWG. NO. box.

412) Click **OK** ✔ from the Note PropertyManager.

413) Right-click **Edit Sheet**.

Fit the drawing to the Graphics window.
414) Press the **f** key. The drawing remains in the Edit Sheet Format mode.

Save the Drawing.
415) Click **Save** 💾.

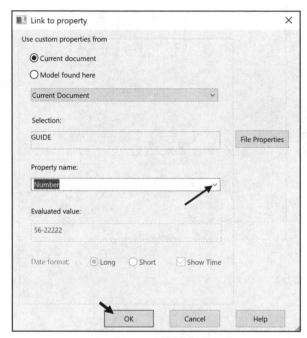

Define Custom Properties for Material and Surface Finish in the GUIDE part. Return to the GUIDE drawing to complete the Title block. The design team decided that the GUIDE part would be fabricated from 304 Stainless Steel. There are numerous types of Stainless steels for various applications. Select the correct material for the application. This is critical!

AISI 304 was set in the part. Open the GUIDE part and create the Material Custom Property. The AISI 304 value propagates to the drawing. Surface Finish is another important property in machining. The GUIDE utilizes an all around 0.80μm high grade machined finish. The numerical value refers to the roughness height of the machined material. Surface finish adds cost and time to the part. Work with the manufacturer to specify the correct part finish for your application.

Create the Material Custom Property.
416) Open the GUIDE part.

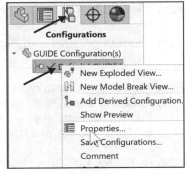

417) Click the **Configuration PropertyManager** icon.

418) Right-click **Default [GUIDE]**.

419) Click **Properties**. The Configuration Properties PropertyManager is displayed.

420) Click the **Custom Properties** button.

421) Click the **Configuration Specific** tab.

Specify the Material.
422) Click inside the **Property Name** box.

423) Select **Material** from the drop-down menu.

424) Click inside the **Value/Text Expression** box.

425) Select **Material** from the drop-down menu.

Determine the correct units before you create Custom Properties linked to Mass, Density, Volume, Center of Mass, Moment of Inertia, etc. Recall Weight = Mass *Gravity. Modify the Title block text to include "g" for gravity, SI units in the Weight box.

Specify the Finish.
426) Click inside the **second row** under the Property Name.

427) Select **Finish** from the drop-down menu.

428) Enter **0.80** for Value/Text Expression as illustrated.

429) Click **OK** from the Summary Information box.

430) Click **OK** ✔ from the Configuration Properties PropertyManager.

Return to the GUIDE drawing.
431) Return to the GUIDE drawing. The Material Custom Property displays AISI 304 in the MATERIAL block and 0.80 in the FINISH block.

432) Right-click **Edit Sheet Format**.

Insert a micrometer symbol.
433) Double-click **0.80**. The Formatting toolbar is displayed.

434) Click **SWGrekc** from the Formatting toolbar for Font.

435) Press the **m** key to display the Greek letter μ. μ is the Greek letter for micro.

436) Click **Century Gothic** for Font. Enter **m** for meter. The text reads μm for micrometer.

437) Click a **position** outside the text box.

Return to the Edit Sheet mode.
438) Right-click in the **Graphics window**.

439) Click **Edit Sheet**. The Title block is displayed in black.

Fit the drawing to the Graphics window.
440) Press the **f** key.

441) Drag the **views** and **annotations** if required for spacing.

Save the GUIDE drawing.
442) Click **Save** ▣.

Establish Custom Properties early in the design process. The MATERIAL and FINISH Custom Properties are established in the part and propagate to the Title block in the drawing.

💡 Create engineering procedures to define the location and values of Custom Properties. The REVISION Custom Property was created in the drawing in the REVISION Table. The REVISION Custom Property can also be established in the part and propagate to the drawing. Does the part or the drawing control the revision? Your company's engineering procedures determine the answer.

Exploded View

Add the GUIDE-ROD Exploded view and Bill of Materials to the drawing. Note: The GUIDE-ROD exploded view assembly was created in Project 3. The Bill of Materials reflects the components of the GUIDE-ROD assembly.

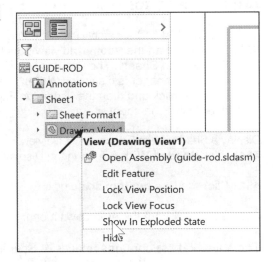

Perform the following steps:

- Create a new drawing from the B-ANSI-MM Drawing Template.

- Display the Exploded view of the assembly.

- Insert the Exploded view of the assembly into the drawing.

- Label each component with Balloon text.

- Create a Bill of Materials.

Activity: Exploded View.

Close all parts and drawings.
443) Click **Windows**, **Close All** from the Menu bar.

Open the GUIDE-ROD assembly.
444) Open the **GUIDE-ROD** assembly from the **ENGDESIGN-W-SOLIDWORKS/PROJECTS** folder. The GUIDE-ROD assembly is displayed in the Graphics window. Note: The GUIDE is updated per the changes to the GUIDE in the drawing.

Create a new drawing.
445) Click **Make Drawing from Part/Assembly** from the Menu bar.

446) Double-click **B-ANSI-MM** from the MY-TEMPLATES tab. The Model View PropertyManager is displayed.

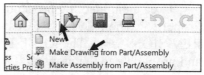

447) If needed, click **Cancel** ✕ from the Model View PropertyManager.

448) Click the **View Palette** tab in the Task Pane.

449) Click **GUIDE-ROD** from the down arrow as illustrated.

Insert the GUIDE-ROD assembly using the View Palette tool.
450) Click and drag the *Isometric view from the View Palette onto Sheet1. The Drawing View1 PropertyManager is displayed. **Note**: You have the option to click and drag the *Isometric Exploded view directly into the sheet.

Display a Shaded With Edges, 1:1 scale view.
451) Click **Shaded With Edges** in the Display Style box.

452) Click the **Use custom scale** box.

453) Select **1:1** from the drop-down menu.

The View Palette contains images of Standard views, Annotation views, Section views, Flat patterns (sheet metal parts) and Isometric Exploded view (if an Exploded configuration is created) of the selected model.

💡 Click the Browse button in the View Palette to locate additional models.

Display an Exploded view.

454) Click **inside** the Isometric view boundary.

455) Right-click **Show in Exploded State**. **Note**: You can also click **Properties**, check **Show in Exploded State**, and click **OK** from the Drawing View Properties dialog box.

456) Click **OK** 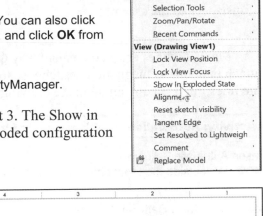 from the Drawing View1 PropertyManager.

Note: The Explode view was created in Project 3. The Show in exploded state option is visible only if an Exploded configuration exists in the assembly.

Fit the drawing to the Graphics window.

457) Press the **f** key.

458) Click **OK** ✔ from the Drawing View1 PropertyManager.

The drawing filename is the same as the assembly filename.

Save the GUIDE-ROD drawing.

459) Click **Save As** 📇.

460) Select **ENGDESIGN-W-SOLIDWORKS\PROJECTS** for folder. GUIDE-ROD is displayed for the File name.

Save the drawing.

461) Click **Save**.

Balloons

Label each component with a unique item number. The item number is placed inside a circle. The circle is called Balloon text. List each item in a Bill of Materials table. Utilize the Auto Balloon tool to apply Balloon text to all components. Utilize the Bill of Materials tool to apply a BOM to the drawing.

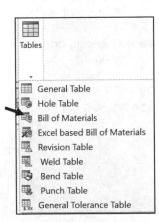

The Circle Split Line option contains the Item Number and Quantity. Item number is determined by the order listed in the assembly FeatureManager. Quantity lists the number of instances in the assembly.

Magnetic lines are a convenient way to align balloons along a line at any angle. You attach balloons to magnetic lines, choose to space the balloons equally or not, and move the lines freely, at any angle in the drawing.

Activity: Auto Balloons with Magnetic Lines.

Insert Automatic Balloons.
462) Click inside the **Isometric view** boundary.

463) Click **Auto Balloon** from the Annotation toolbar. The Auto Balloon PropertyManager is displayed. Accept the default settings.

464) Click and **drag** the balloons. **Note**: Balloon locations will vary.

465) Click **OK** ✔ from the Auto Balloon PropertyManager.

Display Item Number and Quantity.
466) Shift-Select the six **Balloon text**. The Balloon PropertyManager is displayed.

467) Select **Circular Split Line** for Settings from the drop-down menu. View the results in the Graphics window.

Settings
Circular
None
Circular
Triangle
Hexagon
Box
Diamond
Pentagon
Circular Split Line
Flag - Five Sided

468) Click **OK** ✔ from the Balloon PropertyManager.

469) Click and **drag** the circular split line balloons.

Save the drawings.
470) Click **Save** 💾.

The symbol "?" is displayed when the Balloon attachment is not coincident with an edge or face. ASME Y14.2 defines the attachment display as an arrowhead for Edge and a dot for Face.

Arrowhead display is located under Options, Document Properties, Annotations or under the Leaders tab in the Dimension PropertyManager. Note: Edge/vertex attachment is selected by default.

Each Balloon references a component with a SOLIDWORKS File Name. The SOLIDWORKS File Name is linked to the Part Name in the Bill of Materials by default. Customize both the Balloon text and the Bill of Materials according to your company's requirements.

Bill of Materials

The Bill of Materials reflects the components of the GUIDE-ROD assembly. The Bill of Materials is a table in the assembly drawing that contains Item Number, Quantity, Part Number and Description by default. Insert additional columns to customize the Bill of Materials.

Part Number and Description are Custom Properties entered in the individual part documents. Insert additional Custom Properties to complete the Bill of Materials.

Activity: Bill of Materials.

Create a Bill of Materials.
471) Click inside the **Isometric view** in the GUIDE-ROD drawing.

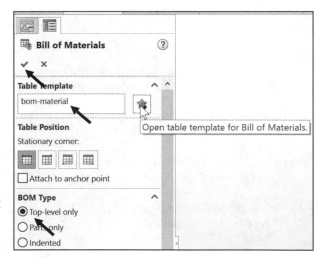

472) Click the **Bill of Materials** tool from the Annotation tab. The Bill of Materials PropertyManager is displayed.

473) Click the **Open table template for Bill of Materials** icon.

474) Double-click **bom-material.sldbomtbt** from the Select BOM Table dialog box.

475) Click **Top-level only**.

476) Click **OK** ✔ from the Bill of Materials PropertyManager.

477) Click a **position** in the upper left corner of Sheet1.

ITEM NO.	PART NUMBER	DESCRIPTION	MATERIAL	QTY.
1	GUIDE	GUIDE SUPPORT		1
2	ROD	ROD 10MM DIA x 100MM		1
3	PLATE	PLATE 56MM x 22MM		1
4	M8-1.25 x 30			2
5	4MMCAPSCREW	CAP SCREW, 4MM		2
6	3MMCAPSCREW	CAP SCREW, 3MM		6

The Bill of Materials requires some editing. The current part file name determines the PART NUMBER values. Material for the GUIDE was defined in the Material Editor and the Material Custom Property.

The current part description determines the DESCRIPTION values. Redefine the PART NUMBER, Description and Material in the BOM. Your BOM values and rows *may vary*. The final BOM should display the below PART NUMBERS, DESCRIPTION, and MATERIAL.

Modify the GUIDE part number in the BOM.

ITEM NO.	PART NUMBER	DESCRIPTION	MATERIAL	QTY.
1	56-A26	GUIDE SUPPORT	AISI 304	1
2	56-A27	ROD	AISI 304	1
3	56-A28	PLATE 56MM x 22MM	AISI 304	1
4	M8-1.25 x 30	FLANGE BOLT M8x1.25x30	AISI 304	2
5	56-333	CAP SCREW, 4MM	AISI 304	2
6	44-4434	CAP SCREW, 3MM	AISI 304	6

478) Right-click on the **GUIDE** in the Isometric view.

479) Click **Open Part**. The Guide FeatureManager is displayed.
Click the GUIDE **ConfigurationManager** tab. Right-click **Default** [**GUIDE**] in the ConfigurationManager.

480) Click **Properties**. The Configuration Properties PropertyManager is displayed.

481) Select **User Specified Name** in the spin box from the Bill of Materials Options.

482) Enter **56-A26** for the Part Number in the Bill of Materials Options.

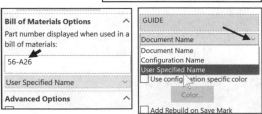

483) Click **OK** ✔ from the Configuration Properties PropertyManager. Default [56-A26] is displayed in the ConfigurationManager.

484) Return to the FeatureManager.

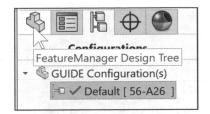

Return to the GUIDE-ROD drawing.
485) Click **Window**, **GUIDE-ROD - Sheet1** from the Menu bar. The part number in the BOM is updated.

Modify the ROD part number.
486) Right-click on the **ROD** in the Isometric view.

487) Click **Open Part**. The ROD FeatureManager is displayed.

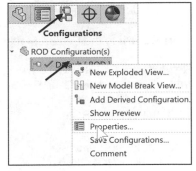

488) Click the ROD **ConfigurationManager** ⬚ tab.

489) Right-click **Default** [**ROD**] from the ConfigurationManager.

490) Click **Properties**.

491) Select **User Specified Name** from the drop-down menu in the Bill of Materials Options.

492) Enter **56-A27** for the Part Number in the Bill of Materials Options.

493) Click **OK** ✔ from the Configuration Properties PropertyManager.

494) Return to the FeatureManager.

Return to the GUIDE-ROD drawing.
495) Click **Window**, **GUIDE-ROD - Sheet1** from the Menu bar. View the updated BOM.

Modify the PLATE part number.
496) Right-click on the **PLATE** in the Isometric view.

497) Click **Open Part**.

498) Click the PLATE **ConfigurationManager** ⬚ tab.

499) Right-click **Default** [**PLATE**] from the ConfigurationManager.

500) Click **Properties**.

ITEM NO.	PART NUMBER
1	56-A26
2	56-A27
3	PLATE

501) Select **User Specified Name** in the spin box from the Bill of Materials Options.

502) Enter **56-A28** for the Part Number in the Bill of Materials Options.

503) Click **OK** ✔ from the Configuration Properties PropertyManager.

504) Return to the FeatureManager.

Return to the GUIDE-ROD drawing.
505) Click **Window, GUIDE-ROD - Sheet1** from the Menu bar. View the updated BOM.

ITEM NO.	PART NUMBER	DESCRIPTION
1	56-A26	GUIDE SUPPORT
2	56-A27	ROD 10MM DIA x 100MM
3	56-A28	PLATE 56MM x 22MM
4	M8-1.25 x 30	
5	4MMCAPSCREW	CAP SCREW, 4MM
6	3MMCAPSCREW	CAP SCREW, 3MM

The flange bolt is a SOLIDWORKS library part. Copy the part with a new name to the ENGDESIGN-W-SOLIDWORKS\PROJECTS folder with the Save As command.

Utilize the flange bolt Part Number for File name.

Modify the flange bolt.
506) Right-click a **flange bolt** in the Isometric view.

507) Click **Open Part**.

Save the flange bolt.
508) Click **Save As** .

509) Select the **ENGDESIGN-W-SOLIDWORKS\PROJECTS** folder.

510) Enter **99-FBM8-1-25** for the File name.

511) Enter **FLANGE BOLT M8x1.25x30** for Description.

512) Click **Save**.

Return to the GUIDE-ROD drawing.
513) Press **Ctrl+Tab** to return to the GUIDE-ROD drawing. View the updated BOM.

ITEM NO.	PART NUMBER	DESCRIPTION
1	56-A26	GUIDE SUPPORT
2	56-A27	ROD 10MM DIA x 100MM
3	56-A28	PLATE 56MM x 22MM
4	M8-1.25 x 30	FLANGE BOLT M8x1.25x30
5	4MMCAPSCREW	CAP SCREW, 4MM
6	3MMCAPSCREW	CAP SCREW, 3MM

Modify ITEM NO. 5 Part Number and ITEM NO. 6 Part Number.

514) Modify ITEM NO. 5 Part Number to **56-333**.

515) Modify ITEM NO. 6 Part Number to **44-4433**.

Add AISI 304 material to all components. First open the part. Second add material to the part. Third enter Customer Properties for the BOM. This will fill (link) the Material cell in the BOM.

516) Open the ROD.

517) Add **AISI 304** Material.

518) Click **File**, **Properties** from the Main menu.

519) Click the **Custom** tab.

520) Select **Material** from the drop-down arrow in the Property Name box as illustrated.

521) Select **Material** from the drop-down arrow in the Value / Text Expression box.

522) Click inside the **Evaluated Value** box. The value is displayed. The material link is displayed in the Value / Text Expression box.

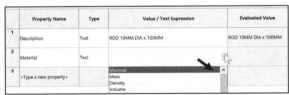

523) Click **OK** from the Properties dialog box. The MATERIAL BOM cell for the ROD displays AISI 304.

524) Repeat the above process for all components in the GUIDE-ROD drawing.

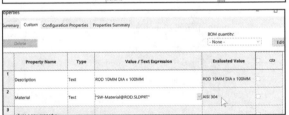

ITEM NO.	PART NUMBER	DESCRIPTION	MATERIAL	QTY.
1	56-A26	GUIDE SUPPORT		1
2	56-A27	ROD 10MM DIA x 100MM	AISI 304	1
3	56-A28	PLATE 56MM x 22MM		1
4	M8-1.25 x 30	FLANGE BOLT M8x1.25x30		2
5	56-333	CAP SCREW, 4MM		2
6	44-4433	CAP SCREW, 3MM		6

525) Click **Save** 💾.

As an exercise, address the Drawing Number, DrawnBy, DrawnDate, etc. as illustrated below.

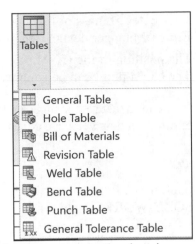

Review of the Parametric Notes, Revision Table and Bill of Materials.

You created a Parametric note in the drawing by inserting a part dimension into the Note text box. The Revision Table was inserted into the drawing to maintain drawing history.

The Bill of Materials listed the Item Number, Part Number, Description, Material and Quantity of components in the assembly.

You developed Custom Properties in the part and utilized the Properties in the drawing and Bill of Materials.

Drawings and Custom Properties are an integral part of the design process. Part, assemblies and drawings all work together.

From your initial design concepts, you created parts and drawings that fulfill the design requirements of your customer.

Refer to Help, Online Tutorial, Lesson3-Drawings exercise and Advanced Drawings exercise for additional information.

Magnetic lines are a convenient way to align balloons along a line at any angle. You attach balloons to magnetic lines, choose to space the balloons equally or not, and move the lines freely, at any angle in the drawing.

Insert a Center of Mass Point into a Drawing

Add a center of mass (COM) point to parts, assemblies and drawings.

COM points added in component documents also appear in the assembly document. In a drawing document of parts or assemblies that contain a COM point you can show and reference the COM point.

Add a COM to a part or assembly document by clicking **Center of Mass** (Reference Geometry toolbar) or **Insert, Reference Geometry, Center of Mass** or check the **Create Center of Mass feature** box in the Mass Properties dialog box.

The center of mass of the model is displayed in the graphics window and in the FeatureManager design tree just below the origin.

The position of the **COM** point ⬩ updates when the model's center of mass changes. The COM point can be suppressed and unsuppressed for configurations.

You can measure distances and add reference dimensions between the COM point and entities such as vertices, edges and faces.

💡 If you want to display a reference point where the CG was located at some particular point in the FeatureManager, you can insert a Center of Mass Reference Point. See SOLIDWORKS Help for additional information.

Add a center of mass (COM) point to a drawing view. The center of mass is a selectable entity in drawings and you can reference it to create dimensions.

In a drawing document, click **Insert**, **Model Items**. The Model Items PropertyManager is displayed. Under Reference Geometry, click the **Center of Mass** icon. Enter any needed additional information. Click **OK** from the Model Items PropertyManager. View the results in the drawing.

The part or assembly **needs to have a COM before** you can view the COM in the drawing. To view the center of mass in a drawing, click **View**, **Hide/Show**, **Center of mass**.

Project Summary

In this project you developed two drawings: GUIDE drawing and the GUIDE-ROD assembly drawing. The drawings contained three standard views, (principle views) and an Isometric view. The drawings utilized a custom Sheet Format containing a Company logo, Title block and Custom Properties. You incorporated the GUIDE part dimensions into the drawing.

The Drawing toolbar contained the Model View tool and the Projected View tool to develop standard views. Additional views were required and utilized the Auxiliary, Detail and Section view tools. Dimensions were inserted from the part and added to the drawing.

You used two major design modes in the drawings: Edit Sheet Format and Edit Sheet.

The detailed GUIDE drawing included annotations and Custom Properties. The GUIDE-ROD assembly drawing incorporated a Bill of Materials and additional Custom Properties.

To show a hidden dimension, click **View**, **Hide/Show**, **Annotations** from the Menu bar.

Questions

1. Describe a Bill of Materials and its contents in a drawing.

2. Name the two major design modes used to develop a drawing in SOLIDWORKS.

3. Identify seven components that are commonly found in a Title block.

4. Describe a procedure to insert an Isometric view into a drawing.

5. In SOLIDWORKS, Drawing file names end with a _____ suffix.

6. In SOLIDWORKS, Part file names end with a _____ suffix.

7. Can a part and drawing have the same name?

8. True or False. In SOLIDWORKS, if a part is modified, the drawing is automatically updated.

9. True or False. In SOLIDWORKS, when a dimension in the drawing is modified, the part is automatically updated.

10. Name three guidelines to create General Notes in a drawing.

11. True or False. Most engineering drawings use the following font: Times New Roman - All small letters.

12. What are Leader lines? Provide an example.

13. Name the two ways that Holes and other circular geometry can be dimensioned.

14. Describe Center Marks. Provide an example.

15. How do you calculate the maximum and minimum variation?

16. Describe the differences between a Drawing Template and a Sheet Format.

17. Describe the key differences between a Detail view and a Section view.

18. Describe a Revision table and its contents.

Exercises

Exercise 4.1: L - BRACKET Drawing

Create the A (ANSI) Landscape - IPS - Third Angle L-BRACKET drawing as illustrated below.

- First create the part from the drawing, then create the drawing.

- Insert the Front, Top, Right and Shaded Isometric view using the View Palette tool from the Task Pane. Hide the Top view.

- Insert dimensions into the Sheet.

- Address the extension line gaps between the feature lines.

	Property Name	Type	Value / Text Expression	Evaluated Value
3	DrawnDate	Text	9-1-2024	9-1-2024
4	CompanyName	Text	D&M ENGINEERING	D&M ENGINEERING
5	Description	Text	L-BRACKET	L-BRACKET
6	Revision	Text	A	A
7	SWFormatSize	Text	8.5in*11in	8.5in*11in
8	PartNo	Text	90990	90990
9	<Type a new property>			

- Think about the proper view and position for your dimensions. Use the default A (ANSI) Landscape Sheet Format/Size.

- Insert Custom Properties: Material, Revision, Number, Description, DrawnBy, DrawnDate and CompanyName.

- Material is 1060 Alloy.

- Insert Company and Third Angle projection icons. The icons are available in the homework folder.

Exercise 4.2: T - SECTION Drawing

Create the A (ANSI) Landscape - IPS - Third Angle T-SECTION drawing as illustrated below.

- First create the part from the drawing, then create the drawing. Use the default A (ANSI) Landscape Sheet Format/Size.

- Insert the Front, Top, Right and Shaded Isometric view using the View Palette tool from the Task Pane.

- Hide the Top view.

- Insert dimensions into the Sheet. Think about the proper view for your dimensions.

- Address the extension line gaps between the feature lines.

- Insert Custom Properties: Material, Revision, Number, Description, DrawnBy, DrawnDate and CompanyName.

- Material is 1060 Alloy.

	Property Name	Type	Value / Text Expression	Evaluated Value
2	DrawnBy	Text	DCP	DCP
3	DrawnDate	Text	9-1-2024	9-1-2024
4	CompanyName	Text	D&M ENGINEERING	D&M ENGINEERING
5	Revision	Text	A	A
6	SWFormatSize	Text	8.5in*11in	8.5in*11in
7	Description	Text	555-001	555-001
8	<Type a new property>			

- Insert Company and Third Angle projection icons. The icons are available in the homework folder.

Exercise 4.3: FLAT BAR - 3HOLE Drawing

Create the A (ANSI) Landscape - IPS - Third Angle
3HOLES drawing as illustrated. Do not display Tangent
Edges. Do not dimension to hidden lines.

- First create the part from the drawing, then create
  the drawing. Use the default A (ANSI) Landscape
  Sheet Format/Size.

- Insert the Front, Top and Shaded Isometric view as
  illustrated.

- Insert dimensions.

- Address the extension line gaps between the feature
  lines.

- Address proper drawing view
  display mode.

- Add a Smart (Linked)
  Parametric note for
  MATERIAL THICKNESS in
  the drawing as illustrated. Hide
  the dimension in the Top view.
  Insert Centerlines. Address
  extension line gaps.

- Modify the Hole dimension
  text to include 3X THRU EQ.
  SP. and 2X as illustrated.

- Insert Custom Properties:
  Material, Revision, Number,
  Description, DrawnBy,
  DrawnDate and
  CompanyName.

- Material is 1060 Alloy.

- Insert Company and Third
  Angle projection icons. The
  icons are available in the
  homework folder.

	Property Name	Type	Value / Text Expression	Evaluated Value
3	DrawnDate	Text	9-1-2024	9-1-2024
4	CompanyName	Text	D&M ENGINEERING	D&M ENGINEERING
5	Description	Text	3HOLES	3HOLES
6	Revision	Text	A	A
7	SWFormatSize	Text	8.5in*11in	8.5in*11in
8	PartNo	Text	556-099	556-099
9				

Exercise 4.4: CYLINDER Drawing

Create the A (ANSI) Landscape - IPS - Third Angle CYLINDER drawing as illustrated. Do not display Tangent Edges. Do not dimension to hidden lines.

- First create the part from the drawing; then create the drawing. Use the default A (ANSI) Landscape Sheet Format/Size.

- Insert drawing views as illustrated. Insert dimensions. Think about the proper view for your dimensions.

- Address the extension line gaps between the feature lines.

- Address proper drawing view display mode.

- Insert Company and Third Angle projection icons. The icons are available in the homework folder.

- Insert Centerlines, Center Marks and Annotations.

- Insert Custom Properties: Material, Revision, Number, Description, DrawnBy, DrawnDate and CompanyName.

- Material is AISI 1020.

Standard sheet size
☑ Only show standard formats

A (ANSI) Landscape
A (ANSI) Portrait
B (ANSI) Landscape
C (ANSI) Landscape
D (ANSI) Landscape
E (ANSI) Landscape
A0 (ANSI) Landscape

	Property Name	Type	Value / Text Expression	Evaluated Value
1	Material	Text	"SW-Material@CylinderPart1.SLDPRT"	AISI 1020
2	Description	Text	CYLINDER	CYLINDER

	Property Name	Type	Value / Text Expression	Evaluated Value
1	DrawnDate	Text	9-1-2024	9-1-2024
2	DrawnBy	Text	DCP	DCP
3	CompanyName	Text	D&M ENGINEERING	D&M ENGINEERING
4	Revision	Text	A	A
5	Number	Text	667-888	667-888
6	SWFormatSize	Text	8.5in*11in	8.5in*11in
7				

Exercise 4.5: PRESSURE PLATE Drawing

Create the A (ANSI) Landscape - IPS - Third Angle PRESSURE PLATE drawing. Do not display Tangent edges. Do not dimension to hidden lines.

- First create the part from the drawing, then create the drawing. Use the default A (ANSI) Landscape Sheet Format/Size.

- Insert the Front and Right view as illustrated. Insert dimensions.

	Property Name	Type	Value / Text Expression	Evaluated Value
2	DrawnDate	Text	9-1-2024	9-1-2024
3	CompanyName	Text	D&M ENGINEERING	D&M ENGINEERING
4	Revision	Text	A	A
5	Description	Text	PRESSURE PLATE	PRESSURE PLATE
6	SWFormatSize	Text	8.5in*11in	8.5in*11in
7	Number	Text	55-568	55-568
8	<Type a new property>			

- Address the extension line gaps between the feature lines.

- Think about the proper view for your dimensions.

- Address proper drawing view display mode.

- Insert Company and Third Angle projection icons. The icons are available in the homework folder.

- Insert Centerlines, Center Marks and Annotations.

- Insert Custom Properties: Material, Revision, Number, Description, DrawnBy, DrawnDate and CompanyName.

- Material is 1060 Alloy.

Exercise 4.6: PLATE-1 Drawing

Create the A (ANSI) Landscape - MMGS - Third Angle
PLATE-1 drawing as illustrated below. Do not display
Tangent edges. Do not dimension to Hidden lines.

- First create the part from the drawing, then create the
  drawing. Use the default A (ANSI) Landscape Sheet
  Format/Size.

- Insert the Front and Right view as
  illustrated. Insert dimensions.

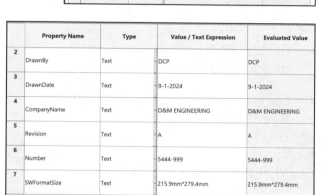

- Address the extension line gaps between
  the feature lines.

- Address proper drawing view
  display mode.

- Insert Company and Third
  Angle projection icons. The
  icons are available in the
  homework folder.

- Insert Centerlines, Center Marks
  and Annotations.

	Property Name	Type	Value / Text Expression	Evaluated Value
2	DrawnBy	Text	DCP	DCP
3	DrawnDate	Text	9-1-2024	9-1-2024
4	CompanyName	Text	D&M ENGINEERING	D&M ENGINEERING
5	Revision	Text	A	A
6	Number	Text	5444-999	5444-999
7	SWFormatSize	Text	215.9mm*279.4mm	215.9mm*279.4mm
8	<Type a new property>			

- Insert Custom
  Properties: Material,
  Revision, Number,
  Description, DrawnBy,
  DrawnDate and
  CompanyName.

- Material is 1060 Alloy.

Exercise 4.7: TUBE-2 Drawing

Create the A (ANSI) Landscape - IPS - Third Angle TUBE-2 drawing as illustrated below. Do not display Tangent edges. Phantom lines are fine.

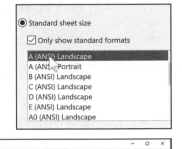

- First create the part from the drawing, then create the drawing. Use the default A (ANSI) Landscape Sheet Format/Size.

- Insert views as illustrated.

- Insert dimensions.

- Address extension line gaps.

- Insert Company and Third Angle projection icons. Icons are available in the homework folder.

- Insert Centerlines, Center Marks and Annotations.

- Address proper drawing view display mode.

- Insert Custom Properties: Material, Revision, Number, Description, DrawnBy, DrawnDate and CompanyName.

- Material is 6061 Alloy.

	Property Name	Type	Value / Text Expression	Evaluated Value
1	CompanyName	Text	D&M ENGINEERING	D&M ENGINEERING
2	DrawnDate	Text	9-1-2024	9-1-2024
3	DrawnBy	Text	DCP	DCP
4	Revision	Text	A	A
5	Number	Text	3334-667A	3334-667A
6	SWFormatSize	Text	8.5in*11in	8.5in*11in
7				

Exercise 4.8: PLATE Drawing

Create the A (ANSI) Landscape - IPS - Third Angle FLAT-PLATE drawing as illustrated below. Do not display Tangent edges. Phantom lines are fine.

- First create the part from the drawing, then create the drawing. Use the default A (ANSI) Landscape Sheet Format/Size.

- Insert the Front, Top, Right and Isometric views as illustrated.

- Insert dimensions.

- Address proper drawing view display mode.

- Address extension line gaps.

- Insert Company and Third Angle projection icons. The icons are available in the homework folder.

- Insert Centerlines, Center Marks and Annotations.

- Insert Custom Properties: Material, Revision, Number, Description, DrawnBy, DrawnDate and CompanyName.

- Material is 1060 Alloy.

	Property Name	Type	Value / Text Expression	Evaluated Value
1	Material	Text	"SW-Material@Figure6.8Part.SLDPRT"	1060 Alloy
2	Description	Text	PLATE	PLATE
3	<Type a new property>			

	Property Name	Type	Value / Text Expression	Evaluated Value
2	DrawnDate	Text	9-1-2024	9-1-2024
3	CompanyName	Text	D&M ENGINEERING	D&M ENGINEERING
4	Description	Text	PLATE	PLATE
5	Revision	Text	A	A
6	Number	Text	2445-990	2445-990
7	SWFormatSize	Text	8.5in*11in	8.5in*11in
8				

Exercise 4.9: **LINKAGE-2 Assembly Drawing**

Create the A (ANSI) Landscape - Third Angle LINKAGE-2 Assembly drawing with a Bill of Materials and balloons.

- Open the LINKAGE-2 assembly from Chapter 4 Homework folder.

- Use the default A (ANSI) Landscape Sheet Format/Size.

- Insert an Isometric Shaded With Edges view.

- Insert a Bill of Materials. Display the top-level sub-assemblies and parts only in the Bill of Materials.

- Select bom-material.sldbomtbt from the Select BOM Table dialog box.

- Resize the text in the Title box and DWG. NO. box.

- Insert Circular Auto-Balloons as illustrated.

- Insert Custom Properties: Revision, Number, Description, DrawnBy, DrawnDate and CompanyName.

- Insert Company and Third Angle projection icons. The icons are available in the homework folder.

	Property Name	Type	Value / Text Expression	Evaluated Value	
1	SWFormatSize	Text	8.5in*11in	8.5in*11in	
2	Revision	Text	A	A	
3	DrawnBy	Text	DCP	DCP	
4	DrawnDate	Text	9-1-2024	9-1-2024	
5	Description	Text	LINKAGE-2	LINKAGE-2	
6	CompanyName	Text	D&M ENGINEERING	D&M ENGINEERING	
7	PartNo	Text	333-223-22	333-223-22	

ITEM NO.	PART NUMBER	DESCRIPTION	MATERIAL	QTY.
1	GIDS-PC-10001	LINEAR ACTUATOR		1
2	GIDS-SC-10017	AXLE ROD	2014 Alloy	3
3	GIDS-SC-10009-9	9HOLES	2014 Alloy	2
4	GIDS-SC-10012-3-16	SHAFT-COLLAR	2014 Alloy	6
5	GIDS-SC-10009-3	FLAT BAR 3 HOLES	6061 Alloy	2

Notes:

Project 5

Extrude and Revolve Features

Below are the desired outcomes and usage competencies based on the completion of Project 5.

Project Desired Outcomes:	Usage Competencies:
• Obtain your customer's requirements for the FLASHLIGHT assembly.	• Ability to incorporate Design Intent into sketches, features, parts and assemblies.

• Two Part Templates:
  o PART-IN-ANSI
  o PART-MM-ISO
	• Aptitude to apply Document Properties in a custom Part Template.

• Four key parts:
  o BATTERY
  o BATTERYPLATE
  o LENS
  o BULB
	• Specific knowledge of the following features: Extruded Boss/Base, Instant3D, Extruded Cut, Revolved Boss/Base, Revolved Cut, Dome, Shell, Circular Pattern and Fillet.

• Core and Cavity Tooling for the BATTERYPLATE.	• Understanding of the Mold tools: Scale, Parting Lines, Parting Surfaces, Shut-off Surfaces, Tooling Split and Draft.

Notes:

Project 5 - Extrude and Revolve Features

Project Objective

Design a FLASHLIGHT assembly according to the customer's requirements. The FLASHLIGHT assembly will be cost effective, serviceable and flexible for future manufacturing revisions.

Design intent is the process in which the model is developed to accept future changes. Build design intent into the FLASHLIGHT sketches, features, parts and assemblies. Create two custom Part Templates. The Part Template is the foundation for the FLASHLIGHT parts.

Create the following parts:

- BATTERY

- BATTERYPLATE

- LENS

- BULB

The other parts for the FLASHLIGHT assembly are addressed in Project 5. Create the Core and Cavity mold tooling required for the BATTERYPLATE.

On the completion of this project, you will be able to:

- Apply design intent to sketches, features, parts and assemblies.

- Select the best profile for a sketch.

- Select the proper Sketch plane.

- Create a template: English and Metric units.

- Set Document Properties.

- Customize the SOLIDWORKS CommandManager toolbar.

- Insert/Edit dimensions.

- Insert/Edit relations.

- Use the following SOLIDWORKS features:
  o Instant3D
  o Extruded Boss/Base
  o Extruded Cut

- o Revolved Boss/Base
- o Revolved Boss Thin
- o Revolved Cut Thin
- o Dome
- o Shell
- o Circular Pattern
- o Fillet
- Use the following Mold tools:
  - o Draft, Scale, Parting Lines, Shut-off Surfaces, Parting Surfaces and Tooling Split

Project Overview

Start the design of the FLASHLIGHT assembly according to the customer's requirements. The FLASHLIGHT assembly will be cost effective, serviceable and flexible for future manufacturing revisions.

A template is the foundation for a SOLIDWORKS document. A template contains document settings for units, dimensioning standards and other properties.

Create two-part templates for the FLASHLIGHT Project:

- PART-IN-ANSI

- PART-MM-ISO

Create four parts for the FLASHLIGHT assembly in this Project:

- BATTERY

- BATTERYPLATE

- LENS

- BULB

Parts models consist of 3D features. Features are the building blocks of a part.

A 2D Sketch Plane is required to create an Extruded feature. Utilize the sketch geometry and sketch tools to create the following features:

- Extruded Boss/Base

- Extruded Cut

Utilize existing faces and edges to create the following features:

- Fillet

- Chamfer

This project introduces you to the Revolved feature.

Create two parts for the FLASHLIGHT assembly in this section:

- LENS

- BULB

A Revolved feature requires a 2D sketch profile and a centerline. Utilize sketch geometry and sketch tools to create the following features:

- Revolved Boss/Base

- Revolved Boss-Thin

- Revolved Cut

Circular pattern of V-shaped cuts.

Utilize existing faces to create the following features:

- Shell

- Dome

- Hole Wizard

Utilize the Extruded Cut feature to create a Circular Pattern.

Utilize the Mold tools to create the Cavity tooling plates for the BATTERYPLATE part.

Design Intent

The SOLIDWORKS definition of design intent is the process in which the model is developed to accept future changes.

Models behave differently when design changes occur. Design for change. Utilize geometry for symmetry, reuse common features and reuse common parts.

BATTERY PLATE

Cavity

Isometric view   Rotated

Mold tools

Build change into the following areas:

1. Sketch

2. Feature

3. Part

4. Assembly

5. Drawing

See the Intelligent modeling techniques Project for additional information.

1. Design Intent in the Sketch

In SOLIDWORKS, relations between sketch entities and model geometry, in either 2D or 3D sketches, are an important means of building in design intent. In this chapter we will address 2D sketches.

Apply design intent in a sketch as the profile is created. A profile is determined from the Sketch Entities. Example: Rectangle, Circle, Arc, Point, Slot, etc.

Develop design intent as you sketch with Geometric relations. Sketch relations are geometric constraints between sketch entities or between a sketch entity and a plane, axis, edge, or vertex. Relations can be added automatically or manually.

A rectangle contains Horizontal, Vertical, and Perpendicular automatic Geometric relations. Apply design intent using added Geometric relations. Example: Horizontal, Vertical, Collinear, Perpendicular, Parallel etc.

Example A: Apply design intent to create a square profile. Sketch a rectangle. Apply the Center Rectangle tool. Note: No construction reference centerline or Midpoint relation is required with the Center Rectangle tool. Insert dimensions to define the square.

Example B: Develop a rectangular profile. Apply the Corner Rectangle tool. The bottom horizontal midpoint of the rectangular profile is located at the Origin. Add a Midpoint relation between the horizontal edge of the rectangle and the Origin. Insert two dimensions to define the width and height of the rectangle as illustrated.

2. Design Intent in the Feature

Build design intent into a feature by addressing symmetry, feature selection, and the order of feature creation.

Example A: The Boss-Extrude1 feature (Base feature) remains symmetric about the Front Plane. Utilize the Mid Plane End Condition option in Direction 1. Modify the depth and the feature remains symmetric about the Front Plane.

Example B: Do you create each tooth separately using the Extruded Cut feature? No. Create a single tooth and then apply the Circular Pattern feature. Create 34 teeth for a Circular Pattern feature. Modify the number of teeth from 32 to 24.

3. Design Intent in the Part

Utilize symmetry, feature order and reusing common features to build design intent into the part.

Example A: Feature order. Is the entire part symmetric? Feature order affects the part. Apply the Shell feature before the Fillet feature and the inside corners remain perpendicular.

4. Design Intent in the Assembly

Utilizing symmetry, reusing common parts and using the Mate relation between parts builds the design intent into an assembly.

Example A: Reuse geometry in an assembly. The assembly contains a linear pattern of holes. Insert one screw into the first hole. Utilize the Component Pattern feature to copy the machine screw to the other holes.

5. Design Intent in the Drawing

Utilize dimensions, tolerance and notes in parts and assemblies to build the design intent into the Drawing.

Example A: Tolerance and material in the drawing.

Insert an outside diameter tolerance +.000/-.002 into the TUBE part. The tolerance propagates to the drawing.

Define the Custom Property MATERIAL in the part. The MATERIAL Custom Property propagates to the drawing.

Project Situation

You work for a company that specializes in providing promotional tradeshow products. The company is expecting a sales order for 100,000 flashlights with a potential for 500,000 units next year. Prototype drawings of the flashlight are required in three weeks.

You are the design engineer responsible for the project. You contact the customer to discuss design options and product specifications. The customer informs you that the flashlights will be used in an international marketing promotional campaign. Key customer requirements:

- Inexpensive reliable flashlight.

- Available advertising space of 10 square inches, 64.5 square centimeters.

- Lightweight semi-indestructible body.

- Self-standing with a handle.

Your company's standard product line does not address the above key customer requirements. The customer made it clear that there is no room for negotiation on the key product requirements.

You contact the salesperson and obtain additional information on the customer and product. This is a very valuable customer with a long history of last minute product changes. The job has high visibility with great future potential.

In a design review meeting, you present a conceptual sketch. Your colleagues review the sketch. The team's consensus is to proceed with the conceptual design.

The first key design decision is the battery. The battery type directly affects the flashlight body size, bulb intensity, case structure integrity, weight, manufacturing complexity and cost.

Review two potential battery options:

- A single 6-volt lantern battery.

- Four 1.5-volt D cell batteries.

The two options affect the product design and specification. Think about it.

A single 6-volt lantern battery is approximately 25% higher in cost and 35% more in weight. The 6-volt lantern battery does provide higher current capabilities and longer battery life.

A special battery holder is required to incorporate the four 1.5 volt D cell configuration. This would directly add to the cost and design time of the FLASHLIGHT assembly.

Time is critical. For the prototype, you decide to use a standard 6-volt lantern battery. This eliminates the requirement to design and procure a special battery holder. However, you envision the four D cell battery model for the next product revision.

Design the FLASHLIGHT assembly to accommodate both battery design options. Battery dimensional information is required for the design. Where do you go? Potential sources are product catalogs, company web sites, professional standards organizations, design handbooks and colleagues.

The team decides to purchase the following parts: 6-volt BATTERY, LENS ASSEMBLY, SWITCH and an O-RING. Model the following purchased parts: BATTERY, LENS assembly, SWITCH and the O-RING. The LENS assembly consists of the LENS and the BULB.

Your company will design, model and manufacture the following parts: BATTERYPLATE, LENSCAP and HOUSING.

Purchased Parts:	Designed Parts:
BATTERY	BATTERYPLATE
LENS assembly	MOLD TOOLING
*SWITCH	*LENSCAP
*O-RING	*HOUSING

*Parts addressed in Project 6.

The BATTERYPLATE, LENSCAP and HOUSING are plastic parts. Review the injection molded manufacturing process and the SOLIDWORKS Mold tools. Modify the part features to eject the part from the mold. Create the MOLD TOOLING for the BATTERYPLATE.

Part Template

Units are the measurement of physical quantities. Millimeter dimensioning and decimal inch dimensioning are the two most common unit types specified for engineering parts and drawings. The FLASHLIGHT project is designed in inch units and manufactured in millimeter units. Inch units are the primary unit and Millimeter units are the secondary unit.

Create two Part templates:

- PART-IN-ANSI

- PART-MM-ISO

Save the Part templates in the **ENGDESIGN-W-SOLIDWORKS\MY-TEMPLATES** folder. System Options, File Locations option controls the file folder location of SOLIDWORKS documents. Utilize the File Locations option to reference your Part templates in the MY-TEMPLATES folder. Add the MY-TEMPLATES folder path name to the Document Templates File Locations list.

Note: If you did not create the templates or the MY-TEMPLATES tab, use a standard SOLIDWORKS template and address the needed document properties and custom properties.

Activity: Create New Two Part Templates.

Create a PART-IN-ANSI Template.

1) Click **New** ⬚ from the Menu bar.

2) Double-click **Part**.

Set Dimensioning Standard.

3) Click **Options** ⚙ ˙ from the Menu bar.

Set Document Properties. Set drafting standard, units, and precision.

4) Click the **Document Properties** tab.

5) Select **ANSI** from the Overall drafting standard drop-down menu.

Set units and precision.

6)   Click **Units**.

7)   Select **IPS** for Unit system.

8)   Select **.123** for Basic unit length decimal place.

9)   Select **millimeters** for Dual dimension length unit.

10)   Select **None** for Basic unit angle decimal place.

11)   Click **OK** from the Document Properties - Units dialog box.

Save the template. Enter name.

12)   Click **Save As** .

13)   Select **Part Templates (*.prtdot)** from the Save as type box.

14)   Select **ENGDESIGN-W-SOLIDWORKS\MY-TEMPLATES** for the Save in folder.

15)   Enter **PART-IN-ANSI** for File name.

16)   Click **Save**.

Utilize the PART-IN-ANSI template to create the PART-MM-ISO template.

17)   Click **Options** ˇ, **Document Properties** tab. Select **ISO** from the Overall drafting standard drop-down menu.

Set units and precision.

18)   Click **Units**. Select **MMGS** for Unit system.

19)   Select **.12** for Basic unit length decimal place.

20)   Select **None** for Basic unit angle decimal place.

21)   Click **OK** from the Document Properties - Units dialog box.

Save the template. Enter name.

22)   Click **Save As** .

23)   Select **Part Templates (*.prtdot)** from the Save as type box. Select **ENGDESIGN-W-SOLIDWORKS\MY-TEMPLATES** for the Save in folder.

24)   Enter **PART-MM-ISO** for File name.

25)   Click **Save**.

If you created the MY-TEMPLATES tab in Project 2, skip this section.
Set System Options - File Locations.

26) Click **Options** 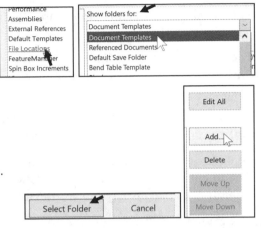 ⁻ from the Menu bar. The System Options General dialog box is displayed.

27) Click the **File Locations** folder.

Add a new tab.

28) Click the **Add** button.

29) Browse to the **ENGDESIGN-W-SOLIDWORKS\MY-TEMPLATES** folder. **Note**: The folder and files were downloaded from the SDC Publications website (www.SDCpublications.com/downloads/978-1-63057-630-1).

30) Click **Select Folder**. A new folder is added.

31) Position the folder at the bottom of the column. Click the **Move Down** button.

32) Click **OK** from the System Options, File Locations dialog box.

33) Click **Yes** to the SOLIDWORKS dialog box. The SOLIDWORKS dialog box displays the new Folder location.

34) Click **Yes**. Click **Yes**.

Close all documents.
35) Click **Windows**, **Close All** from the Menu bar.

Talk to your IT department *before you set* any new locations on a network system.

The document templates and tutorial location will vary depending on your network system setup.

Each folder listed in the System Options, File Locations, Document Templates, Show Folders For option produces a corresponding tab in the New SOLIDWORKS Document dialog box.

The templates used for this project are located in the following folder: ENGDESIGN-W-SOLIDWORKS\MY-TEMPLATES.

The new tab, MY-TEMPLATES is not displayed in the New SOLIDWORKS Document dialog box, **until a template** is added to the folder. Create the PART-MM-ANSI template as an exercise.

The PART-IN-ANSI template contains Document Properties settings for the parts contained in the FLASHLIGHT assembly. Substitute the PART-MM-ISO or PART-MM-ANSI template to create the identical parts in millimeters using a different drafting standard.

The primary units in this project are IPS (inch, pound, second).

The optional secondary units are MMGS (millimeter, gram, second) and are indicated in brackets [ ].

Illustrations are provided in both inch and millimeter. Utilize inch, millimeter or both.

To set dual dimensions, select Options, Document Properties, Dimensions. Check the Dual dimensions display box as illustrated.

To set dual dimensions for an active document, check the Dual Dimension box in the Dimension PropertyManager.

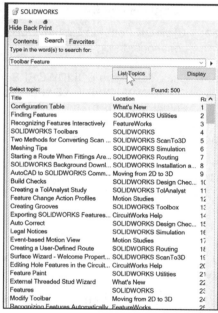

Enter toolbars, features in SOLIDWORKS Help Search category to review the function of each Features toolbar.

Model about the Origin; this provides a point of reference for your dimensions to fully define the sketch.

If you modify a document property from an Overall drafting standard, a modify message is displayed as illustrated.

Review of the Part Templates.

You created two Part templates: PART-IN-ANSI and PART-MM-ISO. The Document Properties Dimensioning Standard, units and decimal places are stored in the Part Templates. **Note**: Additional templates are displayed.

The File Locations System Option, Document Templates option controls the reference to the MY-TEMPLATES folder.

Note: If you did not create the templates or the MY-TEMPLATES tab, use a standard SOLIDWORKS template and address the needed document and custom properties.

BATTERY Part

The BATTERY is a simplified representation of a purchased OEM part. Represent the battery terminals as cylindrical extrusions. The battery dimensions are obtained from the ANSI standard 908D.

A 6-volt lantern battery weighs approximately 1.38 pounds (0.62kg). Locate the center of gravity closest to the center of the battery.

Create the BATTERY part. Use features to create parts. Features are building blocks that add or remove material.

Utilize the Instant3D tool to create the Extruded Boss/Base feature vs. using the Boss-Extrude PropertyManager.

The Extrude Boss/Base features add material. The Base feature (Boss-Extrude1) is the first feature of the part. Note: The default End Condition for Instant3D is Blind.

Apply symmetry. Use the Center Rectangle Sketch tool on the Top Plane. The 2D Sketch profile is centered at the Origin.

Extend the profile perpendicular (⊥) to the Top Plane.

Utilize the Fillet feature to round the four vertical edges.

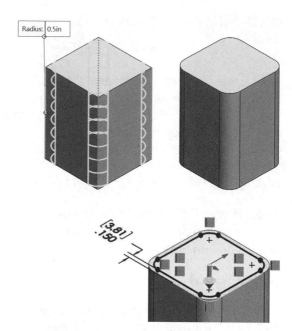

The Extruded Cut feature removes material from the top face. Utilize the top face for the Sketch plane. Utilize the Offset Entity Sketch tool to create the profile.

Utilize the Fillet feature to round the top narrow face.

The Extruded Boss/Base feature adds material. Conserve design time. Represent each of the terminals as a cylindrical Extruded Boss feature.

Think design intent. When do you use the various End Conditions and Geometric sketch relations? What are you trying to do with the design? How does the component fit into the assembly? Design for change and flexibility.

BATTERY Part - Extruded Boss/Base Feature

The Extruded Boss/Base feature requires:

- Sketch plane (Top).

- Sketch profile (Rectangle).

    ○ Geometric relations and dimensions.

- End Condition Depth (Blind) in Direction 1.

Create a new part named BATTERY. Insert an Extruded Boss/Base feature. Extruded features require a Sketch plane. The Sketch plane determines the orientation of the Extruded Base feature. The Sketch plane locates the Sketch profile on any plane or face.

The Top Plane is the Sketch plane. The Sketch profile is a rectangle. The rectangle consists of two horizontal lines and two vertical lines.

Geometric relations and dimensions constrain the sketch in 3D space. The Blind End Condition in Direction 1 requires a depth value to extrude the 2D Sketch profile and to complete the 3D feature.

☀ Alternate between the Features tab and the Sketch tab in the CommandManager to display the available Feature and Sketch tools for the Part document.

☀ CommandManager and FeatureManager tabs and tree folders will vary depending on system setup and Add-ins.

Activity: BATTERY Part - Create the Extruded Base Feature.

Create a new part.

36) Click **New** ⬜ from the Menu bar.

37) Click the **MY-TEMPLATES** tab.

38) Double-click **PART-IN-ANSI**.

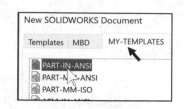

Save the part. Enter part name. Enter part description.

39) Click **Save As** 📄.

40) Select **ENGDESIGN-W-SOLIDWORKS\PROJECTS** for Save in folder.

41) Enter **BATTERY** for File name.

42) Enter **BATTERY, 6-VOLT** for Description.

43) Click **Save**. The Battery FeatureManager is displayed.

Select the Sketch plane.

44) Right-click **Top Plane** from the FeatureManager. This is your Sketch plane.

Sketch the 2D Sketch profile centered at the Origin.

45) Click **Sketch** ✏️ from the Context toolbar. The Sketch toolbar is displayed.

46) Click the **Center Rectangle** 🔲 Sketch tool. The Center Rectangle icon is displayed.

47) Click the **Origin**. This is your first point.

48) Drag and click the **second point** in the upper right quadrant as illustrated. The Origin is located in the center of the sketch profile. The Center Rectangle Sketch tool automatically applies equal relations to the two horizontal and two vertical lines. A midpoint relation is automatically applied to the Origin.

A goal of this book is to expose the new SOLIDWORKS user to various tools, techniques and procedures. The text may not always use the most direct tool or process.

💡 Click **View**, **Hide/Show**, **Sketch Relations** from the Main menu to view sketch relations in the Graphics area.

Dimension the sketch.

49) Click the **Smart Dimension** ↖ Sketch tool.

50) Click the **top horizontal line**.

51) Click a **position** above the horizontal line.

52) Enter **2.700**in, **[68.58]** for width.

53) Click the **Green Check mark** ✔ in the Modify dialog box.

54) Enter **2.700**in, [68.58] for height as illustrated.

55) Click the **Green Check mark** ✔ in the Modify dialog box. The black Sketch status is fully defined.

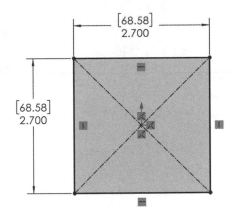

Create a Boss Extrude feature.

56) Click **Extruded Boss/Base** from the Features tab in the CommandManager. Boss-Extrude FeatureManager is displayed.

57) Select **Blind** for End Condition.

58) Enter **4.100in** for depth.

59) Click **OK** ✔ from the Boss-Extrude1 PropertyManager.

Fit the part to the Graphics window.
60)　Press the **f** key.

Rename the Boss-Extrude1 feature.
61)　Rename **Boss-Extrude1** to **Base Extrude**.

Save the BATTERY.
62)　Click **Save** .

The color of the sketch indicates the sketch status.

- **Light Blue** - Currently selected.

- **Blue** - Under defined, requires additional geometric relations and or dimensions.

- **Black** - Fully defined.

- **Red/Orange** - Over defined, requires geometric relations and or dimensions to be deleted or redefined to solve the sketch.

BATTERY Part - Fillet Feature Edge

Fillet features remove sharp edges. Utilize Hidden Lines Visible ⬚ from the Heads-up View toolbar to display hidden edges.

An edge Fillet feature requires:

- A selected edge
- Fillet radius

Select a vertical edge. Select the Fillet feature from the Features toolbar. Enter the Fillet radius. Add the other vertical edges to the Items To Fillet option.

The order of selection for the Fillet feature is not predetermined. Select edges to produce the correct result.

The Fillet feature uses the Fillet PropertyManager. The Fillet PropertyManager provides the ability to select either the Manual or FilletXpert tab.

Fillet

Creates a rounded internal or external face along one or more edges on a solid or surface feature. Fillets can be of constant or variable size. Use the Fillet PropertyManager to select the fillet type and then set other options appropriate to that type.

Each tab has a separate menu and PropertyManager. The Fillet PropertyManager and FilletXpert PropertyManager display the appropriate selections based on the type of fillet you create.

🔅 Use the Fillet tool to create symmetrical conic shaped fillets for parts, assemblies, and surfaces. You can apply conic shapes to *Constant Size*, *Variable Size*, and *Face* fillets.

Activity: BATTERY Part - Fillet Feature Edge.

Display Hidden edges.

63) Click **Hidden Lines Visible** 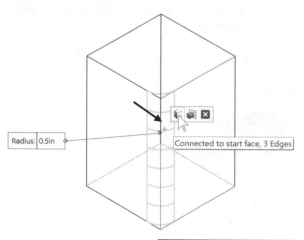 from the Heads-up View toolbar.

Insert a Fillet feature.

64) Click the **Fillet** feature tool. The Fillet PropertyManager is displayed.

65) Click the **Manual** tab.

66) Click **Constant Size Fillet** for type.

67) Click the **front vertical edge** as illustrated to show the mouse pointer edge icon. Edge<1> is displayed in the Items To Fillet box. The fillet option Pop-up toolbar is displayed. Options are model dependent.

68) Select the **Connected to start face, 3 Edges icon**. The four selected edges are displayed in the Edges, Faces, Features, and Loop box.

69) Enter **.500**in, [**12.7**] for Radius. Accept the default settings.

70) Click **OK** from the Fillet PropertyManager. Fillet1 is displayed in the FeatureManager.

Display an Isometric, Shaded with Edges view.

71) Click **Isometric view**.

72) Click **Shaded With Edges** from the Heads-up View toolbar.

Rename the feature.

73) Rename **Fillet1** to **Side Fillets** in the FeatureManager.

Save the BATTERY.

74) Click **Save**.

BATTERY Part - Extruded Cut Feature

An Extruded Cut feature removes material. An Extruded Cut feature requires:

- Sketch plane (Top face).

- Sketch profile (Offset Entities).

- End Condition depth (Blind) in Direction 1.

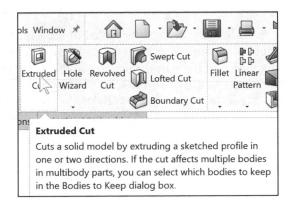

Extruded Cut

Cuts a solid model by extruding a sketched profile in one or two directions. If the cut affects multiple bodies in multibody parts, you can select which bodies to keep in the Bodies to Keep dialog box.

The Offset Entity Sketch tool uses existing geometry, extracts an edge or face and locates the geometry on the current Sketch plane.

Offset the existing Top face for the 2D sketch. Utilize the default Blind End Condition in Direction 1.

Activity: BATTERY Part - Extruded Cut Feature.

Select the Sketch plane.

75) Right-click the **Top face** of the BATTERY in the Graphics window. Base Extruded is highlighted in the FeatureManager.

Create a sketch.

76) Click **Sketch** from the Context toolbar. The Sketch toolbar is displayed.

Offset the existing geometry from the boundary of the Sketch plane.

77) Click the **Offset Entities** Sketch tool. The Offset Entities PropertyManager is displayed.

78) Enter **.150**in, **[3.81]** for the Offset Distance.

79) If needed check the **Reverse** box. The new Offset profile displays ***inside the original profile***.

80) Click **OK** from the Offset Entities PropertyManager.

💡 A leading zero is displayed in the spin box. For inch dimensions less than 1, the leading zero is not displayed in the part dimension in the ANSI standard.

Display an Isometric view, with Hidden Lines Removed.

81) Press the **space bar** to display the Orientation dialog box.

82) Click **Isometric view** .

83) Click **Hidden Lines Removed** from the Heads-up View toolbar.

Insert an Extruded Cut feature. As an exercise, use the Instant3D tool to create the Extruded Cut feature. In this section, the Extruded-Cut PropertyManager is used.

84) Click the **Extruded Cut** feature tool. The Cut-Extrude PropertyManager is displayed.

85) Enter **.200**in, [**5.08**] for Depth in Direction 1. Accept the default settings.

86) Click **OK** from the Cut-Extrude PropertyManager. Cut-Extrude1 is displayed in the FeatureManager.

Rename the feature.

87) Rename **Cut-Extrude1** to **Top Cut** in the FeatureManager.

Save the BATTERY

88) Click **Save** .

The Cut-Extrude PropertyManager contains numerous options. The Reverse Direction option determines the direction of the Extrude. The Extruded Cut feature is valid only when the direction arrow points into material to be removed.

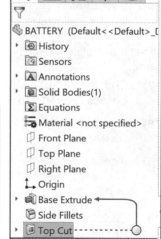

The Flip side to cut option determines if the cut is to the inside or outside of the Sketch profile.

The Flip side to cut arrow points outward. The Extruded Cut feature occurs on the outside of the BATTERY.

Extruded Cut with Flip side to cut option checked

BATTERY Part - Fillet Feature

The Fillet feature tool rounds sharp edges with a constant radius by selecting a face. A Fillet requires a:

- Selected face.

- Fillet radius.

Activity: BATTERY Part - Fillet Feature Face.

Insert a Fillet feature on the top face.

89) Click the **Fillet** feature tool. The Fillet PropertyManager is displayed.

90) Click the **Manual** tab.

91) Click the **top thin face** as illustrated. Note the face icon feedback symbol. Face<1> is displayed in the Items To Fillet box.

92) Click the **Constant Size Fillet** for Fillet Type.

93) Enter **.050**in, **[1.27]** for Radius.

94) Click **OK** from the Fillet PropertyManager. Fillet is displayed in the FeatureManager.

Rename the feature.
95) Rename **Fillet2** to **Top Face Fillet**.

Fit the model to the Graphics window. Display Hidden Lines Visible.
96) Press the **f** key.

97) Click **Hidden Lines Visible**.

Save the BATTERY.
98) Click **Save**.

BATTERY (Default) <<Default>
- History
- Sensors
- Annotations
- Solid Bodies(1)
- Material <not specified>
- Front Plane
- Top Plane
- Right Plane
- Origin
- Base Extrude
- Side Fillets
- Top Cut
- Top Face Fillet

View the mouse pointer for feedback to select Edges or Faces for the fillet.

 Do not select a fillet radius which is larger than the surrounding geometry.

Example: The top edge face width is .150in, [3.81]. The fillet is created on both sides of the face. A common error is to enter a Fillet too large for the existing geometry. A minimum face width of .200in, [5.08] is required for a fillet radius of .100in, [2.54].

The following error occurs when the fillet radius is too large for the existing geometry.

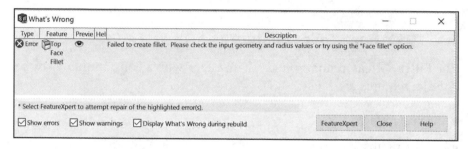

Avoid the fillet rebuild error. Use the FeatureXpert to address a constant radius fillet build error or manually enter a smaller fillet radius size.

BATTERY Part - Extruded Boss/Base Feature

The Extruded Boss feature requires a truncated cone shape to represent the geometry of the BATTERY terminals. The Draft Angle option creates the tapered shape.

Sketch the first circle on the Top face. Utilize the Ctrl key to copy the first circle.

The dimension between the center points is critical. Dimension the distance between the two center points with an aligned dimension. The dimension text toggles between linear and aligned. An aligned dimension is created when the dimension is positioned between the two circles.

An angular dimension is required between the Right Plane and the centerline. Acute angles are less than 90°. Acute angles are the preferred dimension standard. The overall BATTERY height is a critical dimension. The BATTERY height is 4.500in, [114.3].

Calculate the depth of the extrusion: For inches: 4.500in - (4.100in Base-Extrude height - .200in Offset cut depth) = .600in. The depth of the extrusion is .600in.

For millimeters: 114.3mm - (104.14mm Base-Extrude height - 5.08mm Offset cut depth) = 15.24mm. The depth of the extrusion is 15.24mm.

Activity: BATTERY Part - Extruded Boss Feature.

Select the Sketch plane.

99) Right-click the **top face** of the Top Cut feature in the Graphics window. This is your Sketch plane.

Create the sketch.

100) Click **Sketch** from the Context toolbar. The Sketch toolbar is displayed.

Sketch the Close profile.

101) Click the **Circle** Sketch tool. The Circle PropertyManager is displayed.

102) Click the **center point** of the circle coincident to the Origin.

103) Drag and click the **mouse pointer** to the right of the Origin as illustrated.

Add a dimension.

104) Click the **Smart Dimension** Sketch tool.

105) Click the **circumference** of the circle.

106) Click a **position** diagonally to the right.

107) Enter **.500**in, [**12.7**].

108) Click the **Green Check mark** in the Modify dialog box. The black sketch is fully defined.

[12.70]
Ø.500

Copy the sketched circle.

109) Right-click **Select** to de-select the Smart Dimension Sketch tool.

110) Hold the **Ctrl** key down.

111) Click and drag the **circumference** of the circle to the upper left quadrant.

112) Release the **mouse button**.

113) Release the **Ctrl** key. The second circle is selected.

114) Click **OK** ✔ from the Circle PropertyManager.

Add an Equal relation.
115) Hold the **Ctrl** key down.

116) Click the **circumference of the second circle**.

117) Click the **circumference of the first circle**.

118) Release the **Ctrl** key.

119) Right-click **Make Equal** from the Context toolbar.

120) Click **OK** ✔ from the Properties PropertyManager.

Add an aligned dimension.

121) Click the **Smart Dimension** ⟨ Sketch tool.

122) Click the **two center points** of the two circles.

123) Click a **position** off the profile as illustrated.

124) Enter **1.000**in, **[25.4]** for the aligned dimension.

125) Click the **Green Check mark** ✔ in the Modify dialog box.

Insert a centerline.

126) Click the **Centerline** ◌⁻ Sketch tool. The Insert Line PropertyManager is displayed.

127) Sketch a centerline between the **two circle center points** as illustrated.

128) Right-click **Select** to end the line.

 Double-click to end the centerline.

💡 Press the Enter key to accept the value in the Modify dialog box. The Enter key replaces the Green Check mark.

Add an angular dimension.

129) Click the **Smart Dimension** ✎ Sketch tool.

130) Click the **centerline** between the two circles.

131) Click the **Right Plane** (vertical line) from the FeatureManager.

132) Click a **position** between the centerline and the Right Plane, off the profile.

133) Enter **45**.

134) Click **OK** ✅ from the Dimension PropertyManager.

Fit the model to the Graphics window.
135) Press the **f** key.

136) Click **Save** 💾.

💡 Create an angular dimension between three points or two lines. Sketch a centerline/construction line when an additional point or line is required.

Display an Isometric view. Insert an Extruded Boss feature.
137) Click **Isometric view** 🔲.

138) Click the **Extruded Boss/Base** 🔳 feature tool. The Boss-Extrude PropertyManager is displayed. Blind is the default End Condition Type.

139) Enter **.600**in, **[15.24]** for Depth in Direction 1.

140) Click the **Draft ON/OFF** button.

141) Enter **5**deg in the Draft Angle box.

142) Click **OK** ✅ from the Boss-Extrude PropertyManager. The Boss-Extrude2 feature is displayed in the FeatureManager.

Rename the feature and sketch.

143) Rename **Boss-Extrude2** to **Terminals**.

144) Expand Terminals.

145) Rename **Sketch3** to **Sketch-TERMINALS**.

Display Shaded With Edges. Save the model.

146) Click **Shaded With Edges** ⬡ from the Heads-up View toolbar.

147) Click **Save** 🖫.

Each time you create a feature of the same feature type, the feature name is incremented by one. Example: Boss-Extrude1 is the first Extrude feature. Boss-Extrude2 is the second Extrude feature. If you delete a feature, rename a feature or exit a SOLIDWORKS session, the feature numbers will vary from those illustrated in the text.

💡 Utilize the Measure tool to measure distances and add reference dimensions between the COM point and entities such as vertices, edges and faces.

Measure the overall BATTERY height.

148) Click **Front view** 🔲.

149) Click the **Measure** 🔎 tool from the Evaluate tab in the CommandManager. The Measure - BATTERY dialog box is displayed.

150) Click the **Show XYZ Measurements** option. This should be the *only active* option.

151) Click the **top edge** of the battery terminal as illustrated. Click the **bottom edge** of the battery. View the results.

152) Close ✕ the Measure - BATTERY dialog box.

💡 The Measure tool provides the ability to display custom settings. Click **Units/Precision** from the Measure dialog box. View your options. Click **OK**.

The Selection Filter option toggles the Selection Filter toolbar. When Selection Filters are activated, the mouse pointer displays the Filter icon. The Clear All Filters tool removes the current Selection Filters. The Help icon displays the SOLIDWORKS Online Users Guide.

Display a Trimetric view.

153) Click **Trimetric view** from the Heads-up View toolbar.

Save the BATTERY.

154) Click **Save**.

 Review of the BATTERY Part.

The BATTERY utilized a 2D Sketch profile located on the Top Plane. The 2D Sketch profile utilized the Center Rectangle Sketch tool. The Center Rectangle Sketch tool applied equal geometric relations to the two horizontal and two vertical lines. A midpoint relation was added to the Origin.

The Extruded Boss/Base feature was created using the Instant3D tool. Blind was the default End Condition. The Fillet feature rounded sharp edges. All four edges were selected to combine common geometry into the same Fillet feature. The Fillet feature also rounded the top face. The Sketch Offset Entity created the profile for the Extruded Cut feature.

The Terminals were created with an Extruded Boss feature. You sketched a circular profile and utilized the Ctrl key to copy the sketched geometry.

A centerline was required to locate the two holes with an angular dimension. The Draft Angle option tapered the Extruded Boss feature. All feature names were renamed.

Injection Molded Process

Lee Plastics of Sterling, MA is a precision injection molding company. Through the World Wide Web (www.leeplastics.com), review the injection molded manufacturing process.

The injection molding process is as follows:

An operator pours the plastic resin, in the form of small dry pellets, into a hopper. The hopper feeds a large auger screw. The screw pushes the pellets forward into a heated chamber. The resin melts and accumulates into the front of the screw.

At high pressure, the screw pushes the molten plastic through a nozzle, to the gate and into a closed mold (Plates A & B). Plates A and B are the machined plates that you will design in this project.

The plastic fills the part cavities through a narrow channel called a gate.

The plastic cools and forms a solid in the mold cavity. The mold opens (along the parting line), and an ejection pin pushes the plastic part out of the mold into a slide.

Injection Molded Process
(Courtesy of Lee Plastics, Inc.)

BATTERYPLATE Part

The BATTERYPLATE is a critical plastic part. The BATTERYPLATE:

- Aligns the LENS assembly.

- Creates an electrical connection between the BATTERY and LENS.

Design the BATTERYPLATE. Utilize features from the BATTERY to develop the BATTERYPLATE. The BATTERYPLATE is manufactured as an injection molded plastic part. Build Draft into the Extruded Boss/Base features.

Edit the BATTERY features. Create two holes from the original sketched circles. Apply the Instant3D tool to create an Extruded Cut feature.

Modify the dimensions of the Base feature. Add a 3° draft angle.

🔅 A sand pail contains a draft angle. The draft angle assists the sand to leave the pail when the pail is flipped upside down.

Insert an Extruded Boss/Base feature. Offset the center circular sketch.

The Extruded Boss/Base feature contains the LENS. Create an inside draft angle. The draft angle assists the LENS into the Holder.

Insert a Face Fillet and a Multi-radius Edge Fillet to remove sharp edges. Plastic parts require smooth edges. Group Fillet features together into a folder.

Perform a Draft Analysis on the part and create the Core and Cavity mold tooling.

🔅 Group fillets together into a folder to locate them quickly. Features listed in the FeatureManager must be continuous in order to be placed as a group into a folder.

Save As, Delete, Edit Feature and Modify

Create the BATTERYPLATE part from the BATTERY part. Utilize the Save As tool from the Menu bar to copy the BATTERY part to the BATTERYPLATE part.

- The Save As command saves the document to a new file name that becomes the active document without saving the original document.

- The Save as copy and continue command saves the document to a new file name without replacing the active document

- The Save as copy and open command saves the document to a new file name that becomes the active document. The original document remains open. References to the original document are not automatically assigned to the copy.

Reuse existing geometry. Create two holes. Delete the Terminals feature and reuse the circle sketch. Select the sketch in the FeatureManager. Create an Extruded Cut feature from the Sketch-TERMINALS. Blind is the default End Condition. Edit the Boss-Extrude feature. Modify the overall depth. Rebuild the model.

> **Activity: BATTERYPLATE Part - Save As, Delete, Modify, and Edit Feature.**

Apply the Save As tool. Create and save a new part. Enter part name and description.

155) Click **Save As** from the drop-down menu.

156) Select **ENGDESIGN-W-SOLIDWORKS\PROJECTS** for Save In folder.

157) Enter **BATTERYPLATE** for File name.

158) Enter **BATTERY PLATE, FOR 6-VOLT** for Description.

159) Click **Save** from the Save As dialog box. The BATTERYPLATE FeatureManager is displayed. The BATTERY part is closed.

Delete the Terminals feature.
160) Right-click **Terminals** from the FeatureManager.

161) Click **Delete**.

162) Click **Yes** from the Confirm Delete dialog box. Do not delete the two-circle sketch, Sketch-TERMINALS.

Create an Extruded Cut feature from the Sketch-TERMINALS using Instant3D.

163) Activate the Instant 3D tool. If needed, click the

Instant 3D tool ^{Instant3D} icon from the FeatureManager.

164) Click **Sketch-TERMINALS** from the FeatureManager.

165) Click the **circumference** of the center circle as illustrated. An arrow is displayed.

166) Hold the **Alt** key down.

167) Drag the **arrow** downward below the model to create a hole in Direction 1.

168) Release the **mouse button below** the model. This ensures a Through All End Condition.

169) Release the **Alt** key. Boss-Extrude1 is displayed in the FeatureManager.

170) Rename the **Boss-Extrude** feature to **Holes** in the FeatureManager.

Edit the Base Extrude feature.

171) Right-click **Base Extrude** from the FeatureManager.

172) Click **Edit Feature** 🗎 from the Context toolbar. The Base Extrude PropertyManager is displayed.

Modify the overall depth.

173) Enter **.400**in, [**10.16**] for Depth in Direction 1.

174) Click the **Draft ON/OFF** button.

175) Enter **3.00**deg in the Angle box.

176) Click **OK** ✔ from the Base Extrude PropertyManager.

Fit the model to the Graphics window.

177) Press the **f** key.

Save the BATTERYPLATE.

178) Click **Save** 🖫.

BATTERYPLATE Part - Extruded Boss Feature

The Holder is created with a circular Extruded Boss/Base feature. Utilize the Offset Entities ⊏ Sketch tool to create the second circle. Apply a draft angle of 3° in the Extruded Boss feature.

When applying the draft angle to the two concentric circles, the outside face tapers inwards and the inside face tapers outwards.

Draft Angle displayed at 5°

💡 Plastic parts require a draft angle. Rule of thumb: 1° to 5° is the draft angle. The draft angle is created in the direction of pull from the mold. This is defined by geometry, material selection, mold production and cosmetics. Always verify the draft with the mold designer and manufacturer.

Activity BATTERYPLATE Part - Extruded Boss Feature.

Select the Sketch plane.
179) Right-click the **top face** of Top Cut. This is your Sketch plane.

Create the sketch.

180) Click **Sketch** ⌐ from the Context toolbar.

181) Click the **top circular edge** of the center hole.

182) Click the **Offset Entities** ⊏ Sketch tool. The Offset Entities PropertyManager is displayed.

183) Enter **.300**in, **[7.62]** for Offset Distance. Accept the default settings.

184) Click **OK** ✔ from the Offset Entities PropertyManager.

185) Drag the **dimension** off the model.

Create the second offset circle.
186) Click the **offset circle** in the Graphics window.

187) Click the **Offset Entities** ⊏ Sketch tool. The Offset Entities PropertyManager is displayed.

188) Enter **.100**in, **[2.54]** for Offset Distance.

189) Click **OK** ✔ from the Offset Entities PropertyManager. Drag the dimension off the model. Two offset concentric circles define the sketch.

Insert an Extruded Boss/Base feature.

190) Click the **Extruded Boss/Base** 🗔 feature tool. The Boss-Extrude PropertyManager is displayed.

191) Enter **.400**in, **[10.16]** for Depth in Direction 1.

192) Click the **Draft ON/OFF** button.

193) Enter **3**deg in the Angle box.

194) Click **OK** ✔ from the Boss-Extrude PropertyManager. The Boss-Extrude feature is displayed in the FeatureManager.

Rename the feature. Display the model in an Isometric view.

195) Rename the **Boss-Extrude** feature to **Holder** in the FeatureManager.

196) Click **Isometric view** ⬛ .

Save the model.

197) Click **Save** 💾 .

BATTERYPLATE Part - Fillet Features: Full Round and Multiple Radius Options

Use the Fillet 🗔 feature tool to smooth rough edges in a model. Plastic parts require fillet features on sharp edges. Create two Fillets. Utilize different techniques. The current Top Face Fillet produced a flat face. Delete the Top Face Fillet. The first Fillet feature is a Full round fillet. Insert a Full round fillet feature on the top face for a smooth rounded transition.

The second Fillet feature is a Multiple Size fillet. Select a different radius value for each edge in the set. Select the inside and outside edge of the Holder. Select all inside tangent edges of the Top Cut. A Multiple Size fillet is utilized next as an exercise. There are machining instances where radius must be reduced or enlarged to accommodate tooling. Note: There are other ways to create Fillets.

Activity: BATTERYPLATE Part - Fillet Features: Full Round, Multiple Size Options.

Delete the Top Edge Fillet.

198) Right-click **Top Face Fillet** from the FeatureManager.

199) Click **Delete**.

200) Click **Yes** to confirm delete.

201) Drag the **Rollback** bar below Top Cut in the FeatureManager as illustrated.

Create a Full Round fillet feature.

202) Click **Hidden Lines Visible** from the Heads-up View toolbar.

203) Click the **Fillet** feature tool. The Fillet PropertyManager is displayed.

204) Click the **Manual** tab.

205) Click **Full Round Fillet** for Fillet Type.

206) Click the **inside Top Cut face** for Side Face Set 1.

207) Click **inside** the Center Face Set box.

208) Click the **top face** for Center Face Set as illustrated.

Rotate the part.

209) Rotate the model until you can select the outside Base Extrude face.

210) Click **inside** the Side Face Set 2 box.

211) Click the **outside Base Extrude face** for Side Face Set 2. Accept the default settings.

212) Click **OK** from the Fillet PropertyManager. Fillet# is displayed in the FeatureManager.

Rename Fillet1.

213) Rename **Fillet#** to **TopFillet**.

Display an Isometric view with Hidden Lines Removed. Save the BATTERYPLATE.

214) Click **Isometric view**.

215) Click **Hidden Lines Removed** from the Heads-up View toolbar.

216) Drag the **Rollback bar** to the bottom of the FeatureManager.

217) Click **Save**.

Create a Multiple Size fillet feature.

218) Click the **bottom outside circular edge** of the Holder.

219) Click the **Fillet** feature tool. The Fillet PropertyManager is displayed.

220) Click the **Constant Size Fillet** button for Fillet Type.

221) Enter **.050**in, [**1.27**] for Radius.

222) Click the **bottom inside circular edge** of the Top Cut.

223) Click the **inside edge** of the Top Cut.

224) Check the **Tangent propagation** box.

225) Check the **Multiple Radius Fillet** box.

Modify the Fillet values.

226) Click the **Radius** box for the Holder outside edge.

227) Enter **0.060**in, [**1.52**].

228) Click the **Radius** box for the Top Cut inside edge.

229) Enter **0.040**in, [**1.02**].

230) Click **OK** from the Fillet PropertyManager. Fillet is displayed in the FeatureManager.

Rename Fillet#.

231) Rename **Fillet#** to **HolderFillet**.

Display Shaded With Edges.

232) Click **Shaded With Edges** from the Heads-up View toolbar. View the results in the Graphics window.

Group the Fillet features into a new folder.

233) Click **TopFillet** from the FeatureManager.

234) Drag the **TopFillet** feature directly above the HolderFillet feature in the FeatureManager.

235) Click **HolderFillet** in the FeatureManager.

236) Hold the **Ctrl** key down.

237) Click **TopFillet** in the FeatureManager.

238) Right-click **Add to New Folder**.

239) Release the **Ctrl** key.

Rename Folder1.

240) Rename **Folder1** to **Fillet Folder**.

Save the BATTERYPLATE.

241) Click **Save**.

Multi-body Parts and the Extruded Boss/Base Feature

A Multi-body part has separate solid bodies within the same part document.

A wrench consists of two cylindrical bodies. Each extrusion is a separate body. The oval profile is sketched on the right plane and extruded with the Up to Body End Condition option.

The BATTERY and BATTERYPLATE parts consisted of a solid body with one sketched profile. Each part is a single body part.

Multi-body part
Wrench

Refer to **Help**, **Tutorials**, **Advanced Techniques**, **Multibody Parts** for additional information.

 Review of the BATTERYPLATE Part.

The Save As option was utilized to copy the BATTERY part to the BATTERYPLATE part. You created a hole in the BATTERYPLATE using Instant3D and modified features using the PropertyManager.

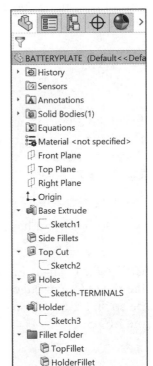

The BATTERYPLATE is a plastic part. The Draft Angle option was added in the Extruded Base (Boss-Extrude1) feature.

The Holder Extruded Boss utilized a circular sketch and the Draft Angle option. The Sketch Offset tool created the circular ring profile. Multi radius Edge Fillets and Face Fillets removed sharp edges.

Similar Fillet features were grouped together into a folder. Features were renamed in the FeatureManager.

The BATTERY and BATTERYPLATE utilized an Extruded Boss/Base feature.

It is considered best practice to fully define all sketches in the model. However, there are times when this is not practical, generally when using the spline tool to create a complex freeform shape.

 Sketch dimensions are displayed in black.

Feature dimensions are displayed in blue.

LENS Part

Create the LENS. The LENS is a purchased part. The LENS utilizes a Revolved Base feature.

Sketch a centerline and a closed profile on the Right Plane. Insert a Revolved Base feature. The Revolved Base feature requires an axis of revolution and an angle of revolution.

Insert the Shell feature. The Shell feature provides uniform wall thickness. Select the front face as the face to be removed.

Utilize the Convert Entities sketch tool to extract the back circular edge for the sketched profile. Insert an Extruded Boss feature from the back of the LENS.

Sketch a single profile. Insert a Revolved Thin feature to connect the LENS to the BATTERYPLATE. The Revolved Thin feature requires thickness.

Insert a Counterbore Hole feature using the Hole Wizard tool. The BULB is located inside the Counterbore Hole.

Insert the front Lens Cover with an Extruded Boss/Base feature. The Extruded Boss/Base feature is sketched on the Front Plane. Add a transparent Lens Shield with the Extruded Boss feature.

LENS Part - Revolved Base Feature

Create the LENS with a Revolved Base feature. The solid Revolved Base feature requires:

- Sketch plane (Right)

- Sketch profile

- Centerline

- Angle of Revolution (360º)

The profile lines reference the Top and Front Planes. Create the curve of the LENS with a 3-point arc.

Activity: LENS Part - Create a Revolved Base Feature.

Create the new part.

242) Click **New** ☐ from the Menu bar.

243) Click the **MY-TEMPLATES** tab.

244) Double-click **PART-IN-ANSI**.

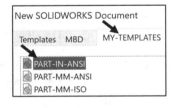

Save the part. Enter part name. Enter part description.

245) Click **Save As** 🖫 .

246) Select **ENGDESIGN-W-SOLIDWORKS\PROJECTS** for Save in folder.

247) Enter **LENS** for File name.

248) Enter **LENS WITH SHIELD** for Description.

249) Click **Save**. The LENS FeatureManager is displayed.

View the Front Plane.

250) Right click **Front Plane** from the FeatureManager.

251) Click **Show** 👁 from the Context toolbar. Hide unwanted planes in the FeatureManager if needed.

Create a sketch.

252) Right-click **Right Plane** from the FeatureManager.

253) Click **Sketch** ⌐ from the Context toolbar. The Sketch toolbar is displayed.

254) Click the **Centerline** ⌐ Sketch tool. The Insert Line PropertyManager is displayed.

255) Sketch a **horizontal centerline** collinear to the Top Plane, through the Origin ↳ as illustrated, approximately 2.0 inches to the left of the Origin. Note the location of the centerline end points.

Sketch the profile. Create three lines.

256) Click the **Line** ╱ Sketch tool. The Insert Line PropertyManager is displayed.

257) Sketch a **vertical line** collinear to the Front Plane coincident with the Origin, approximately 2.0 inches.

258) Sketch a **horizontal line** coincident with the Top Plane, approximately 2 inches

259) Sketch a **vertical line** approximately 1/3 the length of the first line. Right-click **End Chain**.

Create a 3 Point Arc. A 3 Point Arc requires three points.

260) Click the **3 Point Arc** ⌒ Sketch tool from the Consolidated Centerpoint Arc toolbar. Note the mouse pointer feedback icon.

261) Click the **top point** on the left vertical line. This is your first point.

262) Drag the **mouse pointer** to the right.

263) Click the **top point** on the right vertical line. This is your second point.

264) Drag the **mouse pointer** upward.

265) Click a **position** on the arc.

Add an Equal relation.
266) Right-click **Select** to deselect the sketch tool.

267) Click the **left vertical** line.

268) Hold the **Ctrl** key down.

269) Click the **horizontal** line. The Properties PropertyManager is displayed. The selected entities are displayed in the Selected Entities box.

270) Release the **Ctrl** key.

271) Right-click **Make Equal** from the Context toolbar.

272) Click **OK** ✔ from the Properties PropertyManager.

Add dimensions.
273) Click the **Smart Dimension** Sketch tool.

274) Click the **left vertical** line.

275) Click a **position** to the left of the profile.

276) Enter **2.000**in, [**50.8**].

277) Click the **right vertical** line.

278) Click a **position** to the right of the profile.

279) Enter **.400**in, [**10.16**].

280) Click the **arc**.

281) Click a **position** to the right of the profile.

282) Enter **4.000**in, [**101.6**]. The black sketch is fully defined.

Insert the Revolved Base feature.

283) Click the **Revolved Boss/Base** feature tool. The Revolve PropertyManager is displayed.

284) If needed, click the **horizontal centerline** for the axis of revolution. **Note**: The direction arrow points clockwise.

285) Click **OK** ✔ from the Revolve PropertyManager. Revolve1 is displayed in the FeatureManager.

Rename the feature.
286) Rename **Revolve1** to **BaseRevolve**.

Save the model.
287) Click **Save** 📄.

Display the axis of revolution.
288) Click **View**, **Hide/Show**, **Temporary Axes** from the Menu bar.

Revolve features contain an axis of revolution. The axis of revolution utilizes a sketched centerline, edge or an existing feature/sketch or a Temporary Axis. The solid Revolved feature contains a closed profile. The Revolved thin feature contains an open or closed profile.

LENS Part - Shell Feature

The Revolved Base feature is a solid. Utilize the Shell feature to create a constant wall thickness around the front face. The Shell feature removes face material from a solid. The Shell feature requires a face and thickness. Use the Shell feature to create thin-walled parts.

> **Activity: LENS Part - Shell Feature**.

Insert the Shell feature.
289) Click the **front face** of the BaseRevolve feature.

290) Click the **Shell** 🔲 feature tool. The Shell1 PropertyManager is displayed. Face<1> is displayed in the Faces to Remove box.

291) Enter **.250**in, **[6.35]** for Thickness.

Display the Shell feature in an Isometric view.

292) Click **OK** ✔ from the Shell1 PropertyManager. Shell1 is displayed in the FeatureManager.

293) Display an **Isometric view** 🔲.

294) Right-click **Front Plane** from the FeatureManager.

295) Click **Hide** ✎ from the Context toolbar.

Rename the feature.
296) Rename **Shell1** to **LensShell**.

Click **Save** 🖫.

🔆 To insert rounded corners inside a shelled part, apply the Fillet feature before the Shell feature. Select the Multi-thickness option to apply different thicknesses.

Extruded Boss/Base Feature and Convert Entities Sketch tool

Create the LensNeck. The LensNeck houses the BULB base and is connected to the BATTERYPLATE. Use the Extruded Boss/Base feature. The back face of the Revolved Base feature is the Sketch plane.

Utilize the Convert Entities Sketch tool to extract the back circular face to the Sketch plane. The new curve develops an On Edge relation. Modify the back face, and the extracted curve updates to reflect the change. No sketch dimensions are required.

Activity: Extruded Boss Feature and Convert Entities Sketch tool.

Rotate the LENS.

297) **Rotate** the LENS with the middle mouse button to display the back face. The Rotate ↻ icon is displayed.

Sketch the profile.
298) Right-click the **back face** for the Sketch plane. BaseRevolve is highlighted in the FeatureManager. This is your Sketch plane.

299) Click **Sketch** 🖊 from the Context toolbar. The Sketch toolbar is displayed.

300) Click the **Convert Entities** Sketch tool.

Insert an Extruded Boss/Base feature.

301) Click the **Extruded Boss/Base** feature tool. The Boss-Extrude PropertyManager is displayed.

302) Enter **.400**in, **[10.16]** for Depth in Direction 1. Accept the default settings.

303) Click **OK** ✔ from the Boss-Extrude PropertyManager. Boss-Extrude1 is displayed in the FeatureManager.

Display an Isometric view. Rename the feature and save the model.

304) Click **Isometric view** 📦.

305) Rename **Boss-Extrude1** as **LensNeck**.

306) Click **Save** 💾.

LENS Part - Hole Wizard

The LENS requires a Counterbore hole. Apply the Hole Wizard 🕹 feature. The Hole Wizard feature assists in creating complex and simple holes. The Hole Wizard Hole type categories are *Counterbore, Countersink, Hole, Straight Tap, Tapered Tap, Legacy Hole* (holes created before SOLIDWORKS 2000), *Counterbore Slot, Countersink Slot,* and *Slot*.

Specify the user parameters for the custom Counterbore hole. The parameters are *Description, Standard, Screw Type, Hole Size, Fit, Counterbore diameter, Counterbore depth* and *End Condition*. Select the face or plane to locate the hole profile.

Insert a Coincident relation to position the hole center point. Dimensions for the Counterbore hole are provided in both inches and millimeters.

Activity: LENS Part - Hole Wizard Counterbore Hole Feature.

Create the Counterbore hole.

307) Click **Front view** 📦.

308) Click the **Hole Wizard** 🕹 feature tool. The Hole Specification PropertyManager is displayed. Type is the default tab.

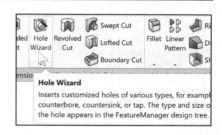

309) Click the **Counterbore** icon as illustrated.

Note: For a metric hole, skip the next few steps.

For inch Counterbore hole:
310) Select **ANSI Inch** for Standard. Select **Hex Bolt** for Type.

311) Select ½ for Size. Check the **Show custom sizing** box.

312) Click inside the **Counterbore Diameter** value box.

313) Enter **.600**in. Click inside the **Counterbore Depth** value box.

314) Enter **.200**in.

315) Select **Through All** for End Condition.

316) Click the **Position** tab.

317) Click the small **inside back face** of the LensShell feature as illustrated. Do not select the Origin. LensShell is highlighted in the FeatureManager.

The Point tool ✏️ icon is displayed.

318) Click the **origin**. A Coincident icon relation is displayed.

Deselect the Point tool.
319) Right-click **Select** in the Graphics window.

Note: For an inch hole, skip the next few steps to address millimeter.

For millimeter Counterbore hole:
320) Select **ANSI Metric** for Standard. Enter **Hex Bolt** for Type. Select **M5** for Size. Click **Through All** for End Condition. Check the **Show custom sizing** box.

321) Click inside the **Counterbore Diameter** value box.

322) Enter **15.24**. Click inside the **Counterbore Depth** value box.

323) Enter **5**. Click the **Position** tab.

324) Click the small **inside back face** of the LensShell feature as illustrated. Do not select the Origin. LensShell is highlighted in the FeatureManager. The Point tool icon is displayed.

325) Click the **origin**. A Coincident relation is displayed.

Deselect the Point tool.
326) Right-click **Select** in the Graphics window.

Return to the Type tab.
327) Click the **Type** tab.

Add the new hole type to your Favorites list.
328) Expand the Favorites box.

329) Click the **Add or Update Favorite** ✦ icon.

330) Enter **CBORE for ½ Hex Head Bolt**.

331) Click **OK** from the Add or Update a Favorite dialog box.

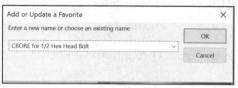

332) Click **OK** ✔ from the Hole Specification PropertyManager.

Expand the Hole feature.
333) Expand the CBORE feature in the FeatureManager. Note: Sketch3 and Sketch4 created the CBORE feature.

Display an Isometric Section view.
334) Click **Right Plane** from the FeatureManager.

335) Click **Section view** ⊞ from the Heads-up View toolbar.

336) Click **OK** ✔ from the Section View PropertyManager.

337) Click **Isometric view** ▢.

Return to a Full view.
338) Click **Section view** ⊞ from the Heads-up View toolbar.

Rename the feature and save the model.
339) Rename **CBORE for ½ Hex Head Bolt1** to **BulbHole**.

340) Click **Save** 💾.

LENS Part - Revolved Boss Thin Feature

Create a Revolved Boss Thin feature. Rotate an open sketched profile around an axis. The sketch profile must be open and cannot cross the axis. A Revolved Boss Thin feature requires:

- Sketch plane (Right Plane)

- Sketch profile (Center point arc)

- Axis of Revolution (Temporary axis)

- Angle of Rotation (360)

- Thickness .100in [2.54]

A Revolved feature produces silhouette edges in 2D views. A silhouette edge represents the extent of a cylindrical or curved face.

Select the Temporary Axis for Axis of Revolution. Select the Revolved Boss feature. Enter .100in [2.54] for Thickness in the Revolve PropertyManager. Enter 360° for Angle of Revolution.

Activity: LENS Part - Revolved Boss Thin Feature.

Create a sketch.
341) Right-click **Right Plane** from the FeatureManager.

342) Click **Sketch** ⌙ from the Context toolbar.

343) Click **Right view** ⬚.

344) Zoom in on the LensNeck.

345) Click the **Centerpoint Arc** ⌒ Sketch tool.

346) Click the **top horizontal edge** of the LensNeck. Do not select the midpoint of the silhouette edge.

347) Click the **top right corner** of the LensNeck.

348) Drag the **mouse pointer** counterclockwise to the left.

349) Click a **position** directly above the first point as illustrated. If needed, insert a Vertical relation between the end third and first point.

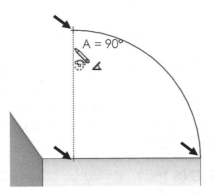

Add a dimension.
350) Click the **Smart Dimension** ⟋ Sketch tool.

351) Click the **arc**.

352) Click a **position** to the right of the profile.

353) Enter **.100**in, **[2.54]**. The Sketch is fully defined.

Insert a Revolved Thin feature.

354) Click the **Revolved Boss/Base** feature tool. The Revolve PropertyManager is displayed.

355) Select **Mid-Plane** for Revolve Type in the Thin Feature box.

356) Enter **.050**in, [**1.27**] for Direction1 Thickness.

357) Click the **Temporary Axis** for Axis of Revolution.

358) Click **OK** ✔ from the Revolve PropertyManager.

Rename the feature.
359) Rename **Revolve-Thin1** to **LensConnector**.

Fit the model to the Graphics window and save.
360) Press the **f** key.

361) Click **Save** 💾.

A Revolved sketch that remains open results in a Revolve Thin feature. A Revolved sketch that is automatically closed results in a line drawn from the start point to the end point of the sketch. The sketch is closed and results in a non- Revolve Thin feature .

LENS Part - Extruded Boss/Base Feature and Offset Entities

Use the Extruded Boss/Base feature tool to create the front LensCover. Utilize the Offset Entities Sketch tool to offset the outside circular edge of the Revolved feature. The Sketch plane for the Extruded Boss feature is the front circular face.

The Offset Entities Sketch tool requires an Offset Distance and direction. Utilize the Bi-direction option to create a circular sketch in both directions. The extrude direction is away from the Front Plane.

Activity: LENS Part - Extruded Boss Feature and Offset Entities.

Create the Sketch.

362) Click **Hidden Lines Removed** from the Heads-up View toolbar.

363) Click **Front view** .

364) Right-click the **front circular face** for the Sketch plane.

365) Click **Sketch** from the Context toolbar.

Offset the selected edge.

366) Click the **outside circular edge** of the LENS in the Graphics window.

367) Click the **Offset Entities** Sketch tool. The Offset Entities PropertyManager is displayed.

368) Check the **Bi-directional** box.

369) Enter **.250**in, [**6.35**] for Offset Distance. Accept the default settings.

370) Click **OK** from the Offset Entities PropertyManager.

Display an Isometric view with Shaded With edges.

371) Click **Isometric view** from the Heads-up View toolbar.

372) Click **Shaded With Edges** from the Heads-up View toolbar.

Insert an Extruded Boss/Base feature.

373) Click the **Extruded Boss/Base** feature tool. The Boss-Extrude PropertyManager is displayed.

374) Enter **.250**in, [**6.35**] for Depth in Direction 1. Accept the default settings.

375) Click **OK** from the Boss-Extrude PropertyManager. Boss-Extrude# is displayed in the PropertyManager.

Verify the position of the extruded feature.

376) Click the **Top view** . View the extruded feature.

Rename the feature. Display an Isometric view. Save the model.

377) Rename **Boss-Extrude#** to **LensCover**.

378) Click **Isometric view** .

379) Click **Save** .

LENS Part - Extruded Boss Feature and Transparency

Apply the Extruded Boss/Base feature to create the LensShield. Utilize the Convert Entities Sketch tool to extract the inside circular edge of the LensCover and place it on the Front plane.

Apply the Transparent Optical property to the LensShield to control the ability for light to pass through the surface. Transparency is an Optical Property found in the Color PropertyManager.

Activity: LENS Part - Extruded Boss Feature and Transparency.

Create the sketch.

380) Right-click **Front Plane** from the FeatureManager. This is your Sketch plane.

381) Click **Sketch** from the Context toolbar. The Sketch toolbar is displayed.

382) Click **Isometric view** .

383) Click the **front inner circular edge** of the LensCover (Boss-Extrude2) as illustrated.

384) Click the **Convert Entities** ⬡ Sketch tool.

Insert an Extruded Boss feature.

385) Click the **Extruded Boss/Base** ⬡ feature tool. The Boss-Extrude FeatureManager is displayed.

386) Enter **.100**in, **[2.54]** for Depth in Direction 1.

387) Click **OK** ✔ from the Boss-Extrude PropertyManager. Boss-Extrude3 is displayed in the FeatureManager.

Rename the feature. Save the model.
388) Rename **Boss-Extrude3** to **LensShield**.

389) Click **Save** 💾.

Change Transparency of the LensShield.
390) Right-click **LensShield** in the FeatureManager.

391) Click **Change Transparency**.

392) Click **inside** the Graphics area. View the result.

Hide the axis of revolution. Display Shaded with Edges.
393) Click **View**, **Hide/Show**, and uncheck **Temporary Axes** from the Menu bar.

394) Click **Shaded with Edges** ⬡ from the Heads-up toolbar.

Suppress the LensShield. Save the model.
395) Right-click **LensShield** from the FeatureManager.

396) Click **Suppress**.

397) Click **Save** 💾.

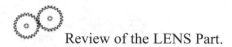 Review of the LENS Part.

The LENS feature utilized a Revolved Base feature. A Revolved feature required an axis, profile and an angle of revolution. The Shell feature created a uniform wall thickness.

You utilized the Convert Entities Sketch tool to create the Extruded Boss feature for the LensNeck. The Counterbore hole was created using the Hole Wizard feature.

The Revolved Thin feature utilized a single 3 Point Arc. Geometric relations were added to the silhouette edge to define the arc. The LensCover and LensShield utilized existing geometry to Offset and Convert the geometry to the sketch.

Color and Optics determined the LensShield transparency.

BULB Part

The BULB fits inside the LENS. Use the Revolved feature as the Base feature for the BULB.

Insert the Revolved Base feature from a sketched profile on the Right Plane.

Insert a Revolved Boss feature using a Spline sketched profile. The profile utilizes a complex curve called a Spline (Non-Uniform Rational B-Spline or NURB). Draw Splines with control points.

Insert a Revolved Cut Thin feature at the base of the BULB. A Revolved Cut Thin feature removes material by rotating an open sketch profile about an axis.

Insert a Dome feature at the base of the BULB. A Dome feature creates spherical or elliptical shaped geometry. Use the Dome feature to create the Connector feature of the BULB. The Dome feature requires a face and a height value.

Insert a Circular Pattern feature from an Extruded Cut feature.

BULB Part - Revolved Base Feature

Create the new part, BULB. The BULB utilizes a solid Revolved Base feature.

The solid Revolved Base feature requires a:

- Sketch plane (Right Plane)

- Sketch profile (Lines)

- Axis of Revolution (Centerline)

- Angle of Rotation (360°)

Utilize the centerline to create a diameter dimension for the profile. The flange of the BULB is located inside the Counterbore hole of the LENS. Align the bottom of the flange with the Front Plane. The Front Plane mates against the Counterbore face.

Activity: BULB Part - Revolved Base Feature.

Create a new part.

398) Click **New** ⬜ from the Menu bar.

399) Click the **MY-TEMPLATES** tab.

400) Double-click **PART-IN-ANSI**.

Save the part. Enter name. Enter description.

401) Click **Save As** 🖫 .

402) Select **ENGDESIGN-W-SOLIDWORKS\PROJECTS** for Save in folder.

403) Enter **BULB** for File name.

404) Enter **BULB FOR LENS** for Description.

405) Click **Save**. The BULB FeatureManager is displayed.

Select the Sketch plane.
406) Right-click **Right Plane** from the FeatureManager. This is your Sketch plane.

Create the sketch.
407) Click **Sketch** ⌐ from the Context toolbar. The Sketch toolbar is displayed.

Sketch a centerline.

408) Click the **Centerline** ✐ Sketch tool. The Insert Line PropertyManager is displayed.

409) Sketch a **horizontal centerline** through the Origin ⌐ as illustrated.

Create six profile lines.

410) Click the **Line** ✐ Sketch tool. The Insert Line PropertyManager is displayed.

411) Sketch a **vertical line** to the left of the Front Plane.

412) Sketch a **horizontal line** with the endpoint coincident to the Front Plane.

413) Sketch a short **vertical line** towards the centerline, collinear with the Front Plane.

414) Sketch a **horizontal line** to the right.

415) Sketch a **vertical line** with the endpoint collinear with the centerline.

416) Sketch a **horizontal line** to the first point to close the profile.

Add dimensions.

417) Click the **Smart Dimension** ✐ Sketch tool.

418) Click the **centerline**.

419) Click the **top right horizontal line** as illustrated.

420) Click a **position** below the centerline and to the right.

421) Enter **.400**in, [**10.016**].

[10.16]
.400

🔆 Click **View**, **Hide/Show**, **Sketch Relations** from the Menu bar to display the relations of the model in the Graphics window.

422) Click the **centerline**.

423) Click the **top left horizontal line**.

424) Click a **position** below the centerline and to the left.

425) Enter **.590**in, **[14.99]**.

426) Click the **top left horizontal line**.

427) Click a **position** above the profile.

428) Enter **.100**in, **[2.54]**.

429) Click the **top right horizontal line**.

430) Click a **position** above the profile.

431) Enter **.500**in, **[12.7]**.

Fit the model to the Graphics window.
432) Press the **f** key.

Insert a Revolved Base feature.

433) Click the **Revolved Boss/Base** feature tool. The Revolve PropertyManager is displayed. Accept the default settings.

434) Click **OK** from the Revolve PropertyManager. Revolve1 is displayed in the FeatureManager.

Display an Isometric view. Save the model.

435) Click **Isometric view** . Click **Save** .

BULB Part - Revolved Boss Feature and Spline Sketch Tool

The BULB requires a second solid Revolved feature. The profile utilizes a complex curve called a Spline (Non-Uniform Rational B-Spline or NURB). Draw Splines with control points. Adjust the shape of the curve by dragging the control points. Note: The complex curve (sketch) does not need to be fully defined.

Activity: BULB Part - Revolved Boss Feature and Spline Sketch Tool.

Display the Temporary Axes. Create the sketch.

436) Click **View**, **Hide/Show**, check **Temporary Axes** from the Menu bar.

437) Right-click **Right Plane** from the FeatureManager for the Sketch plane.

438) Click **Sketch** from the Context toolbar.

Sketch the profile.

439) Click the **Spline** Sketch tool.

440) Click the **left vertical edge** of the Base feature for the Start point.

441) Drag the **mouse pointer** to the left.

442) Click a **position** above the Temporary Axis for the Control point.

443) Click the **Temporary Axis.**

444) Right-click **Select** to end the Spline.

445) Click the **Line** Sketch tool.

446) Sketch a **horizontal line** from the Spline endpoint to the left edge of the Revolved feature.

447) Sketch a **vertical line** to the Spline start point, collinear with the left edge of the Revolved feature. Dimensions are not required to create a feature.

Insert a Revolved Boss feature. Display an Isometric view. Save the part.

448) De-select the Line Sketch tool. Right-click **Select**.

449) Click the **Temporary Axis** from the Graphics window as illustrated.

450) Click the **Revolved Boss/Base** feature tool. The Revolve PropertyManager is displayed. Accept the default settings.

451) Click **OK** from the Revolve PropertyManager.

452) Click **Isometric view** .

453) Click **Save** .

The points of the Spline dictate the shape of the Spline. Edit the control points in the sketch to produce different shapes for the Revolved Boss feature.

BULB Part - Revolved Cut Thin Feature

A Revolved Cut Thin feature removes material by rotating an open sketch profile around an axis. Sketch an open profile on the Right Plane. Add a Coincident relation to the silhouette and vertical edge. Insert dimensions.

Sketch a Centerline to create a diameter dimension for a revolved profile. The Temporary axis does not produce a diameter dimension.

Note: If lines snap to grid intersections, uncheck Tools, Sketch Settings, Enable Snapping for the next activity.

Activity: BULB Part - Revolved Cut Thin Feature.

Create a sketch.
454) Right-click **Right Plane** from the FeatureManager.

455) Click **Sketch** from the Context toolbar.

456) Click the **Line** Sketch tool.

457) Click the **midpoint** of the top silhouette edge.

458) Sketch a **line** downward and to the right as illustrated.

459) Sketch a **horizontal line** to the right vertical edge.

460) De-select the Line Sketch tool. Right-click **Select**.

If needed, add a Coincident relation.
461) Click the **endpoint** of the line.

462) Hold the **Ctrl** key down.

463) Click the right **vertical edge**.

464) Release the **Ctrl** key.

465) Click **Coincident** .

466) Click **OK** from the Properties PropertyManager.

Sketch a Centerline.
467) Click **View**, **Hide/Show**, and un-check **Temporary Axes** from the Menu bar.

468) Click the **Centerline** Sketch tool.

469) Sketch a **horizontal centerline** through the Origin.

Add dimensions.
470) Click the **Smart Dimension** Sketch tool.

471) Click the **horizontal centerline**.

472) Click the **short horizontal line**.

473) Click a **position** below the profile to create a diameter dimension.

474) Enter **.260**in, [**6.6**].

475) Click the **short horizontal line**.

476) Click a **position** above the profile to create a horizontal dimension. Enter **.070**in, [**1.78**]. The Sketch is fully defined and is displayed in black.

For Revolved features, the ∅ symbol is not displayed in the part. The ∅ symbol is displayed when inserted into the drawing.

Insert the Revolved Cut Thin feature.
477) De-select the Smart Dimension Sketch tool. Right-click **Select**. Click the **centerline** in the Graphics window.

478) Click the **Revolved Cut** feature tool. The Cut-Revolve PropertyManager is displayed.

479) Click **No** to the Warning Message, "Would you like the sketch to be automatically closed?" The Cut-Revolve PropertyManager is displayed.

480) Check the **Thin Feature** box.

481) Enter **.150**in, [**3.81**] for Thickness.

482) Click the **Reverse Direction** box.

483) Click **OK** ✔ from the Cut-Revolve PropertyManager. Cut-Revolve-Thin1 is displayed in the FeatureManager.

Save the model.

484) Click **Save** 🖫.

BULB Part - Dome Feature

The Dome ⬭ feature creates spherical or elliptical shaped geometry. Use the Dome feature to create the Connector feature of the BULB. The Dome feature requires a face and a height/distance value.

Activity: BULB Part - Dome Feature.

Insert the Dome feature.

485) Click the **back circular face** of Revolve1. Revolve1 is highlighted in the FeatureManager.

486) Click **Insert**, **Features**, **Dome** ⬭ from the Main menu. The Dome PropertyManager is displayed. Face1 is displayed in the Parameters box.

487) Enter **.100**in, [**2.54**] for Distance.

488) Click **OK** ✔ from the Dome PropertyManager. Dome1 is displayed in the FeatureManager.

Display an Isometric view. Save the model.
489) Click **Isometric view** 📦.

490) Click **Save** 💾.

💡 Before creating sketches that use Geometric relations, check the Enable Snapping option in the Document Properties dialog box.

BULB Part - Circular Pattern Feature

A Pattern feature creates one or more instances of a feature or a group of features. The Circular Pattern feature places the instances around an axis of revolution.

Seed   Pattern

The Circular Pattern 🔄 feature requires a seed feature. The seed feature is the first feature in the pattern. The seed feature in this section is a V-shaped Extruded Cut feature.

Activity: BULB Part - Circular Pattern Feature.

Create the Seed Cut feature.
491) Right-click the **front face** of the Base feature, Revolve1 in the Graphics window for the Sketch plane as illustrated. Revolve1 is highlighted in the FeatureManager.

492) Click **Sketch** 📝 from the Context toolbar. The Sketch toolbar is displayed.

493) Click the **outside circular edge** of the BULB.

494) Click the **Convert Entities** ⬭ Sketch tool.

495) Zoom in on the top half of the BULB.

Sketch a centerline.

496) Click the **Centerline** ⟋ Sketch tool.

497) Sketch a **vertical centerline** coincident with the top and bottom circular circles and coincident with the Right plane.

Converted outside circular edge

Centerline endpoints coincident with circular edges

Sketch a V-shaped line.

498) Click **Tools**, **Sketch Tools**, **Dynamic Mirror** from the Menu bar. The Mirror PropertyManager is displayed.

499) Click the **centerline** from the Graphics window.

500) Click the **Line** ⟋ Sketch tool.

501) Click the **midpoint** of the centerline.

502) Click the coincident **outside circle edge to the right** of the centerline.

Deactivate the Dynamic Mirror tool.

503) Click **Tools**, **Sketch Tools**, **Dynamic Mirror** from the Menu bar.

Trim unwanted geometry.

504) Click the **Trim Entities** ⟊ Sketch tool. The Trim PropertyManager is displayed.

505) Click **Power trim** from the Options box.

506) Click a **position** in the Graphics window and drag the mouse pointer until it intersects the circle circumference.

507) Click **OK** ✔ from the Trim PropertyManager.

Add a Perpendicular relation.

508) Click the **left V shape** line.

509) Hold the **Ctrl** key down.

510) Click the **right V shape** line. The Properties PropertyManager is displayed.

511) Release the **Ctrl** key.

512) Click **Perpendicular** ⊥ from the Add Relations box.

513) Click **OK** ✔ from the Properties PropertyManager. The sketch is fully defined.

Create an Extruded Cut feature.

514) Click the **Extruded Cut** 📄 feature tool. The Cut-Extrude PropertyManager is displayed.

515) Click **Through All** for End Condition in Direction 1. Accept the default settings.

516) Click **OK** ✔ from the Cut-Extrude PropertyManager. The Cut-Extrude1 feature is displayed in the FeatureManager.

Display an Isometric view.

517) Click **Isometric view** 🧊.

Fit the drawing to the Graphics window.

518) Press the **f** key.

Save the model.

519) Click **Save** 💾.

🔅 Reuse Geometry in the feature. The Cut-Extrude1 feature utilized the centerline, Mirror Entity, and geometric relations to create a sketch with no dimensions.

The Cut-Extrude1 feature is the seed feature for the pattern. Create four copies of the seed feature. A copy of a feature is called an Instance. Modify the four copies of Instances to eight.

Insert the Circular Pattern feature.

520) Click the **Cut-Extrude1** feature from the FeatureManager.

521) Click the **Circular Pattern** feature tool. The Circular Pattern PropertyManager is displayed. Cut-Extrude1 is displayed in the Features to Pattern box.

522) Click **View**, **Hide/Show**, and check **Temporary Axes** from the Menu bar.

523) Click the **Temporary Axis** from the Graphics window. Axis<1> is displayed in the Pattern Axis box.

524) Enter **4** in the Number of Instances box.

525) Check the **Equal spacing** box.

526) Enter **360deg**.

527) Check the **Geometry pattern** box. Accept the default settings.

528) Click **OK** from the Circular Pattern PropertyManager. CirPattern1 is displayed in the FeatureManager.

Edit the Circular Pattern feature.
529) Right-click **CirPattern1** from the FeatureManager.

530) Click **Edit Feature** from the Context toolbar. The CirPattern1 PropertyManager is displayed.

531) Enter **8** in the Number of Instances box.

532) Click **OK** from the CirPattern1 PropertyManager.

Rename the feature.
533) Rename **Cut-Extrude1** to **SeedCut**.

Hide the reference geometry.
534) **Hide** all reference geometry.

Save the model.
535) Click **Save**.

Rename the seed feature of a pattern to locate it quickly for future assembly.

Customizing Toolbars and Short Cut Keys

The default toolbars contain numerous icons that represent basic functions. Additional features and functions are available that are not displayed on the default toolbars.

You have utilized the z key for Zoom In/Out, the f key for Zoom to Fit and Ctrl-C/Ctrl-V to Copy/Paste. Short Cut keys save time.

Assign a key to execute a SOLIDWORKS function. Create a Short Cut key for the Temporary Axis.

Activity: Customizing Toolbars and Short Cut Keys.

Customize the toolbar.
536) Click **Tools**, **Customize** from the Menu bar. The Customize dialog box is displayed.

Place the Freeform icon on the Features toolbar.
537) Click the **Commands** tab.

538) Click **Features** from the category text box.

539) Drag the **Freeform** feature icon into the Features toolbar. The Freeform feature option is displayed.

Customize the keyboard for
the Temporary Axes.

540) Click the **Keyboard** tab
from the Customize
dialog box.

541) Select **View** for
Categories.

542) Select **Temporary
Axes** for Commands.

543) Enter **R** for Shortcut(s)
key.

544) Click **OK**.

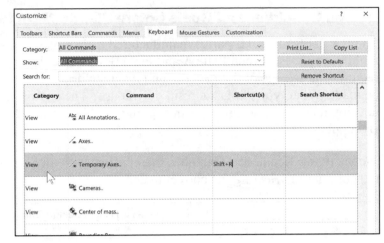

545) Press the **R** key to toggle the display of the Temporary Axes.

 Test the proposed Short cut key before you customize your keyboard. Refer to the
default Keyboard Short cut table in the Appendix.

Review of the BULB Part.

The Revolved Base feature utilized a sketched profile on
the Right plane and a centerline. The Revolved Boss
feature utilized a Spline sketched profile. A Spline is a
complex curve. A complex curve (sketch) does not need
to be fully defined.

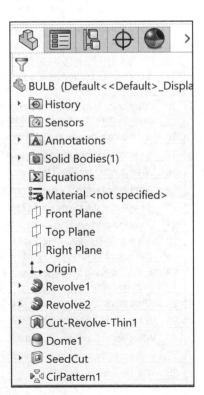

You created the Revolved Cut Thin feature at the base of
the BULB to remove material. A centerline was inserted
to add a diameter dimension. The Dome feature was
inserted on the back face of the BULB. The Circular
Pattern feature was created from an Extruded Cut feature.
The Extruded Cut feature utilized existing geometry and
required no dimensions.

Toolbars and keyboards were customized to save time.
Always verify that a Short cut key is not predefined in
SOLIDWORKS.

Design Checklist and Goals before Plastic Manufacturing

The BATTERYPLATE part is manufactured from a plastic resin. A plastic part requires mold base plates, named Plate A and Plate B, to create the mold. A plastic part also requires knowledge of the material and manufacturing process. Critical decisions need to be made. What type of plastic resin do you use? The answer is derived from the development team: designer, resin supplier, mold maker and mold base, (Plate A & Plate B) supplier.

Ticona (www.ticona.com) supplies their customers with a designer's checklist to assist the designer in selecting the correct material.

TICONA-Celanese Typical Design Checklist
1.   What is the function of the part?
2.   What is the expected lifetime of the part?
3.   What agency approvals are required? (UL, FDA, USDA, NSF, USP, SAE, MIL spec)
4.   What electrical characteristics are required and at what temperatures?
5.   What temperature will the part see and for how long?
6.   What chemicals will the part be exposed to?
7.   Is moisture resistance necessary?
8.   How will the part be assembled? Can parts be combined into one plastic part?
9.   Is the assembly going to be permanent or one time only?
10.   Will adhesives be used? Some resins require special adhesives.
11.   Will fasteners be used? Will threads be molded in?
12.   Does the part have a snap fit? Glass filled materials will require more force to close the snap fit but will deflect less.
13.   Will the part be subjected to impact? If so, radius the corners.
14.   Is surface appearance important? If so, beware of weld lines, parting lines, ejector location and gate location.
15.   What color is required for the part? Is a specific match required or will the part be color-coded? Some glass or mineral filled materials do not color as well as unfilled materials.
16.   Will the part be painted? Is primer required? Will the part go through a high temperature paint oven?
17.   Is weathering or UV exposure a factor?
18.   What are the required tolerances? Can they be relaxed to make molding more economical?
19.   What is the expected weight of the part? Will it be too light (or too heavy)?
20.   Is wear resistance required?
21.   Does the part need to be sterilized? With what methods (chemical, steam, radiation)?
22.   Will the part be insert molded or have a metal piece press fit in the plastic part? Both methods result in continuous stress in the part.
23.   Is there a living hinge designed in the part? Be careful with living hinges designed for crystalline materials such as acetal.
24.   What loading and resulting stress will the part see? And, at what temperature and environment.
25.   Will the part be loaded continuously or intermittently? Will permanent deformation or creep be an issue?
26.   What deflections are acceptable?
27.   Is the part moldable? Are there undercuts? Are there sections that are too thick or thin?
28.   Will the part be machined?
29.   What is the worst possible situation the part will be in? Worst-case environment.

In "Designing With Plastic - The Fundamentals," Ticona lists three design goals for creating injection molded parts:

Goal 1: Maximize Functionality

Mold bases are costly. Design functionality into the part. A single plastic chassis replaces several sheet metal components.

Reduce assembly time and part weight whenever it is appropriate.

Sheet Metal Assembly

Injection Molded Thermoplastic

Goal 2: Optimize Material Selection

For material selection, consider the following elements: part functionality, operating environment, cost/price constraint and any special cosmetic requirements.

Several materials should be selected with a developed list of advantages and disadvantages for review.

For maximum product performance at a competitive cost in this project, use Celanese® (Nylon 6/6) (P/A 6/6), where:

- Celanese® is the registered trademark name.

- Nylon 6/6 is the common plastic name.

- P/A 6/6 (Polyhexamethyleneadipamide) is the chemical name.

Goal 3: Minimize Material Use

Optimize part wall thickness for cost and performance.

The minimum volume of plastic that satisfies the structural, functional, appearance and molding requirements of the application is usually the best choice.

Machined Valve Body

Injected Molded Valve Body

Mold Base

Designing a custom mold base is expensive and time consuming. An example of a mold base supplier is Progressive Components (www.procomps.com).

The mold base positions the mold cavities. The mold base plates are machined to create the mold cavities. The mold base is designed to withstand high pressure, heating and cooling sequences during the fabrication process.

The mold base assembly is composed of the following two plates and a variety of support plates, ejector plates, pins and bushings:

- PLATE A

- PLATE B

PLATE A

PLATE B

Courtesy of Progressive Components
Wauconda, IL USA

Applying SOLIDWORKS Features
for Mold Tooling Design

SOLIDWORKS features such as Draft, Fillet and Shell assist in the design of plastic parts. Utilize the Draft tool to analyze the part before creating the Mold Tooling, PLATE A and PLATE B.

Most molded parts require Draft. Plastic molded injection parts require a draft angle to eject the part from the mold. To properly eject the part, design parts with a draft angle in the direction of the mold movement. A draft angle of 1° - 3° is the design rule for most injection molded parts. There are exceptions based on material type and draw depth. For illustration purposes, a draft angle of 1° is used in the BATTERYPLATE.

The draft angle is an option in both the Extruded Boss/Base and Extruded Cut features. The Draft feature adds a specified Draft Angle to faces of an existing feature.

Use the Fillet feature to remove sharp edges from plastic parts. For thin walled parts, ensure that the inside sharp edges are removed. Sketch Arcs and 2D Fillets in the 2D profile to remove sharp edges. The Shell feature, Extruded Cut feature or Extruded Thin feature all provide wall thickness. Select the correct wall thickness for a successful part.

If the wall thickness is too thin, the part will structurally fail. If the wall thickness is too thick, the part will be overweight and too costly to manufacture. This will increase cycle time and decrease profits.

There are numerous types of plastic molds. SOLIDWORKS contains a variety of Mold tools to assist the designer, mold maker and mold manufacturer. For a simple Cavity Mold follow the following steps:

- *Draft analysis*. A draft analysis examines the faces of the model for sufficient draft, to ensure that the part ejects properly from the tooling.

- *Undercut detection*. Identifies trapped areas that prevent the part from ejecting.

- *Parting Lines*. This tool has two functions:

    o Verifies that you have draft on your model, based on the angle you specify.

    o Creates a parting line from which you create a parting surface. The Parting Lines tool includes the option to select an edge and have the system Propagate to all edges.

- *Shut-off Surfaces*. Creates surface patches to close up through holes in the molded part.

- *Parting Surfaces*. Extrude from the parting line to separate mold cavity from core. You can also use a parting surface to create an interlock surface.

- *Ruled Surface*. Adds draft to surfaces on imported models. You can also use the Ruled Surface tool to create an interlock surface.

- *Tooling Split*. Creates the core and cavity bodies, based on the steps above.

Activity: SOLIDWORKS Features for Mold Tooling Design.

Create the new part.

546) Click **New** from the Menu bar. Click the **MY-TEMPLATES** tab.

547) Double-click **PART-IN-ANSI**, [**PART-MM-ISO**].

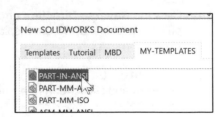

Save the part. Enter part name. Enter part description.

548) Click **Save As** . Select **PROJECTS** for Save in folder.

549) Enter **MOLD-TOOLING** for File name.

550) Enter **MOLD FEATURES FOR TOOLING** for Description. Click **Save**. The MOLD-TOOLING FeatureManager is displayed.

Insert a part.
551) Click **Insert**, **Part** from the Menu bar.

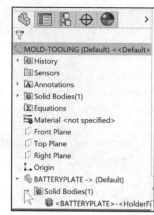

552) Double-click **BATTERYPLATE** from the PROJECTS folder. The Insert Part PropertyManager is displayed. If needed, Click **Yes**.

553) Click **OK** ✔ from the Insert Part PropertyManager. The BATTERYPLATE part is the first feature in the MOLD-TOOLING part.

554) Click **OK**.

💡 Utilize the Insert Part option to maintain the original part and a simplified FeatureManager. The Insert Part maintains an External reference to the original part ->. Modify the original part and the Insert Part updates on a Rebuild. Apply the Insert Part option to SOLIDWORKS parts and imported parts.

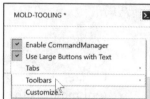

Display the Mold toolbar.
555) Right-click in the **gray area** of the CommandManager toolbar.

556) Click **Toolbars**, click **Mold Tools**. The Mold toolbar is displayed. Click and drag the **Mold** toolbar into the Graphics window.

Perform a Draft Analysis.
557) Click the **Draft** 🔲 tool from the Mold toolbar. The Draft PropertyManager is displayed.

558) Click the **DraftXpert** tab. Click the **Add** tab.

Create the Direction of Pull.
559) **Rotate** the part and click the **bottom face**. Face<1> is displayed in the Neutral Plane box.

560) Click the **Reverse Direction** arrow. The direction of pull arrow points upwards. Display an **Isometric view**.

Select a face to draft.
561) **Zoom in** and click the **inside hole face** of the center hole.

562) **Zoom in** and click the **inside hole face** of the small hole. The selected faces are displayed.

563) Enter **1**deg for Draft angle. Check the **Auto Paint** box.

564) Click the **Apply** button.

565) Rotate the **part** to display the faces that require a draft. The faces are displayed in different colors. View the results in the Graphics window.

566) Click **OK** ✔ from the DraftXpert PropertyManager. Draft1 is displayed.

The Top Cut, Extruded Cut, and the Holes Extruded Cut requires a draft. The bottom flat surface will become the Parting surface.

Modify the BATTERYPLATE part In-Context of the MOLD-TOOLING part. Both parts are open during the modification session.

Modify the BATTERYPLATE In-Context.
567) Right-click **BATTERYPLATE** in the FeatureManager.

568) Click **Edit In Context**. The BATTERYPLATE FeatureManager is displayed.

Insert Draft into the Top Cut.
569) Right-click **Top Cut** in the BATTERYPLATE FeatureManager.

570) Click **Edit Feature** . The Top Cut PropertyManager is displayed. If needed, click the **Draft on/off** button. Enter **1**deg for Draft Angle.

571) Click **OK** ✔ from the Top Cut PropertyManager.

Insert a draft into the holes.
572) Right-click **Holes** in the BATTERYPLATE FeatureManager.

573) Click **Edit Feature** . The Holes PropertyManager is displayed. If needed, click the **Draft on/off** button.

574) Enter **1**deg for Draft Angle.

575) Click **OK** ✔ from the Holes PropertyManager.

Return to the MOLD-TOOLING part.
576) Press **Ctrl-Tab**. The MOLD TOOLING FeatureManager is displayed.

Scale:

The Scale feature increases or decreases the part's volume by a specified factor. The hot thermoplastic utilized in this part cools during the molding process and hardens and shrinks. Increase the part with the Scale feature to compensate for shrinkage.

Insert the Scale feature.

577) Click **Scale** from the Mold toolbar. The Scale PropertyManager is displayed.

578) Enter **1.2** for Scale Factor.

579) Click **OK** from the Scale PropertyManager. Scale1 is displayed in the FeatureManager.

Parting Lines:

The Parting Lines are the boundary between the core and cavity surfaces. The Parting Lines create the edges of the molded part. The Parting Lines consist of the bottom eight edges.

Insert the Parting Lines feature.

580) Click **Parting Lines** from the Mold toolbar. The Parting Line PropertyManager is displayed.

Display the Temporary Axis.
581) Click **View**, **Hide/Show**, **Temporary Axes** from the Menu bar.

582) Click the **Temporary Axis** displayed through the center hole for the Direction of Pull.

583) Click **Reverse Direction** if required. The direction arrow points upward.

584) Enter **1**deg for Draft Angle.

585) Click the **Draft Analysis** button. View the results.

586) Click **OK** from the Parting Line PropertyManager. Parting Line1 is displayed in the FeatureManager.

Hide the Temporary Axes.

587) Click **View**, **Hide/Show**, uncheck **Temporary Axes** from the Menu bar.

Shut-off Surfaces:

Holes and windows that penetrate through an entire part require separate mold tooling. A shut-off area in a mold is where two pieces of the mold tooling contact each other. These areas require a Shut-off Surface. There are two Thru Holes in the BATTERYPLATE.

Insert the Shut-off Surfaces feature.

588) Click **Shut-off Surfaces** from the Mold toolbar. The Shut-Off Surface PropertyManager is displayed.

589) Select **All Tangent** for both edges. Note the direction arrows. If needed, check the **Knit** box.

590) Click **OK** from the Shut-Off Surface PropertyManager. A surface closes off the opening of each hole. Shut-Off Surface1 is displayed in the FeatureManager.

591) Rotate the model and view the closed surface.

Parting Surface:

A Parting Surface is utilized to separate the mold tooling components. A Parting Surface extends radially, a specified distance from the Parting Lines. Usually, the surface is normal to the direction of pull for simple molds. The boundary of the Parting Surface must be larger than the profile for the Tooling Split.

Insert the Parting Suface feature.

592) Click **Parting Surfaces** from the Mold toolbar. The Parting Surface PropertyManager is displayed.

593) Click **Top view**.

594) Enter **1.00**in, **[25.4]** for Distance. Accept the default settings.

595) Click **OK** from the Parting Surface PropertyManager. A Parting Surface (Radiate Surface) is created from the Parting Lines and the Distance value.

Fit the model to the Graphics window.
596) Press the **f** key.

Save the model.
597) Click **Save** 🖫.

Tooling Split:

The Tooling Split separates the solid bodies from the parting surfaces. The solid bodies become the mold tooling and the molded part. The Tooling Split for the BATTERYPLATE creates two mold base plates named Tooling Split1 and Tooling Split2.

Insert the Tooling Split feature.
598) Click **Parting Surface1** in the FeatureManager.

599) Click **Tooling Split** 🖾 from the Mold toolbar.

600) Click the **front face** of Parting Surface1.

601) Click the **Center Rectangle** Sketch tool.

602) Click the **Origin**.

603) Sketch a **rectangle** as illustrated about the Origin.

De-select the Center Rectangle Sketch tool.
604) Right-click **Select**.

Add an Equal relation.
605) Click the **top horizontal line** of the rectangle.

606) Hold the **Ctrl key** down.

607) Click the **left vertical line, right vertical line,** and **bottom horizontal line** of the rectangle.

608) Release the **Ctrl key**.

609) Click **Equal** =.

610) Click **OK** ✔ from the Properties PropertyManager.

Add a dimension.

611) Click the **Smart Dimension** Sketch tool.

612) Click the **top horizontal** line.

613) Enter **4.000**in, **[101.60]**.

614) Click **Exit Sketch**. The Tooling Split PropertyManager is displayed.

Enter values for the Block size.
615) Click **Front view**.

616) Enter **1.00**in, **[25.4]** for Depth in Direction 1.

617) Enter **.500**in, **[12.7]** for Depth in Direction 2.

618) Click **Isometric view** .

619) Click **OK** from the Tooling Split PropertyManager. Tooling Split1 is displayed in the FeatureManager.

The Mold tools produce Surface Bodies and Solid Bodies. The Solid Bodies are utilized in an assembly for the actual mold base.

The Surface Bodies were utilized as interim features to create the mold tooling parts. Hide the Surface Bodies to display the solid mold tooling parts.

Hide the Surface Bodies.
620) Expand the Surface Bodies folder in the FeatureManager. View the created bodies.

621) Right-click the **Surface Bodies** folder.

622) Click **Hide**.

Display the Solid Bodies.
623) Expand the Solid Bodies folder in the FeatureManager.

View each entry.
624) Click each **entry** in the Solid Bodies folder.

Separate Tooling Split1.
625) Click **Tooling Split1[1]** from the FeatureManager.

626) Click **Insert**, **Features**, **Move/Copy** from the Menu bar. The Move/Copy Body PropertyManager is displayed.

627) Enter **3.000in** in the Y box as illustrated.

628) Click **OK** from the PropertyManager. Body-Move/Copy1 is displayed in the FeatureManager.

Separate Tooling Split1[2].

629) Click **Tooling Split1[2]** from the FeatureManager.

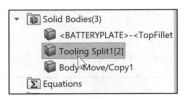

630) Click **Insert**, **Features**, **Move/Copy** from the Menu bar.

631) Click and drag the **vertical axis** of the Triad upward approximately 4in, [100mm] to display the BATTERYPLATE part.

632) Click **OK** from the PropertyManager. Body-Move/Copy2 is displayed in the FeatureManager.

Save the MOLD-TOOLING part.

633) Click **Save** .

634) Click **Yes** to save the referenced part (BATTERYPLATE).

Manufacturing Design Issues

For the experienced mold designer, the complete mold base requires additional mold tooling, lines and fittings. Utilize the Solid Bodies, Insert into New Part option to create the MOLD-TOOLING-CAVITY part from Body2 in the Solid Bodies folder.

The mold designer works with a material supplier and mold manufacturer. A wrong material selection is a costly mistake.

As a designer, you work with the material supplier to determine the best material for the project. For example, a material supplier application engineer asks the following questions:

- What is the plastic material utilized in this application?

- Will the material be corrosive or abrasive?

- Will the mold be textured?

- How important is the surface finish?

- What are the quantities of the part to be produced?

- Are part tolerances held within close limits?

- Have you inserted fillets on all sharp corners?

- Did you utilize adequate wall thickness and overall dimensions?

You send an eDrawing of your MOLD-TOOLING part for advice to the application engineer. Take the time early in the design process to determine the tooling material.

Project Summary

You are designing a FLASHLIGHT assembly that is cost effective, serviceable and flexible for future design revisions. The FLASHLIGHT assembly consists of various parts. The BATTERY, BATTERYPLATE, LENS and BULB parts were modeled in this project.

This Project concentrated on the Extruded Boss/Base feature and the Revolved Boss/Base feature. The Extruded Boss/Base feature required a Sketch plane, Sketch profile and End Condition (Depth). The BATTERY and BATTERYPLATE parts incorporated an Extruded Boss/Base (Boss-Extrude1) feature. A Revolved feature requires a Sketch plane, Sketch profile, Axis of Revolution and an Angle of Rotation.

You also utilized the Extruded Boss/Base, Extruded Cut, Fillet, Chamfer, Revolved Cut, Revolved Thin-Cut, Shell, Hole Wizard, Dome and Circular Pattern features.

The Hole Wizard provides the ability to define the end condition of a hole to the depth of the drill tip or to the depth of the shoulder. The options are available for all Hole Wizard features (including the Hole Wizard Assembly feature) and Advanced Hole types with the following end conditions: Blind, Up to Vertex, Up to Surface, and Offset to Surface. Previously, the end condition of a hole was calculated only up to the full diameter of the shoulder.

You addressed the following Sketch tools in this project: Sketch, Smart Dimension, Line, Rectangle, Circle, Tangent Arc and Centerline. You addressed additional Sketch tools that utilized existing geometry: Add Relations, Display/Delete Relations, Mirror Entities, Convert Entities and Offset Entities. Geometric relations were utilized to build symmetry into the sketches.

The BATTERYPLATE utilized various Mold tools to develop the Core and Cavity tooling plates required in the manufacturing process. Plastic parts require Draft. The Draft Analysis tool determined areas of positive and negative draft. Features were modified to accommodate the manufacturing process.

Practice these concepts with the project exercises. The other parts for the FLASHLIGHT assembly are addressed in Project 6.

Questions

1. Provide a short explanation of the following features:

Fillet	Extruded Cut	Draft	Scale
Extruded Boss	Revolved Base	Shell	Move/Copy
Revolved Cut Thin	Circular Pattern	Hole Wizard	Parting Surface

2. True or False. Design intent exists only in the sketch.

3. Describe a Symmetric relation.

4. Describe an Angular dimension.

5. Describe a draft angle? When do you apply a draft angle in a model?

6. When do you apply the Mirror feature? Utilize the SOLIDWORKS Online Users Guide to determine the difference between the Sketch Mirror and the Dynamic Mirror Sketch tool.

7. Describe the Spline Sketch tool.

7.a. Identify the required information for a Circular Pattern Feature.

8. Describe the procedure for adding a feature icon to the default Feature toolbar.

9. Describe the procedure to create a Short Cut key to Show/Hide Planes.

10. For a simple mold, describe the procedure to create the core and cavity mold tooling.

Exercise 5.1: FLAT PLATE

- Create the ANSI FLAT PLATE part on the Top Plane.

- The FLAT PLATE is machined from 0.09in, 6061 Alloy stock.

- Primary Unit system: IPS.

- The 8.69in, x 5.69in, FLAT PLATE contains a Linear Pattern of ∅.190in, Thru Holes.

- The holes are equally spaced, .50in apart.

- Utilize the Geometric Pattern option for the Linear Pattern feature.

DETAIL A
SCALE 1 : 1

Exercise 5.2: IM15-MOUNT Part

Create the ANSI IM15-MOUNT part as illustrated on the Right plane.

The IM15-MOUNT part is machined from 0.09in, AISI 304 flat stock.

Primary Unit system: IPS.

Exercise 5.3: ANGLE BRACKET Part

Create the ANSI ANGLE BRACKET part. The Boss-Extrude1 feature is sketched with an L-Shaped profile on the Right Plane.

The ANGLE BRACKET part is machined from 0.060in, [1.5mm] AISI 304 flat stock.

Primary Unit system: IPS.

- Create the Boss-Extrude1 feature.

- Apply the Mid Plane option for Direction 1 End Condition.

- Create the first Cut-Extrude feature on the Top view for the hole.

- Use the Linear Pattern feature to create six additional holes along the ANGLE BRACKET.

- Add Fillets to the four edges, .25in.

- Create the second Cut-Extrude feature.

- Apply the Linear Pattern feature.

- Create the third Cut-Extrude feature for the center slot.

Exercise 5.4

Create the illustrated ANSI part. All edges of the model are not located on perpendicular planes. Think about the steps required to build the model. Insert two features: Extruded Boss/Base and Extruded Cut.

Note the location of the Origin. Select the Right Plane as the Sketch plane. Apply construction geometry. Insert the required geometric relations and dimensions for Sketch1. Calculate the overall mass of the part.

Given:
A = 3.00, B = 1.00
Material: 6061 Alloy
Density = .097 lb/in^3
Units: IPS
Decimal place: 2

There are numerous ways to build the models in this section.

Exercise 5.5

Create the illustrated ANSI part. Think about the required steps to build this part. Insert four features: Boss-Extrude1, Cut-Extrude1, Cut-Extrude2 and Fillet. Note the location of the Origin.

Select the Right Plane as the Sketch plane. The part Origin is located in the lower left corner of the sketch. Calculate the overall mass of the part.

Given:
A = 4.00
B = R.50
Material: 6061 Alloy
Density = .0975 lb/in^3
Units: IPS
Decimal place: 2

Exercise 5.6

Create the illustrated ANSI part. Think about the required steps to build this part. Insert three features: Boss-Extrude1, Boss-Extrude2, and Mirror. Three holes are displayed with a 1.00in diameter.

Note the location of the Origin.

Create Sketch1. Select the Top Plane as the Sketch plane. Apply the Tangent Arc and Line Sketch tool. Insert the required Geometric relations and dimensions. Calculate the overall mass of the part.

Given:
A = Ø1.00
All Thru Holes
Material: Brass
Density = .307 lb/in^3
Units: IPS
Decimal place: 2

There are numerous ways to build the models in this section.

Exercise 5.7

Create the illustrated ANSI part. Think about the required steps to build this part. Insert a Revolved Base feature and an Extruded Cut feature to build this part.

Given:
A = Ø12
Material: Cast Alloy Steel
Density = .0073 g/mm^3
Units: MMGS

Note the location of the Origin. Select the Front Plane as the Sketch plane.

Apply the Centerline Sketch tool for the Revolve1 feature.

Insert the required geometric relations and dimensions for Sketch1.

Sketch1 is the profile for the Revolve1 feature. Calculate the overall mass of the part.

Exercise 5.8

Create the illustrated ANSI part. Think about the required steps to build this part.

Insert two features: Boss-Extrude1 and Revolve1.

Note the location of the Origin.

Select the Top Plane as the Sketch plane.

Apply construction geometry.

Apply the Tangent Arc and Line Sketch tool.

Insert the required geometric relations and dimensions. Calculate the overall mass of the part.

Given:
A = 60, B = 40, C = 8
Material: Cast Alloy Steel
Density = .0073 g/mm^3
Units: MMGS

Exercise 5.9

Create the illustrated ANSI part. Think about the required steps to build the part. Apply the Extruded Boss/Base, Extruded Cut and Shell features. View the provided dimensions.

Units: MMGS.

Decimal place: 2

Part Origin is arbitrary in this example.

Material: Plain Carbon Steel.

Calculate the overall mass of the part in grams.

Mass properties of Exercise 5-9
   Configuration: Default
   Coordinate system: -- default --

Density = 0.01 grams per cubic millimeter

Mass = 1484.11 grams

Volume = 190271.14 cubic millimeters

Surface area = 49308.51 square millimeters

Center of mass: (millimeters)
   X = 77.43
   Y = 13.26
   Z = 0.00

Notes:

Project 6

Swept, Lofted and Additional Features

Below are the desired outcomes and usage competencies based on the completion of Project 6.

Project Desired Outcomes:	Usage Competencies:
<ul><li>Four FLASHLIGHT parts:<ul><li>O-RING</li><li>SWITCH</li><li>LENSCAP</li><li>HOUSING</li></ul></li></ul>	<ul><li>Ability to apply the following features: Extruded Boss/Base, Extruded Cut, Swept Boss/Base, Shell, Lofted Boss/Base, Draft, Rib, Linear Pattern, Circular Pattern, Plane, Helix/Spiral, Mirror, Dome and Revolved Cut.</li></ul>
<ul><li>Three O-RING configurations:<ul><li>Small, Medium and Large</li></ul></li></ul>	<ul><li>Create and edit a O-RING Design Table for three configurations.</li></ul>
<ul><li>Four assemblies:<ul><li>LENSANDBULB assembly</li><li>CAPANDLENS assembly</li><li>BATTERYANDPLATE assembly</li><li>FLASHLIGHT assembly</li></ul></li></ul>	<ul><li>Skill to combine multiple features to create components.</li></ul><ul><li>Knowledge of the Bottom-up assembly modeling method using Coincident, Concentric and Distance mates.</li></ul>

Notes:

Project 6 - Swept, Lofted, and Additional Features

Project Objective

Create four parts: O-RING, SWITCH, LENSCAP and HOUSING of the FLASHLIGHT assembly. Create four assemblies: LENSANDBULB, CAPANDLENS, BATTERYANDPLATE and the final FLASHLIGHT assembly.

On the completion of this project, you will be able to:

- Utilize the following SOLIDWORKS features:
  - Extruded Boss/Base, Extruded Cut, Swept Boss/Base, Shell, Lofted Boss/Base, Draft, Rib, Linear Pattern, Circular Pattern, Plane, Helix/Spiral, Mirror, Dome and Revolved Cut.

- Performance part Evaluation.

- Create and edit a Design Table for an O-RING with three configurations.

- Create an inch and metric Custom Assembly template with Custom document properties.

- Apply Bottom-up assembly modeling techniques.

- Insert the following Standard Mate types: Coincident, Concentric and Distance.

- Utilize the following Assembly tools: Insert Component, Fix, Float, Hide/Show, Suppress, Move Component, Rotate Component, Exploded View and Interference Detection.

- Apply Selection Filters.

- Export SOLIDWORKS files.

- Use the Pack and Go option to save an assembly with references.

- Create a SOLIDWORKS eDrawing.

Project Overview

Project 6 introduces the Swept Boss/Base and Lofted Boss/Base features. The O-RING utilizes a simple Swept Base feature.

The SWITCH utilizes the Lofted Base feature. The LENSCAP and HOUSING utilize the Swept Boss and Lofted Boss feature.

A simple Swept Base feature requires a *path* and *profile*. Sketch the path and enter the circular profile diameter. The profile follows the path to create the following Swept features:

- Swept Base

- Swept Boss

The Lofted feature requires a minimum of two profiles sketched on different planes. The profiles are blended together to create the following Lofted features:

- Lofted Base

- Lofted Boss

Utilize existing features to create the Rib, Linear Pattern and Mirror features.

Utilize existing faces to create the Draft and Dome features.

The LENSCAP and HOUSING combines the Extruded Boss/Base, Extruded Cut, Revolved Cut Thin, Shell and Circular Pattern with the Swept and Lofted features.

Features in the SWITCH, LENSCAP and HOUSING have been simplified for educational purposes.

Create four assemblies in this project:

1. LENSANDBULB assembly

2. CAPANDLENS assembly

3. BATTERYANDPLATE assembly

4. FLASHLIGHT assembly

Create an inch and metric Assembly template.

- ASM-IN-ANSI

- ASM-MM-ISO

Develop an understanding of assembly modeling techniques.

Combine the LENSANDBULB assembly, CAPANDLENS assembly, BATTERYANDPLATE assembly, HOUSING part and SWITCH part to create the FLASHLIGHT assembly.

Create the following Standard Mate types: Coincident, Concentric and Distance.

Utilize the following tools: Insert Component, Hide/Show, Suppress, Mate, Move Component, Rotate Component, Exploded View and Interference Detection.

Project Situation

Communication is a major component to a successful development program. Provide frequent status reports to the customer and to your team members.

Communicate with suppliers. Ask questions. Check on details. What is the delivery time for the BATTERY, LENS and SWITCH parts?

Talk to colleagues to obtain useful manufacturing suggestions and ideas. Your team decided that plastic injection molding is the most cost-effective method to produce large quantities of the desired part.

Investigate surface finishes that support the customer's advertising requirement. You have two fundamental choices:

- Adhesive label

- Silk-screen

There are time, quantity and cost design constraints. For the prototype, use an adhesive label. In the final product, create a silk screen.

Investigate the options on O-Ring material. Common O-Ring materials for this application are Buna-N (Nitrile®) and Viton®. Be cognizant of compatibility issues between O-RING materials and lubricants.

The LENSCAP encloses the LENS. The HOUSING protects the BATTERY. How do you design the LENSCAP and HOUSING to ease the transition from development to manufacturing? Answer: Review the fundamental design rules behind the plastic injection manufacturing process:

- Maintain a constant wall thickness. Inconsistent wall thickness creates stress.

- Create a radius on all corners. No sharp edges. Sharp edges create vacuum issues when removing the mold. Filleting inside corners on plastic parts will strengthen the part and improve plastic flow characteristics around corners in the mold.

- Understand the Draft feature. The Draft feature is commonly referred to as plus draft or minus draft. Plus draft adds material. Minus draft removes material.

- Allow a minimum draft angle of 1°. Draft sides and internal ribs. Draft angles assist in removing the part from the mold.

- Ribs are often added to improve strength. Ribs must also be added to improve plastic flow to entirely fill the cavity and to avoid a short shot - not enough material to fill the mold.

- Generally, use a 2:3 ratio for rib thickness compared to overall part thickness.

- Shrinkage of the cooling plastic requires that the cavity is larger than the finished part. Shrinkage around the core will cause the part to bind onto the core. The Draft feature helps eject the part from the mold. Shrinkage on the cavity side will cause the part to pull from the mold. Less draft is needed on the cavity faces than on the core faces.

- Sink marks occur where there is an area with thicker material than the rest of the part wall thickness. A problem may occur where a tall rib with draft meets a face of the part.

- Mold cavities should be vented to allow trapped gases to escape through the channels.

Obtain additional information on material and manufacturing from material suppliers or mold manufacturers. Example: GE Plastics of Pittsfield, MA provides design guidelines for selecting raw materials and creating plastic parts and components.

O-RING Part - Swept Base Feature

The O-RING is positioned between the LENSCAP and the LENS.

Path(Sketch-path)

A Swept Boss/Base feature can be simple or complex. The O-RING is classified as a simple Swept Base feature. A complex Swept Boss feature utilizes 3D curves and Guide Curves.

Create the O-RING with the Swept Base 🌀 feature. The Swept Base feature requires two sketches or a sketch path and a circular profile.

Utilize the PART-IN-ANSI template created in the previous Project for inch units. Utilize the PART-MM-ISO template for millimeter units. Millimeter dimensions are provided in brackets [x].

🔅 For non-circular sketch profiles, create the sketch profile on a perpendicular plane to the path and use a pierce relation to locate the profile on the path.

Activity: O-RING Part - Swept Base Feature.

Create a new part. Enter part name. Enter part description.

1)   Click **New** ▯ from the Menu bar.

2)   Click the **MY-TEMPLATES** tab.

3)   Double-click **PART-IN-ANSI**.

Templates   MBD   MY-TEMPLATES

PART-IN-ANSI
PART-MM-ANSI
PART-MM-ISO

4)   Click **Save As** 🖫 from the drop down menu.

5)   Select **ENGDESIGN-W-SOLIDWORKS\PROJECTS** for the Save in folder.

6)   Enter **O-RING** for File name.

7)   Enter **O-RING FOR LENS** for Description.

8)   Click **Save**. The O-RING FeatureManager is displayed.

Create the Swept path.

9)   Right-click **Front Plane** from the FeatureManager for the Sketch plane. This is your Sketch plane.

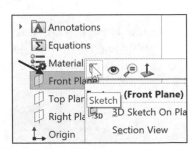

10)   Click **Sketch** 🖉 from the Context toolbar. The Sketch toolbar is displayed.

11)   Click the **Circle** ⊙ Sketch tool. The Circle PropertyManager is displayed.

12) Sketch a circle centered at the **Origin** .

Add a dimension.

13) Click the **Smart Dimension** Sketch tool.

14) Click the **circumference** of the circle.

15) Click a **position** off the profile.

16) Enter **4.350**in, [**110.49**] as illustrated.

Close the sketch. Rename the sketch.

17) **Rebuild** the model. Sketch1 is displayed in the FeatureManager.

[110.49]
⌀ 4.350

18) **Rename** Sketch1 to Sketch-path.

Insert the Swept Base feature.

19) Click the **Swept Boss/Base** feature tool. The Sweep PropertyManager is displayed.

20) **Expand** O-RING from the fly-out FeatureManager.

21) Click the **Circular Profile** box.

22) Click **Sketch-path** from the fly-out FeatureManager. Sketch-path is displayed.

23) Enter **0.125**in for Diameter.

24) Click **OK** from the Sweep PropertyManager. Sweep1 is displayed in the FeatureManager.

Rename the feature.

25) **Rename** Sweep1 to Base-Sweep.

Display an Isometric view. Save the model.

26) Click **Isometric view** .

27) Click **Save** .

O-RING Part - Design Table

A Design Table is a spreadsheet used to create multiple configurations in a part or assembly. The Design Table controls the dimensions and parameters in the part. Utilize the Design Table to modify the overall path diameter and profile diameter of the O-RING.

Create three configurations of the O-RING:

- Small

- Medium

- Large

The O-RING contains two dimension names in the Design Tables. Parts contain hundreds of dimensions and values. Rename dimension names for clarity.

The part was initially designed in inches. You are required to manufacture the configurations in millimeters. Modify the part units to millimeters.

Activity: O-RING Part - Design Table.

Modify the Primary Units.

28) Click **Options** ⚙ , **Document Properties** tab from the Menu bar. Select **Units**.

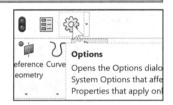

29) Select **MMGS**. Select **.12** for Basic unit length decimal place.

30) Click **OK** from the Documents Properties - Units box.

Insert a Design Table.

31) Click **Insert**, **Tables**, **Excel Design Table** from the Menu bar. The Auto-create option is selected. Accept the default settings.

32) Click **OK** ✔ from the Excel Design Table PropertyManager.

33) Click **D1@Sketch-path**. Hold the **Ctrl** key down.

34) Click **D3@BaseSweep**. Release the **Ctrl** key.

35) Click **OK** from the Dimensions dialog box.

Note: The dimension variable name will be different if sketches or features were deleted.

The input dimension names and default values are automatically entered into the Excel Design Table. The value Default is entered in Cell A3.

The values for the O-RING are entered in Cells B3 through C6. The sketch-path diameter is controlled in Column B. The sketch-profile diameter is controlled in Column C.

Enter the three configuration names.
36) Double click **Cell A4**. Enter **Small**.

37) Double click **Cell A5**. Enter **Medium**.

38) Double click **Cell A6**. Enter **Large**.

Enter the dimension values for the Small configuration.

39) Double click **Cell B4**. Enter **100**.

40) Double click **Cell C4**. Enter **3**.

Enter the dimension values for the Medium configuration.

41) Double click **Cell B5**. Enter **150**.

42) Double click **Cell C5**. Enter **4**.

Enter the dimension values for the Large configuration.

43) Double click **Cell B6**. Enter **200**.

44) Double click **Cell C6**. Enter **5**.

	A	B	C	D
1	Design Table for: O-RING			
2		D1@Sketch-path	D3@Base-Sweep	
3	Default	110.49	3.175	
4	Small	100	3	
5	Medium	150	4	
6	Large	200	5	
7				
8				
9				
10				
11				
12				
13				
14				

Sheet1 +

Build the three configurations.

45) Click a **position** inside the Graphics window.

46) Click **OK** to generate the three configurations. The Design Table icon is displayed in the FeatureManager.

47) **Rebuild** 🔘 the model.

View the three configurations.

48) Double-click **Small**. Double-click **Medium**.

49) Double-click **Large**. Double-click **Default**.

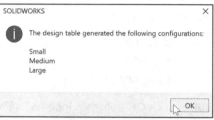

SOLIDWORKS ✕

ℹ The design table generated the following configurations:

Small
Medium
Large

OK

Return to the FeatureManager.

50) Click the **FeatureManager** 🔖 tab.

Save the model.

51) Click **Save** 💾.

Configurations

🔷 O-RING Configuration(s) (D
   📋 Tables
      📊 Excel Design Table
      ✔ Default [O-RING]
      — Large [O-RING]
      — Medium [O-RING]
      — Small [O-RING]

The O-RING is classified as a simple Swept feature. A complex Swept feature utilizes 3D curves and Guide Curves. Investigate additional Swept features later in the project.

An example of a complex Swept is a Thread. A Thread requires a Helical/Spiral curve for the path and a circular profile.

Another example of a complex Swept is a violin body. The violin body requires Guide Curves to control the Swept geometry.

Thread

Without Guide Curves the profile follows the straight path to produce a rectangular shape.

Review of the O-RING Part.

The O-RING utilized the Swept feature. The Swept feature required a sketch path and a sketch profile. The path was a circle sketched on the Front Plane. The profile was the diameter of the circular profile.

For non-circular sketch profiles, create the sketch profile on a perpendicular plane to the path and use a pierce relation to locate the profile on the path.

Swept features can be simple or complex.

Small, Medium and Large configurations of the O-RING were developed with an Excel Design Table. The Excel Design Table icon displayed in the ConfigurationManager is an Excel based spread sheet utilized to create variations of dimensions and features.

Note: If you did not create the templates or the MY-TEMPLATES tab in the book, use a standard SOLIDWORKS template and address the needed Document properties and Custom properties.

💡 The Sketch toolbar contains two areas: Sketch Entities and Sketch Tools. There are numerous ways to access these areas: Select the Sketch ⬚ icon from the Sketch toolbar, right-click and select the Sketch icon from the Context toolbar or click Tools, Sketch Entities or Sketch Tools from the Menu bar.

SWITCH Part-Lofted Base Feature

The SWITCH is a purchased part. The SWITCH is a complex assembly. Create the outside casing of the SWITCH as a

simplified part. Create the SWITCH with the Lofted Base feature.

The orientation of the SWITCH is based on the position in the assembly. The SWITCH is comprised of three cross section profiles. Sketch each profile on a different plane.

The first plane is the Top Plane. Create two reference planes parallel to the Top Plane.

Sketch one profile on each plane. The design intent of the sketch is to reduce the number of dimensions.

Utilize symmetry, construction geometry and Geometric

relations to control three sketches with one dimension.

Insert a Lofted feature. Select the profiles to create the Loft feature.

Insert the Dome feature to the top face of the Loft and modify the Loft Base feature. If the Dome feature is not displayed in the Feature CommandManager, click Insert, Features, Dome from the Main menu.

Activity: Switch Part - Loft Base Feature.

Create a new part. Enter part name. Enter part description.

52) Click **New** from the Menu bar.

53) Click the **MY-TEMPLATES** tab. Additional templates are displayed.

54) Double-click **PART-IN-ANSI.**

55) Click **Save As** .

56) Select **ENGDESIGN-W-SOLIDWORKS\PROJECTS** for the Save in folder.

57) Enter **SWITCH** for File name.

58) Enter **BUTTON STYLE** for Description.

59) Click **Save**. The SWITCH FeatureManager is displayed.

Display the Top Plane.
60) Right-click **Top Plane** from the FeatureManager.

61) Click **Show** .

Display an Isometric view.
62) Click **Isometric view** from the Heads-up View toolbar.

Insert two Reference planes.
63) Hold the **Ctrl** key down.

64) Click and drag the **Top Plane** upward. The Plane PropertyManager is displayed.

65) Release the **mouse button**.

66) Release the **Ctrl** key.

67) Enter **.500**in, [**12.7**] for Offset Distance.

68) Enter **2** for # of Planes to Create.

69) Click **OK** from the Plane PropertyManager. Plane1 and Plane2 are displayed in the FeatureManager.

70) Click **Front view** to display Plane1 and Plane2 offset from the Top plane.

> Hold the Ctrl key down. Drag the Top Plane upward. Pick an edge, not the handles.

Display a Top view.

71)   Click **Top view**.

Insert Sketch1. Sketch1 is a square on the Top Plane centered about the Origin.

72)   Right-click **Top Plane** from the FeatureManager.

73)   Click **Sketch** from the Context toolbar.

74)   Click the **Center Rectangle** tool from the Consolidated Sketch toolbar.

75)   Click the **Origin**.

76)   Click a **point** in the upper right hand of the window as illustrated. The Center Rectangle Sketch tool automatically inserts a midpoint relation about the Origin.

77)   Click **OK** from the Rectangle PropertyManager.

Add an Equal relation between the four edges.

78)   Click the **left vertical** line.

79)   Hold the **Ctrl** key down.

80)   Click the **top horizontal** line.

81)   Click the **right vertical** line.

82)   Click the **bottom horizontal** line.

83)   Release the **Ctrl** key.

84)   Click **Equal =** .

Add a dimension.

85)   Click the **Smart Dimension** Sketch tool.

86)   Click the **top horizontal** line.

87)   Click a **position** above the profile.

88)   Enter **.500**in, [**12.7**].

Close, fit and rename the sketch.

89)   **Rebuild** the model.

90)   Press the **f** key to fit the sketch to the Graphics area.

91)   **Rename** Sketch1 to Sketch1-lower.

Save the model.

92)   Click **Save**.

Display a Top view. Insert Sketch2 on Plane1.

93)   Click **Top view** .

94)   Right-click **Plane1** from the FeatureManager.

95)   Click **Sketch** from the Context toolbar.

96)   Click the **Circle** Sketch tool. The Circle PropertyManager is displayed.

97)   Sketch a **Circle** centered at the Origin as illustrated.

Add a Tangent relation.

98)   Right-click **Select** to deselect the Circle Sketch tool.

99)   Click the **circumference** of the circle.

100)   Hold the **Ctrl** key down.

101)   Click the **top horizontal Sketch1-lower line**. The Properties Propert

102)   Release the **Ctrl** key.

103)   Click **Tangent** .

104)   Click **OK** from the Properties PropertyManager.

Close and rename the sketch.

105)   Click **Exit Sketch**.

106)   **Rename** Sketch2 to Sketch2-middle.

Display an Isometric view.

107)   Click **Isometric view** .

Display a Top view. Insert Sketch3 on Plane2.

108)   Click **Top view** .

109)   Right-click **Plane2** from the FeatureManager.

110)   Click **Sketch** from the Context toolbar. The Sketch toolbar is displayed.

111)   Click the **Centerline** Sketch tool.

112) Sketch a **centerline** coincident with the Origin upward up to the midpoint as illustrated.

113) Click the **Point** Sketch tool.

114) Click the **midpoint** of the right diagonal centerline. The Point PropertyManager is displayed.

115) Click **Circle** from the Sketch toolbar.

116) Sketch a **Circle** centered at the Origin to the midpoint of the diagonal centerline.

Close and rename the sketch.
117) Click **Exit Sketch**.

118) **Rename** Sketch3 to Sketch3-upper.

Display an Isometric view. Insert the Lofted Base feature.

119) Click the **Lofted Boss/Base** feature tool. The Loft PropertyManager is displayed.

120) Click **Isometric view** from the Heads-up View toolbar.

121) Right-click in the **Profiles box**.

122) Click **Clear Selections**.

123) Click **Sketch1-lower** from the fly-out FeatureManager.

124) Click **Sketch2-middle** from the fly-out FeatureManager.

125) Click **Sketch3-upper** from the fly-out FeatureManager. The selected sketch entities are displayed in the Profiles box.

126) Click **OK** from the Loft PropertyManager. Loft1 is displayed in the FeatureManager.

Rename the feature.
127) Rename **Loft1** to **Base Loft**.

Hide the planes.
128) Click **View**, **Hide/Show,** uncheck **Planes** from the Menu bar.

Save the part.

129) Click **Save**.

The system displays a preview curve and loft as you select the profiles. Use the Up button and Down button in the Loft PropertyManager to rearrange the order of the profiles.

🔅 Redefine incorrect selections efficiently. Right-click in the Graphics window and click Clear Selections to remove selected profiles. Select the correct profiles.

SWITCH Part - Dome Feature

Insert the Dome feature on the top face of the Lofted Base feature. The Dome feature forms the top surface of the SWITCH. Note: You can specify an elliptical dome for cylindrical or conical models and a continuous dome for polygonal models. A continuous dome's shape slopes upwards, evenly on all sides.

Activity: SWITCH Part - Dome Feature

Insert the Dome feature.

130) Click the **top face** of the Base Loft feature in the Graphics window.

131) Click the **Dome** 🔵 feature tool. The Dome PropertyManager is displayed. Face<1> is displayed in the Faces to Dome box. Note: If the Dome feature is not displayed in the Feature CommandManager, click Insert, Features, Dome from the Main menu.

Enter distance.

132) Enter .**20**in, [**5.08**] for Distance.

133) **View** the dome feature in the Graphics window.

134) Click **OK** ✔ from the Dome PropertyManager.

In the next section, modify the offset distance between the Top Plane and Plane1.

Modify the Loft Base feature.

135) **Expand** the Base Loft feature.

136) Right-click on **Annotations** in the FeatureManager.

137) Click **Show Feature Dimensions**. Move all dimensions off the model as illustrated.

138) Click on the Plane1 offset dimension, **.500**in, [**12.700**].

139) Enter **.125**in, [**3.180**]. Click **Rebuild**.

140) Click **OK** ✔ from the Dimension PropertyManager.

Hide Feature dimensions.

141) Right-click on **Annotations** in the FeatureManager.

142) Un-check **Show Feature Dimensions**.

Display Performance Evaluation.

143) Click **Performance Evaluation** from the Evaluate tab in the CommandManager. View the results.

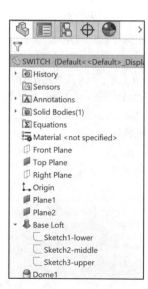

Performance Evaluation

SWITCH
Features 7, Solids 1, Surfaces 0
Total rebuild time in seconds: 0.11

Feature Order	Time %	Time(s)
Dome1	71.82	0.08
Sketch3-upper	14.55	0.02
Base Loft	13.64	0.01
Plane1	0.00	0.00
Plane2	0.00	0.00
Sketch1-lower	0.00	0.00
Sketch2-middle	0.00	0.00

Close SWITCH Performance Evaluation.

144) Click **Close** from the Performance Evaluation dialog box.

145) Click **Save** .

The Dome feature created the top for the SWITCH. The Performance Evaluation report displays the rebuild time for the Dome feature and the other SWITCH features. As feature geometry becomes more complex, the rebuild time increases.

 Review of the SWITCH Part.

The SWITCH utilized the Lofted Base and Dome feature. The Lofted Base feature required three planes: Sketch1-lower, Sketch2-middle and Sketch3-upper. A profile was sketched on each plane. The three profiles were combined to create the Lofted Base feature.

The Dome feature created the final feature for the SWITCH.

🔆 The SWITCH utilized a simple Lofted Base feature. Lofts become more complex with additional Guide Curves. Complex Lofts can contain hundreds of profiles.

Four Major Categories of Solid Features

The LENSCAP and HOUSING combine the four major categories of solid features:

- Extrude: Requires one profile.

- Revolve: Requires one profile and axis of revolution.

- Swept: Requires one profile and one path.

- Lofted: Requires two or more profiles sketched on different planes.

Identify the simple features of the LENSCAP and HOUSING. Extrude and Revolve are simple features. Only a single sketch profile is required. Swept and Loft are more complex features. Two or more sketches are required.

Example: The O-RING was created as a Swept.

Could the O-RING utilize an Extruded feature?

Answer: No. Extruding a circular profile produces a cylinder.

Can the O-RING utilize a Revolved feature?
Answer: Yes. Revolving a circular profile about a centerline creates the O-RING.

Revolved feature   Sweep feature

A Swept feature is required if the O-RING contained a non-circular path. Example: A Revolved feature does not work with an elliptical path or a more complex curve as in a paper clip. Combine the four major features and additional features to create the LENSCAP and HOUSING.

LENSCAP Part

The LENSCAP is a plastic part used to position the LENS to the HOUSING. The LENSCAP utilizes an Extruded Boss/Base, Extruded Cut, Extruded Thin Cut, Shell, Revolved Cut and Swept features.

The design intent for the LENSCAP requires that the Draft Angle be incorporated into the Extruded Boss/Base and Revolved Cut feature. Create the Revolved Cut feature by referencing the Extrude Base feature geometry. If the Draft angle changes, the Revolved Cut also changes.

Insert an Extruded Boss/Base 🗔 feature with a circular profile on the Front Plane. Use a Draft option in the Boss-Extrude PropertyManager. Enter 5deg for Draft angle.

Insert an Extruded Cut feature. The Extruded Cut feature should be equal to the diameter of the LENS Revolved Base feature.

Insert a Shell feature. Use the Shell feature for a constant wall thickness.

Insert a Revolved Cut feature on the back face. Sketch a single line on the Silhouette edge of the Extruded Base.

Utilize the Thin Feature option in the Cut-Revolve PropertyManager.

Utilize a Swept feature for the thread. Insert a new reference plane for the start of the thread. Insert a Helical Curve for the path. Sketch a trapezoid for the profile.

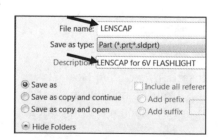

LENSCAP Part - Extruded Boss/Base, Extruded Cut and Shell Features

Create the LENSCAP. Review the Extruded Boss/Base, Extruded Cut and Shell features introduced in Project 5. The first feature is a Boss-Extrude feature. Select the Front Plane for the Sketch plane. Sketch a circle centered at the Origin for the profile. Utilize a Draft angle of 5deg.

Create an Extruded Cut feature on the front face of the Base feature. The diameter of the Extruded Cut equals the diameter of the Revolved Base feature of the LENS.

The Shell feature removes the front and back face from the solid LENSCAP.

Activity: LENSCAP Part - Extruded Base, Extruded Cut, and Shell Features.

Create a new part. Enter part name. Enter part description.

146) Click **New** from the Menu bar.

147) Click the **MY-TEMPLATES** tab.

148) Double-click **PART-IN-ANSI**.

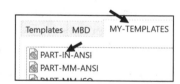

149) Click **Save As** .

150) Select **ENGDESIGN-W-SOLIDWORKS\PROJECTS** for the Save in folder.

151) Enter **LENSCAP** for File name.

152) Enter **LENSCAP for 6V FLASHLIGHT** for Description.

153) Click **Save**. The LENSCAP FeatureManager is displayed.

Create the sketch for the Extruded Base feature.

154) Right-click **Front Plane** from the FeatureManager.

155) Click **Sketch** from the Context toolbar.

156) Click the **Circle** Sketch tool. The Circle PropertyManager is displayed.

157) Sketch a **circle** centered at the Origin.

Add a dimension.

158) Click the **Smart Dimension** Sketch tool.

159) Click the **circumference** of the circle.

160) Click a **position** off the profile. Enter **4.900**in, [**124.46**].

Insert an Extruded Boss/Base feature.

161) Click the **Extruded Boss/Base** feature tool. The Boss-Extrude PropertyManager is displayed. Blind is the default End Condition in Direction 1.

162) Click the **Reverse Direction** box.

163) Enter **1.725**in, [**43.82**] for Depth in Direction 1.

164) Click the **Draft On/Off** button.

165) Enter **5**deg for Angle. Click the **Draft outward** box.

166) Click **OK** from the Boss-Extrude PropertyManager. Boss-Extrude1 is displayed in the FeatureManager.

167) **Rename** Boss-Extrude1 to Base Extrude.

168) Click **Save**.

Create the sketch for the Extruded Cut feature.

169) Right-click the **front face** for the Sketch plane.

170) Click **Sketch** from the Context toolbar. The Sketch toolbar is displayed.

171) Click the **Circle** Sketch tool. The Circle PropertyManager is displayed.

172) Sketch a **circle** centered at the Origin.

Add a dimension.

173) Click the **Smart Dimension** Sketch tool.

174) Click the **circumference** of the circle. Click a **position** off the profile.

175) Enter **3.875**in, [**98.43**].

Insert an Extruded Cut feature.

176) Click the **Extruded Cut** feature tool. The Cut-Extrude PropertyManager is displayed. Blind is the default End Condition.

177) Enter .275in, [6.99] for Depth in Direction 1.

178) Click the **Draft On/Off** button.

179) Enter 5deg for Angle. Accept the default settings.

180) Click **OK** ✔ from the Cut-Extrude PropertyManager. Cut-Extrude1 is displayed in the FeatureManager.

181) **Rename** Cut-Extrude1 to Front-Cut.

Insert the Shell feature.

182) Click the **Shell** feature tool. The Shell1 PropertyManager is displayed.

183) Click the **front face** of the Front-Cut as illustrated.

184) **Rotate** the model to view the back face.

185) Click the **back face** of the Base Extrude.

186) Enter .150in, [3.81] for Thickness.

187) Click **OK** ✔ from the Shell1 PropertyManager. Shell1 is displayed in the FeatureManager.

Display an Isometric view.

188) Press the **space bar** to display the Orientation dialog box.

189) Click **Isometric view** 🔲.

Display the inside of the Shell. Save the model.

190) Click **Right view** 🔲.

191) Click **Hidden Lines Visible** 🔲 from the Heads-up View toolbar.

192) Click **Save** 🔲.

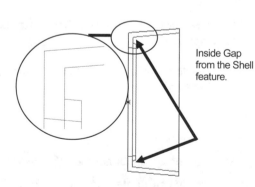

Inside Gap from the Shell feature.

💡 Use the inside gap created by the Shell feature to seat the O-RING in the assembly.

LENSCAP Part - Revolved Thin Cut Feature

The Revolved Thin Cut 🔀 feature removes material by rotating a sketched profile around a centerline.

The Right Plane is the Sketch plane. The design intent requires that the Revolved Cut maintains the same Draft angle as the Extruded Base feature.

Utilize the Convert Entities Sketch tool to create the profile. Small thin cuts are utilized in plastic parts. Utilize the Revolved Thin Cut feature for cylindrical geometry in the next activity.

Utilize a Swept Cut for non-cylindrical geometry. The semi-circular Swept Cut profile is explored in the project exercises.

Sweep Cut Example

Activity: LENSCAP Part - Revolved Thin Cut Feature.

Create the sketch.
193) Right-click **Right Plane** from the FeatureManager. This is your Sketch plane.

194) Click **Sketch** 🖰 from the Context toolbar. The Sketch toolbar is displayed.

Sketch a centerline.
195) Click the **Centerline** ✏ Sketch tool. The Insert Line PropertyManager is displayed.

196) Sketch a **horizontal centerline** through the Origin as illustrated.

Create the profile.
197) Right-click **Select** to deselect the Centerline Sketch tool.

198) Click the **top silhouette outside edge** of Base Extrude as illustrated.

199) Click the **Convert Entities** 🗍 Sketch tool.

200) Click and drag the **left endpoint 2/3** towards the right endpoint.

201) Release the **mouse button**.

Add a dimension.
202) Click the **Smart Dimension** ✎ Sketch tool.

203) Click the **line**. Do not click the midpoint. The aligned dimension arrows are parallel to the profile line.

204) Drag the **text upward** and to the right.

205) Enter .250in, [6.35].

Insert a Revolved Cut feature.

206) Click **Revolved Cut** from the Features toolbar. Do not close the Sketch. The warning message states, "The sketch is currently open."

207) Click **No**. The Cut-Revolve PropertyManager is displayed.

208) Click the **Reverse Direction** box in the Thin Feature box. The arrow points counterclockwise. Enter **.050**in, **[1.27]** for Direction 1 Thickness.

209) Click **OK** ✔ from the Cut-Revolve PropertyManager. Cut-Revolve-Thin1 is displayed in the FeatureManager.

Display the Revolved Thin Cut feature.
210) **Rotate** the part to view the back face.

211) Click **Isometric view** 🔲.

212) Click **Shaded With Edges** 🔲.

Rename and save the feature.
213) **Rename** Cut-Revolve-Thin1 to BackCut.

214) Click **Save** 💾.

LENSCAP Part - Thread, Swept Feature and Helix/Spiral Curve

Utilize the Swept feature 🪱 to create the required threads. The thread requires a spiral path. This path is called the ThreadPath. The thread requires a Sketched profile. This cross-section profile is called the ThreadSection.

The plastic thread on the LENSCAP requires a smooth lead in. The thread is not flush with the back face. Use an Offset plane to start the thread. There are numerous steps required to create a thread:

- Create a new plane for the start of the thread.

- Create the Thread path. Utilize Convert Entities and Insert, Curve, Helix/Spiral.

- Create a large thread cross section profile for improved visibility.

- Insert the Swept feature.

- Reduce the size of the thread cross section.

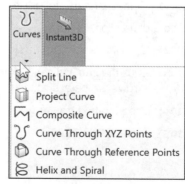

Activity: LENSCAP Part - Thread, Swept Feature, and Helix/Spiral Curve.

Create the offset plane.

215) **Rotate** and **Zoom to Area** on the back face of the LENSCAP.

216) Click the **narrow back face** of the Base Extrude feature. Note the mouse feedback icon.

217) Click **Insert**, **Reference Geometry, Plane** from the Menu bar. The Plane PropertyManager is displayed.

218) Enter **.450**in, **[11.43]** for Distance.

219) Click the **Flip offset** box.

220) Click **OK** ✔ from the Plane PropertyManager. Plane1 is displayed in the FeatureManager.

221) **Rename** Plane1 to ThreadPlane.

Display the Isometric view with Hidden Lines Removed.

222) Click **Isometric view** .

223) Click **Hidden Lines Removed** .

Save the model.

224) Click **Save** .

Utilize the Convert Entities Sketch tool to extract the back circular edge of the LENSCAP to the ThreadPlane.

Create the Thread path.

225) Right-click **ThreadPlane** from the FeatureManager.

226) Click **Sketch** from the Context toolbar.

227) Click **Isometric view** .

228) Click the **back inside circular edge** of the Shell as illustrated.

229) Click the **Convert Entities** Sketch tool.

230) Click **Top view** . The circular edge is displayed on the ThreadPlane.

Access the Plane tool from the Consolidated Reference Geometry toolbar.

Insert the Helix/Spiral curve path.

231) Click **Insert, Curve, Helix/Spiral** from the Menu bar. The Helix/Spiral PropertyManager is displayed.

232) Enter **.250**in, **[6.35]** for Pitch.

233) Check the **Reverse** direction box.

234) Enter **2.5** for Revolutions.

235) Enter **0**deg for Starting angle. The Helix start point and end point are Coincident with the Top Plane.

236) Click the **Clockwise** box.

237) Click the **Taper Helix** box.

238) Enter **5**deg for Angle.

239) Uncheck the **Taper outward** box.

240) Click **OK** ✔ from the Helix/Spiral PropertyManager.

Rename the feature.
241) **Rename** Helix/Spiral1 to ThreadPath.

Save the model.
242) Click **Save** 💾.

The Helix tapers with the inside wall of the LENSCAP. Position the Helix within the wall thickness to prevent errors in the Swept.

Sketch the profile on the Top Plane. Position the profile to the Top right of the LENSCAP in order to pierce to the ThreadPath in the correct location.

If required, Show the ThreadPlane.
243) Right-click **ThreadPlane** from the FeatureManager.

244) Click **Show** from the Context toolbar.

245) Click **Hidden Lines Removed** ⬚ from the Heads-up View toolbar.

Sketch to the Top right ➡

Select the Plane for the Thread.
246) Right-click **Top Plane** from the FeatureManager.

Sketch the profile.

247) Click **Sketch** from the Context toolbar.

248) Click the **Centerline** Sketch tool.

249) Create a short **vertical centerline** off to the upper top area of the ThreadPath feature as illustrated.

250) Create a second **centerline** horizontal from the Midpoint to the left of the vertical line.

251) Create a third centerline coincident with the left horizontal endpoint. Drag the **centerline upward** until it is approximately the same size as the right vertical line.

252) Create a fourth **centerline** coincident with the left horizontal endpoint. Drag the **centerline** downward until it is approximately the same size as the left vertical line as illustrated.

Add an Equal relation.

253) Right-click **Select** to de-select the Centerline Sketch tool.

254) Click the **right vertical** centerline.

255) Hold the **Ctrl** key down.

256) Click the **two left vertical** centerlines.

257) Release the **Ctrl** key. The selected sketch entities are displayed in the Selected Entities box.

258) Click **Equal** =

259) Click **OK** ✔ from the Properties PropertyManager.

🔅 Utilize centerlines and construction geometry with geometric relations to maintain relationships with minimal dimensions.

🔅 Check **View**, **Hide/Show**, **Sketch Relations** from the Menu bar to show/hide sketch relation symbols.

Add a dimension.

260) Click the **Smart Dimension** Sketch tool.

261) Click the two **left vertical endpoints**. Click a **position** to the left.

262) Enter **.500**in, **[12.7]**.

Sketch the profile. The profile is a trapezoid.

263) Click the **Line** Sketch tool.

264) Click the **endpoints** to create the trapezoid as illustrated.

265) Right-click **Select** to de-select the Line Sketch tool.

Add an Equal relation.
266) Click the **left vertical** line.

267) Hold the **Ctrl** key down.

268) Click the **top** and **bottom lines** of the trapezoid.

269) Release the **Ctrl** key.

270) Click **Equal** .

271) Click **OK** from the Properties PropertyManager.

Click and drag the sketch to a position above the top right corner of the LENSCAP.

Add a Pierce relation.
272) Click the **left midpoint** of the trapezoid.

273) Hold the **Ctrl** key down.

274) Click the **starting left back edge** of the ThreadPath.

275) Release the **Ctrl** key.

276) Click **Pierce** from the Add Relations box. The sketch is fully defined.

Selected Entities

Edge<1>
Point4

277) Click **OK** from the Properties PropertyManager.

Select edge on the left side

Display an Isometric view.

278) Click **Isometric view** .

Modify the dimension.
279) Double-click the **.500** dimension text.

280) Enter **.125**in, [**3.18**].

Close the sketch.

281) Rebuild the model.

Rename the sketch.
282) Rename Sketch# to ThreadSection.

Save the model.

283) Click **Save** .

Insert the Swept feature.

284) Click the **Swept Boss/Base** feature tool. The Sweep PropertyManager is displayed.

285) Click **Sketch Profile**.

286) Click **ThreadSection** for the fly-out FeatureManager.

287) Click inside the **Path** box.

288) Click **ThreadPath** from the fly-out FeatureManager. ThreadPath is displayed.

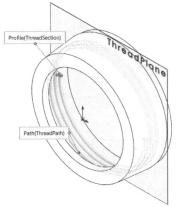

289) Click **OK** from the Sweep PropertyManager. Sweep1 is displayed in the FeatureManager.

Rename the feature. Display Shaded With Edges. Hide all planes.
290) Rename **Sweep1** to **Thread**.

291) Click **Shaded With Edges** from the Heads-up View toolbar.

292) Hide ThreadPath.

Save the model.

293) Click **Save** .

Swept geometry cannot intersect itself. If the ThreadSection geometry intersects itself, the cross section is too large. Reduce the cross section size and recreate the Swept feature.

The Thread feature is composed of the following: ThreadSection and ThreadPath.

The ThreadPath contains the circular sketch and the helical curve.

Most threads require a beveled edge or smooth edge for the thread part start point. A 30° Chamfer feature can be utilized on the starting edge of the trapezoid face. This action is left as an exercise.

Create continuous Swept features in a single step. Pierce the cross-section profile at the start of the swept path for a continuous Swept feature.

Un-suppress the Pattern feature to resolve both the Pattern feature and the seed feature at the same time.

The LENSCAP is complete. Review the LENSCAP before moving onto the last part of the FLASHLIGHT.

Review of the LENSCAP Part.

The LENSCAP utilized the Extruded Base feature with the Draft Angle option. The Extruded Cut feature created an opening for the LENS. You utilized the Shell feature with constant wall thickness to remove the front and back faces.

The Revolved Thin Cut feature created the back cut with a single line. The line utilized the Convert Entities tool to maintain the same draft angle as the Extruded Base feature.

You utilized a Swept feature with a Helical Curve and Thread profile to create the thread.

HOUSING Part

The HOUSING is a plastic part utilized to contain the BATTERY and to support the LENS. The HOUSING utilizes an Extruded Boss/Base, Lofted Boss, Extruded Cut, Draft, Swept, Rib, Mirror and Linear Pattern features.

Insert an Extruded Boss/Base (Boss-Extrude1) feature centered at the Origin.

Insert a Lofted Boss feature. The first profile is the converted circular edge of the Extruded Base. The second profile is a sketch on the BatteryLoftPlane.

Insert the second Extruded Boss/Base (Boss-Extrude2) feature. The sketch is a converted edge from the Loft Boss. The depth is determined from the height of the BATTERY.

Insert a Shell feature to create a thin-walled part.

Insert the third Extruded Boss (Boss-Extrude3) feature. Create a solid circular ring on the back-circular face of the Boss-Extrude1 feature. Insert the Draft feature to add a draft angle to the circular face of the HOUSING. The design intent for the Boss-Extrude1 feature requires you to maintain the same LENSCAP draft angle.

Insert a Swept feature for the Thread. Insert a Swept feature for the Handle. Reuse the Thread profile from the LENSCAP part.

Insert an Extruded Cut feature to create the hole for the SWITCH.

Insert the Rib feature on the back face of the HOUSING.

Insert a Linear Pattern feature to create a row of Ribs.

Insert a Rib feature along the bottom of the HOUSING.

Utilize the Mirror feature to create the second Rib.

💡 Reuse geometry between parts. The LENSCAP thread is the same as the HOUSING thread. Copy the ThreadSection from the LENSCAP to the HOUSING.

💡 Reuse geometry between features. The Linear Pattern and Mirror Pattern utilized existing features.

Activity: HOUSING Part - Extruded Base Feature.

Create a new part. Enter part name. Enter part description.

294) Click **New** ⬜ from the Menu bar.

295) Click the **MY-TEMPLATES** tab.

296) Double-click **PART-IN-ANSI**.

297) Click **Save As** 🖫 . Select **ENGDESIGN-W-SOLIDWORKS\PROJECTS** for the Save in folder.

298) Enter **HOUSING** for File name.

299) Enter **HOUSING FOR 6VOLT FLASHLIGHT** for Description.

300) Click **Save**. The HOUSING FeatureManager is displayed.

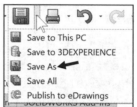

Create the Base sketch.

301) Right-click **Front Plane** from the FeatureManager. This is your Sketch plane. Click **Sketch** ⬕ from the Context toolbar.

302) Click the **Circle** ⊙ Sketch tool. The Circle PropertyManager is displayed.

303) Sketch a circle centered at the **Origin** ↳ as illustrated.

Add a dimension.

304) Click the **Smart Dimension** ✎ Sketch tool.

305) Click the **circumference**.

306) Enter **4.375**in, [**111.13**].

Insert an Extruded Boss/Base feature. Display an Isometric view.

307) Click the **Extruded Boss/Base** 🗔 feature tool. The Boss-Extrude PropertyManager is displayed.

308) Enter **1.300**, [**33.02**] for Depth in Direction 1. Accept the default settings.

309) Click **OK** ✔ from the Boss-Extrude1 PropertyManager.

310) Click **Isometric view** 🧊 .

Rename the feature. Save the model.

311) **Rename** Boss-Extrude1 to Base Extrude.

312) Click **Save** 💾.

HOUSING Part - Lofted Boss Feature

The Lofted Boss feature is composed of two profiles. The first sketch is named Sketch-Circle. The second sketch is named Sketch-Square.

Create the first profile from the back face of the Extruded feature. Utilize the Convert Entities sketch tool to extract the circular geometry to the back face.

Create the second profile on an Offset Plane. The FLASHLIGHT components must remain aligned to a common centerline. Insert dimensions that reference the Origin and build symmetry into the sketch. Utilize the Mirror Entities Sketch tool.

Activity: HOUSING Part - Lofted Boss Feature.

Create the first profile.
313) Right-click the **back face** of the Base Extrude feature. This is your Sketch plane.

314) Click **Sketch** ✏️ from the Context toolbar. The Sketch toolbar is displayed.

315) Click the **Convert Entities** 🗋 Sketch tool to extract the face to the Sketch plane.

Close and rename the sketch.
316) Right-click **Exit Sketch**.

317) **Rename** Sketch2 to SketchCircle.

Create an offset plane.
318) Click the **back face** of the Base Extrude feature.

319) Click **Plane** from the Consolidated Reference Geometry Features toolbar. The Plane PropertyManager is displayed.

320) Enter **1.300**in, [**33.02**] for Distance.

Verify the Plane position.
321) Click **Top view** 📐. View the Graphics window.

322) Click **OK** ✔️ from the Plane PropertyManager. Plane1 is displayed in the FeatureManager.

Rename Plane1. Rebuild and save the model.
323) **Rename** Plane1 to BatteryLoftPlane.

324) Click **Save** 💾.

Create the second profile.

325) Right-click **BatteryLoftPlane** in the FeatureManager.

326) Click **Sketch** from the Context toolbar. The Sketch toolbar is displayed.

327) Click **Back view**.

328) Click the **circumference** of the circle.

329) Click the **Convert Entities** Sketch tool.

330) Click the **Centerline** Sketch tool.

331) Sketch a **vertical centerline** coincident to the Origin and to the top edge of the circle as illustrated.

332) Click the **Line** Sketch tool.

333) Sketch a **horizontal line** to the right side of the centerline as illustrated.

334) Sketch a **vertical line** down to the circumference as illustrated.

335) Click the **Sketch Fillet** Sketch tool. The Sketch Fillet PropertyManager is displayed. If needed clear the Entities to Fillet box.

336) Click the **horizontal line** in the Graphics window.

337) Click the **vertical line** in the Graphics window.

338) Enter **.1**in, [**2.54**] for Radius.

339) Click **OK** from the Sketch Fillet PropertyManager.

340) Click **OK** from the Sketch Fillet PropertyManager. View the Sketch Fillet in the Graphics window.

Mirror the profile.

341) Click the **Mirror Entities** Sketch tool. The Mirror PropertyManager is displayed.

342) Click the **horizontal line**, **fillet** and **vertical line**. The selected entities are displayed in the Entities to mirror box.

343) Click inside the **Mirror about** box.

344) Click the **centerline** from the Graphics window.

345) Click **OK** from the Mirror PropertyManager.

Trim unwanted geometry.

346) Click the **Trim Entities** Sketch tool. The Trim PropertyManager is displayed.

347) Click **PowerTrim** from the Options box.

348) Click a **position** to the far right of the circle.

349) Drag the **mouse pointer** to intersect the circle.

350) Perform the same **actions** on the left side of the circle.

351) Click **OK** from the Trim PropertyManager.

Add dimensions.

352) Click the **Smart Dimension** Sketch tool.

Create a horizontal dimension.
353) Click the **left vertical** line.

354) Click the **right vertical** line.

355) Click a **position** above the profile.

356) Enter **3.100**in, **[78.74]**. View the results.

Create a vertical dimension.

357) Click the **Origin** .

358) Click the **top horizontal** line.

359) Click a **position** to the right of the profile.

360) Enter **1.600**in, **[40.64]**.

Modify the fillet dimension.

361) Double-click the **.100** fillet dimension.

362) Enter **.500**in, **[12.7]**.

Remove all sharp edges.

363) Click the **Sketch Fillet** Sketch tool. The Sketch Fillet PropertyManager is displayed.

364) Enter **.500**in, **[12.7]** for Radius.

365) Click the **lower left corner point** as illustrated.

366) Click the **lower right corner point** as illustrated.

367) Click **OK** from the Sketch Fillet PropertyManager.

368) Click **OK** from the PropertyManager.

Close and rename the sketch.

369) Right-click **Exit Sketch**.

370) **Rename** the Sketch3 to SketchSquare.

Save the model.

371) Click **Save**.

The Loft feature is composed of the SketchSquare and the SketchCircle. Select two individual profiles to create the Loft. The Isometric view provides clarity when selecting Loft profiles.

Display an Isometric view.

372) Click **Isometric view**.

Insert a Lofted Boss feature.

373) Click the **Lofted Boss/Base** feature tool. The Loft PropertyManager is displayed.

374) Click the **upper right side** of the SketchCircle as illustrated.

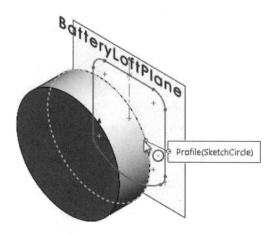

375) Click the **upper right side** of the SketchSquare as illustrated or from the Fly-out FeatureManager. The selected entities are displayed in the Profiles box.

376) Click **OK** ✅ from the Loft PropertyManager.

Rename Loft1. Save the model.
377) **Rename** Loft1 to Boss-Loft1.

378) Click **Save** 💾.

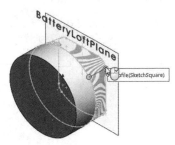

💡 Organize the FeatureManager to locate Loft profiles and planes. Insert the Loft reference planes directly before the Loft feature. Rename the planes, profiles and guide curves with clear descriptive names.

HOUSING Part - Second Extruded Boss/Base Feature

Create the second Extruded Boss/Base feature from the square face of the Loft. How do you estimate the depth of the second Extruded Boss/Base feature? Answer: The Boss-Extrude1 feature of the BATTERY is 4.100in, [104.14mm].

Ribs are required to support the BATTERY. Design for Rib construction. Ribs add strength to the HOUSING and support the BATTERY. Use a 4.400in, [111.76mm] depth as the first estimate. Adjust the estimated depth dimension later if required in the FLASHLIGHT assembly.

The Extruded Boss/Base feature is symmetric about the Right Plane. Utilize Convert Entities to extract the back face of the Loft Base feature. No sketch dimensions are required.

Activity: HOUSING Part - Second Extruded Boss/Base Feature.

Select the Sketch plane.
379) **Rotate** the model to view the back.

380) Right-click the **back face** of Boss-Loft1. This is your Sketch plane.

Create the sketch.

381) Click **Sketch** 📝 from the Context toolbar. The Sketch toolbar is displayed.

382) Click the **Convert Entities** 🗊 Sketch tool.

Insert the second Extruded Boss/Base feature.

383) Click the **Extruded Boss/Base** feature tool. The Boss-Extrude PropertyManager is displayed.

384) Enter **4.400**in, [**111.76**] for Depth in Direction 1.

385) Click the **Draft On/Off** box.

386) Enter **1**deg for Draft Angle.

387) Click **OK** from the Boss-Extrude PropertyManager.

Display a Right view.

388) Click **Right view**.

Rename Boss-Extrude2. Save the model.
389) **Rename** Boss-Extrude2 to Boss-Battery.

390) Click **Save**.

HOUSING Part - Shell Feature

The Shell feature removes material. Use the Shell feature to remove the front face of the HOUSING. In the injection-molded process, the body wall thickness remains constant.

A dialog box is displayed if the thickness value is greater than the Minimum radius of Curvature.

SOLIDWORKS   ✕

⚠ The Thickness value is greater than the Minimum radius of Curvature. The shell may succeed, but could cause undesirable results, such as bad geometry. To find the Minimum radius of Curvature, use Tools, Check.

☐ Don't show again in this session   OK   Cancel

Activity: HOUSING Part - Shell Feature.

Display an Isometric view. Insert the Shell feature.

391) Click **Isometric view**. If needed **Show** the BatteryLoftPlane from the FeatureManager.

392) Click the **Shell** feature tool. The Shell1 PropertyManager is displayed.

393) Click the **front face** of the Base Extrude feature as illustrated.

394) Enter **.100**in, [**2.54**] for Thickness.

395) Click **OK** ✓ from the Shell1 PropertyManager. Shell1 is displayed in the FeatureManager.

💡 The Shell feature position in the FeatureManager determines the geometry of additional features. Features created before the Shell contained the wall thickness specified in the Thickness option. Position features of different thickness such as the Rib feature and Thread Swept feature after the Shell. Features inserted after the Shell remain solid.

HOUSING Part - Third Extruded Boss/Base Feature

The third Extruded Boss/Base feature creates a solid circular ring on the back circular face of the Boss-Extrude1 feature. The solid ring is a cosmetic stop for the LENSCAP and provides rigidity at the transition of the HOUSING. Design for change. The Extruded Boss/Base feature updates if the Shell thickness changes.

Utilize the Front plane for the sketch. Select the inside circular edge of the Shell. Utilize Convert Entities to obtain the inside circle. Utilize the Circle Sketch tool to create the outside circle. Extrude the feature towards the front face.

Activity: HOUSING Part - Third Extruded Boss Feature.

Select the Sketch plane.
396) Right-click **Front Plane** from the FeatureManager. This is your Sketch plane.

Create the sketch.

397) Click **Sketch** ⌐ from the Context toolbar. The Sketch toolbar is displayed.

398) Zoom-in and click the **front inside circular edge** of Shell1 as illustrated.

399) Click the **Convert Entities** Sketch tool.

Display the Front view. Create the outside circle.

400) Click **Front view** ▱.

401) Click the **Circle** ⊙ Sketch tool. The Circle PropertyManager is displayed.

402) Sketch a **circle** centered at the Origin.

Add a dimension.

403) Click the **Smart Dimension** ✎ Sketch tool. Click the **circumference** of the circle.

404) Enter **5.125**in, [**130.18**].

Insert the third Extruded Boss/Base feature.

405) Click the **Extruded Boss/Base** feature tool. The Boss-Extrude PropertyManager is displayed.

406) Enter **.100**in, [**2.54**] for Depth in Direction 1. The Extrude arrow points to the front. Click **OK** ✔ from the Boss-Extrude PropertyManager.

Display an Isometric view.

407) Click **Isometric view** ⬚.

Rename Boss-Extrude3. Hide the BatteryLoftPlane. Save the model.

408) **Rename** Boss-Extrude3 to Boss-Stop. If needed, **Hide** the BatteryLoftPlane.

409) Click **Save** 💾.

HOUSING Part - Draft Feature

The Draft feature tapers selected model faces by a specified angle by utilizing a Neutral Plane or Parting Line. The Neutral Plane option utilizes a plane or face to determine the pull direction when creating a mold.

The Parting Line option drafts surfaces around a parting line of a mold. Utilize the Parting Line option for non-planar surfaces. Apply the Draft feature to solid and surface models.

A 5deg draft is required to ensure proper thread mating between the LENSCAP and the HOUSING. The LENSCAP Extruded Boss/Base (Boss-Extrude1) feature has a 5deg draft angle.

The outside face of the Extruded Boss/Base (Boss-Extrude1) feature HOUSING requires a 5° draft angle. The inside HOUSING wall does not require a draft angle. The Extruded Boss/Base (Boss-Extrude1) feature has a 5° draft angle. Use the Draft feature to create the draft angle. The front circular face is the Neutral Plane. The outside cylindrical surface is the face to draft.

You created the Extruded Boss/Base and Extruded Cut features with the Draft Angle option. The Draft feature differs from the Extruded feature, Draft Angle option. The Draft feature allows you to select multiple faces to taper.

In order for a model to eject from a mold, all faces must draft away from the parting line which divides the core from the cavity. Cavity side faces display a positive draft and core side faces display a negative draft. Design specifications include a minimum draft angle, usually less than 5deg.

For the model to eject successfully, all faces must display a draft angle greater than the minimum specified by the Draft Angle. The Draft feature, Draft Analysis Tools and DraftXpert utilize the draft angle to determine what faces require additional draft base on the direction of pull.

🔆 You can apply a draft angle as a part of an Extruded Boss/Base or Extruded Cut feature.

The Draft PropertyManager provides the ability to select either the *Manual* or *DraftXpert* tab. Each tab has a separate menu and option selections. The Draft PropertyManager displays the appropriate selections based on the type of draft you create.

🔆 The DraftXpert PropertyManager provides the ability to manage the creation and modification of all Neutral Plane drafts. Select the draft angle and the references to the draft. The DraftXpert manages the rest.

Activity: HOUSING Part - Draft Feature.

Insert the Draft feature.

410) Click the **Draft** 🔲 feature tool. The Draft PropertyManager is displayed.

411) Click the **Manual** tab.

412) Click **Neutral plane** for Type of Draft.

413) Click inside the **Neutral Plane** box.

414) **Zoom-in** and click the thin **front circular face** of Base Extrude.

415) Click inside the **Faces to draft** box.

416) Click the **outside cylindrical face**. Note: The face feedback icon.

417) Enter **5**deg for Draft Angle.

418) Click **OK** ✔ from the Draft PropertyManager. Draft1 is displayed in the FeatureManager.

Display the draft angle and the straight interior.

419) Click **Right view** ▥.

420) Click **Hidden Lines Visible** ▥ from the Heads-up View toolbar.

Save the model.

421) Click **Save** 💾.

Draft Angle

Straight

The order of feature creation is important. Apply threads after the Draft feature for plastic parts to maintain a constant thread thickness.

HOUSING Part - Thread with Swept Feature

The HOUSING requires a thread. Create the threads for the HOUSING on the outside of the Draft feature. Create the thread with the Swept feature. The thread requires two sketches: ThreadPath and ThreadSection.

The LENSCAP and HOUSING Thread utilize the same technique. Create a ThreadPlane. Utilize Convert Entities to create a circular sketch referencing the HOUSING Extruded Boss/Base (Boss-Extrude1) feature.

Insert a Helix/Spiral curve to create the path.

Reuse geometry between parts. The ThreadSection is copied from the LENSCAP and is inserted into the HOUSING Top Plane.

Activity: HOUSING Part - Thread with Swept Feature.

Insert the ThreadPlane.

422) Click **Isometric view** from the Heads-up View toolbar.

423) Click **Hidden Lines Removed** from the Heads-up View toolbar.

424) Click the **thin front circular face**, Base Extrude.

425) Click **Plane** from the Consolidated Reference Geometry toolbar. The Plane PropertyManager is displayed.

426) Check the **Flip offset** box.

427) Enter **.125**in, [**3.18**] for Distance. Accept the default settings.

428) Click **OK** from the Plane PropertyManager. Plane2 is displayed in the FeatureManager.

Rename Plane2. Save the model.

429) Rename Plane2 to ThreadPlane.

430) Click **Save**.

Insert the ThreadPath.

431) Right-click **ThreadPlane** from the FeatureManager. This is your Sketch plane.

432) Click **Sketch** from the Context toolbar. The Sketch toolbar is displayed.

433) Click the **front outside circular edge** of the Base Extrude feature.

434) Click the **Convert Entities** Sketch tool. The circular edge is displayed on the ThreadPlane.

Insert the Helix/Spiral curve.

435) Click the **Helix and Spiral** tool from the Consolidated Curves toolbar as illustrated. The Helix/Spiral PropertyManager is displayed.

436) Enter .**250**in, [**6.35**] for Pitch.

437) Click the **Reverse direction** box.

438) Enter **2.5** for Revolution.

439) Enter **180** in the Start angle box. The Helix start point and end point are Coincident with the Top Plane.

440) Click the **Taper Helix** box.

441) Enter **5**deg for Angle.

442) Check the **Taper outward** box.

443) Click **OK** ✔ from the Helix/Spiral PropertyManager. HexliSpiral1 is displayed in the FeatureManager.

444) **Rename** Helix/Spiral1 to ThreadPath.

Display an Isometric view.

445) Click **Isometric view** ▣.

Save the model.

446) Click **Save** 🖫.

Copy the LENSCAP ThreadSection.
447) **Open** the LENSCAP part. The LENSCAP FeatureManager is displayed.

448) **Expand** the Thread feature from the FeatureManager.

449) Click the **ThreadSection** sketch. ThreadSection is highlighted.

450) Click **Edit**, **Copy** from the Menu bar.

451) **Close** the LENSCAP.

Open the HOUSING.

452) **Return** to the Housing.

Paste the LENSCAP ThreadSection.

453) Click **Top Plane** from the HOUSING FeatureManager.

454) Click **Edit**, **Paste** from the Menu bar. The ThreadSection is displayed on the Top Plane. The new Sketch7 name is added to the bottom of the FeatureManager.

455) **Hide** ThreadPlane.

Rename Sketch7. Save the model.

456) **Rename** Sketch7 to Thread Section.

457) Click **Save** .

Add a Pierce relation.

458) Right-click **ThreadSection** from the FeatureManager.

459) Click **Edit Sketch**.

460) Click **ThreadSection** from the HOUSING FeatureManager.

461) **Zoom in** on the Midpoint of the ThreadSection.

462) Click the **Midpoint** of the ThreadSection.

463) Click **Isometric view** .

464) Hold the **Ctrl** key down.

465) Click the **right back edge of the ThreadPath**. Note: Do not click the end point. The Properties PropertyManager is displayed. The selected entities are displayed in the Selected Entities box.

466) Release the **Ctrl** key.

Click **Pierce** from the Add Relations box.

467) Click **OK** from the Properties PropertyManager.

Caution: Do not click the front edge of the Thread path. The Thread is then created out of the HOUSING.

Midpoint of the ThreadSection

Pierce to the back edge of ThreadPath.

Close the sketch.

468) Click **Exit Sketch**. ThreadSection is fully defined.

Insert the Swept feature.

469) Click the **Swept Boss/Base** feature tool. The Swept PropertyManager is displayed.

470) Select **Sketch Profile**.

471) **Expand** HOUSING from the fly-out FeatureManager.

472) Click **ThreadSection** from the fly-out FeatureManager.

473) Click **ThreadPath** from the fly-out FeatureManager.

474) Click **OK** ✔ from the Sweep PropertyManager. Sweep1 is displayed in the FeatureManager.

Rename Sweep1. Save the model.
475) **Rename** Sweep1 to Thread.

476) Click **Save** 🖫.

💡 Creating a ThreadPlane provides flexibility to the design. The ThreadPlane allows for a smoother lead. Utilize the ThreadPlane offset dimension to adjust the start of the thread.

HOUSING Part - Handle with Swept Feature

Create the handle with the Swept feature. The Swept feature consists of a sketched path and cross section profile. Sketch the path on the Right Plane. The sketch uses edges from existing features. Sketch the profile on the back circular face of the Boss-Stop feature.

Activity: HOUSING Part - Handle with Swept Feature.

Create the Swept path sketch.

477) Right-click **Right Plane** from the FeatureManager.

478) Select **Sketch** ⌐ from the Context toolbar. The Sketch toolbar is displayed.

479) Click **Right view** ⬚.

480) Click **Hidden Lines Removed** ⬚ from the Heads-up View toolbar.

481) Click the **Line** ✏ Sketch tool.

482) Sketch a **vertical lin**e up from the top corner as illustrated.

483) Sketch a **horizontal line** over to the left below the top of the Boss Stop. Add any needed relations.

Insert a 2D Fillet.

484) Click the **Sketch Fillet** ⌐ Sketch tool.

485) Click the **right top corner** of the sketch lines as illustrated.

486) Enter **.500**in, [**12.7**] for Radius.

487) Click **OK** ✔ from the Sketch Fillet PropertyManager.

488) Click **OK** ✔ from the PropertyManager.

Add an Intersection relation.

489) Click the **bottom end point** of the vertical line.

490) Hold the **Ctrl** key down.

491) Click the **right vertical edge** of the Housing.

492) Click the **horizontal edge** of the Housing.

493) Release the **Ctrl** key.

494) Click **Intersection** ⤬ from the Add Relations box.

495) Click **OK** ✔ from the Properties PropertyManager.

Add a dimension.

496) Click the **Smart Dimension** ✧ Sketch tool.

497) Click the **Origin**.

498) Click the **horizontal** line.

499) Click a **position** to the right.

500) Enter **2.500**in, [**63.5**].

Close and rename the sketch.

501) Click **Exit Sketch**.

502) **Rename** Sketch8 to HandlePath.

Save the model.

503) Click **Save** 💾.

Create the Swept Profile.

504) Click **Back view** 🔲.

505) Right-click the **back circular face** of the Boss-Stop feature as illustrated.

506) Click **Sketch** ⌐ from the Context toolbar. The Sketch toolbar is displayed.

💡 The book is designed to expose the new user to various methods in creating sketches and features. In the next section create a slot **without using** the Slot Sketch tool.

507) Click the **Centerline** Sketch tool.

508) Sketch a **vertical centerline** collinear with the Right Plane, coincident to the Origin.

509) Sketch a **horizontal centerline**. The left end point of the centerline is coincident with the vertical centerline on the Boss-Stop feature. Do not select existing feature geometry.

510) **Zoom in** on the top of the Boss-Stop.

511) Click the **Line** Sketch tool.

512) Sketch a **line** above the horizontal centerline as illustrated.

513) Click the **Tangent Arc** Sketch tool.

514) Sketch a **90° arc**.

515) Right-click **Select** to exit the Tangent Arc tool.

Add an Equal relation.
516) Click the **horizontal** centerline.

517) Hold the **Ctrl** key down.

518) Click the **horizontal** line.

519) Release the **Ctrl** key.

520) Click **Equal** = from the Add Relations box.

521) Click **OK** from the Properties PropertyManager.

Add a Horizontal relation.
522) Click the **right end point** of the tangent arc.

523) Hold the **Ctrl** key down.

524) Click the **arc center point**.

525) Click the **left end point** of the centerline.

526) Release the **Ctrl** key.

527) Click **Horizontal** from the Add Relations box.

528) Click **OK** from the Properties PropertyManager.

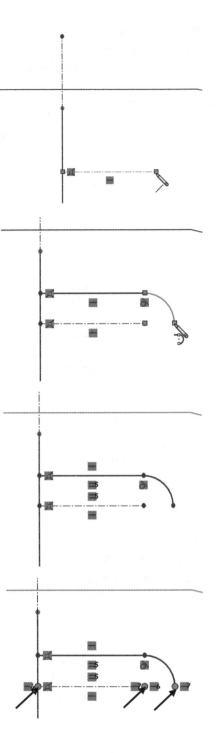

Mirror about the horizontal centerline.

529) Click the **Mirror Entities** ⊢⊣ Sketch tool. The Mirror PropertyManager is displayed.

530) Clear the Entities to mirror box.

531) Click the **horizontal** line as illustrated.

532) Click the **90° arc**. The selected entities are displayed in the Entities to mirror box.

533) Click inside the **Mirror about** box.

534) Click the **horizontal** centerline.

535) Click **OK** ✔ from the Mirror PropertyManager.

Mirror about the vertical centerline.

536) Click the **Mirror Entities** ⊢⊣ Sketch tool. The Mirror PropertyManager is displayed.

537) Window-Select the **two horizontal lines**, the **horizontal centerline** and the **90° arc** for Entities to mirror. The selected entities are displayed in the Entities to mirror box.

538) Click inside the **Mirror about** box.

539) Click the **vertical** centerline.

540) Click **OK** ✔ from the Mirror PropertyManager.

Add dimensions.

541) Click the **Smart Dimension** ↖ Sketch tool.

542) Enter **1.000**in, **[25.4]** between the arc center points.

543) Enter .**100**in, **[2.54]** for Radius.

Add a Pierce relation.

544) Right-click **Select**.

545) Click **Isometric view** .

546) Click the **top midpoint** of the Sketch profile.

547) Hold the **Ctrl** key down.

548) Click the **line** from the Handle Path.

549) Release the **Ctrl** key.

550) Click **Pierce** from the Add Relations box. The sketch is fully defined.

551) Click **OK** from the Properties PropertyManager.

Close and rename the sketch.

552) Click **Exit Sketch**.

553) **Rename** Sketch8 to HandleProfile.

554) **Hide** ThreadPlane and BatteryLoftPlane if needed.

Insert the Swept feature.

555) Click the **Swept Boss/Base** feature tool. The Sweep PropertyManager is displayed.

556) Select **Sketch Profile**.

557) **Expand** HOUSING from the fly-out FeatureManager.

558) Click inside the **Profile** box.

559) Click **HandleProfile** from the fly-out FeatureManager.

560) Click **HandlePath** from the fly-out FeatureManager.

561) Click **OK** from the Sweep PropertyManager. Sweep# is displayed in the FeatureManager.

Fit the profile to the Graphics window.

562) Press the **f** key.

Display Shaded With Edges.

563) Click **Shaded With Edges** from the Heads-up View toolbar.

Rename Sweep. Save the model.
564) Rename **Sweep#** to **Handle**.

565) Click **Save** 💾.

How does the Handle Swept feature interact with other parts in the FLASHLIGHT assembly? Answer: The Handle requires an Extruded Cut to insert the SWITCH.

HOUSING Part - Extruded Cut Feature with Up To Surface

Create an Extruded Cut in the Handle for the SWITCH. Utilize the top face of the Handle for the Sketch plane. Create a circular sketch centered on the Handle.

Utilize the Up To Surface End Condition in Direction 1. Select the inside surface of the HOUSING for the reference surface.

> **Activity: HOUSING Part - Extruded Cut Feature with Up To Surface**.

Select the Sketch plane.
566) Right-click the **top face** of the Handle. Handle is highlighted in the FeatureManager. This is your Sketch plane.

Create the sketch.
567) Click **Sketch** 🖉 from the Context toolbar. The Sketch toolbar is displayed.

568) Click **Top view** 🗗.

569) Click **Circle** ⊙ from the Sketch toolbar. The Circle PropertyManager is displayed.

570) Sketch a **circle** on the Handle near the front as illustrated.

Deselect the circle sketch tool.
571) Right-click **Select**.

Add a Vertical relation.
572) Click the **Origin**.

573) Hold the **Ctrl** key down.

574) Click the **center point** of the circle. The Properties PropertyManager is displayed. The selected entities are displayed in the Selected Entities box.

575) Release the **Ctrl** key.

576) Click **Vertical** │.

577) Click **OK** ✔ from the Properties PropertyManager.

Add dimensions.

578) Click the **Smart Dimension** Sketch tool.

579) Enter **.510**in, [**12.95**] for diameter.

580) Enter **.450**in, [**11.43**] for the distance from the Origin.

Insert an Extruded Cut feature.
581) **Rotate** the model to view the inside Shell1.

582) Click the **Extruded Cut** feature tool. The Cut-Extrude PropertyManager is displayed.

583) Select the **Up To Surface** End Condition in Direction 1.

584) Click the **top inside face** of the Shell1 as illustrated.

585) Click **OK** from the Cut-Extrude PropertyManager. The Cut-Extrude1 feature is displayed in the FeatureManager.

Rename Cut-Extrude1. Display an Isometric view.
586) **Rename** the Cut-Extrude1 feature to SwitchHole.

587) Click **Isometric view** from the Heads-up View toolbar.

Save the model.
588) Click **Save**.

HOUSING Part - First Rib and Linear Pattern Feature

The Rib feature adds material between contours of existing geometry. Use Ribs to add structural integrity to a part.

A Rib requires:

- A sketch

- Thickness

- Extrusion direction

The first Rib profile is sketched on the Top Plane. A 1° draft angle is required for manufacturing. Determine the Rib thickness by the manufacturing process and the material.

Rule of thumb states that the Rib thickness is ½ the part wall thickness. The Rib thickness dimension is .100 inches, [2.54mm] for illustration purposes.

The HOUSING requires multiple Ribs to support the BATTERY. A Linear Pattern feature creates multiple instances of a feature along a straight line. Create the Linear Pattern feature in two directions along the same vertical edge of the HOUSING.

The Instance to Vary option in the Linear Pattern PropertyManager allows you to vary the dimensions and locations of instances in a feature pattern *after it is created*. You can vary the dimensions of a series of instances, so that each instance is larger or smaller than the previous one. You can also change the dimensions of a single instance in a pattern and change the position of that instance relative to the seed feature of the pattern. For linear patterns, you can change the spacing between the columns and rows in the pattern.

Activity: HOUSING Part - First Rib and Linear Pattern Feature.

Display all hidden lines.

589) Click **Hidden Lines Visible** from the Heads-up View toolbar.

Create the sketch.

590) Right-click **Top Plane** from the FeatureManager. This is your Sketch plane.

591) Click **Sketch** from the Context toolbar.

592) Click **Top view**.

593) Click the **Line** Sketch tool.

594) Sketch a **horizontal line**. The endpoints are located on either side of the Handle.

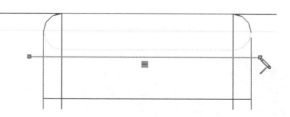

Add a dimension.

595) Click the **Smart Dimension** Sketch tool.

596) Click the **inner back** edge.

597) Click the **horizontal** line.

598) Click a **position** to the right off the profile.

599) Enter **.175**in, [**4.45**].

Insert the Rib feature.

600) Click the **Rib** feature tool. The Rib PropertyManager is displayed.

601) Click the **Both Sides** button.

602) Click the **Draft On/Off** box.

603) Enter **1**deg for Draft Angle.

604) Enter **.100**in, [**2.54**] for Rib Thickness.

605) Click the **Parallel to Sketch** button. The Rib direction arrow points to the back. Flip the material side if required. Select the Flip material side check box if the direction arrow does not point towards the back.

606) Click **Front view** .

607) Click the **Rib** sketch inside the HOUSING.

608) Click **OK** from the Rib PropertyManager. Rib1 is displayed in the FeatureManager.

Display an Isometric view. Save the model.

609) Click **Isometric view** .

610) Click **Save** .

Existing geometry defines the Rib boundaries. The Rib does not penetrate through the wall.

Insert the Linear Pattern feature.

611) **Zoom to Area** on Rib1.

612) Click **Rib1** from the FeatureManager.

613) Click the **Linear Pattern** feature tool. Rib1 is displayed in the Features to Pattern box.

Rib pointing inwards

614) Click inside the Direction 1 Pattern Direction box.

615) Click the **hidden upper back vertical edge** of Shell1 in the Graphics window. The direction arrow points upward. Click the Reverse direction button if required.

616) Enter **.500**in, [**12.7**] for Spacing.

617) Enter **3** for Number of Instances.

618) Click inside the **Direction 2 Pattern** Direction box.

619) Click the hidden **lower back vertical edge** of Shell1 in the Graphics window. The direction arrow points downward. Click the Reverse direction button if required.

620) Enter **.500**in, [**12.7**] for Spacing.

621) Enter **3** for Number of Instances.

622) Click the **Pattern seed only** box.

623) Drag the Linear Pattern **Scroll bar** downward to display the Options box.

624) Check the **Geometry pattern** box. Accept the default values.

625) Click **OK** ✔ from the Linear Pattern PropertyManager. LPattern1 is displayed in the FeatureManager.

Display an Isometric view. Save the model.

626) Click **Isometric view** 🔲.

627) Click **Save** 💾.

💡 Utilize the Geometry pattern option to efficiently create and rebuild patterns. Know when to check the Geometry pattern.

Check Geometry pattern. You require an exact copy of the seed feature. Each instance is an exact copy of the faces and edges of the original feature. End conditions are not calculated. This option saves rebuild time.

Uncheck Geometry pattern. You require the end condition to vary. Each instance will have a different end condition. Each instance is offset from the selected surface by the same amount.

Surface for end condition

💡 Suppress Patterns when not required. Patterns contain repetitive geometry that takes time to rebuild. Pattern features also clutter the part during the model creation process. Suppress patterns as you continue to create more complex features in the part. Unsuppress a feature to restore the display and load into memory for future calculations. Hide features to improve clarity. Show feature to display hidden features.

Rib sketches are not required to be fully defined. The Linear Rib option blends sketched geometry into existing contours of the model.

Example: Create an offset reference plane from the inside back face of the HOUSING.

Sketch two under defined arcs. Insert a Rib feature with the Linear option. The Rib extends to the Shell walls.

HOUSING Part - Second Rib Feature

The Second Rib feature supports and centers the BATTERY. The Rib is sketched on a reference plane created through a point on the Handle and parallel with the Right Plane. The Rib sketch references the Origin and existing geometry in the HOUSING. Utilize an Intersection and Coincident relation to define the sketch.

Activity: HOUSING Part - Second Rib Feature.

Insert a Reference plane for the second Rib feature. Create a Parallel Plane at Point.

628) Click **Wireframe** ⊞ from the Heads-up View toolbar.

629) **Zoom to Area** 🔍 on the back right side of the Handle.

630) Click **Plane** from the Features toolbar. The Plane PropertyManager is displayed.

631) Click **Right Plane** from the fly-out FeatureManager.

632) Click the **vertex** (point) at the back right of the handle as illustrated. Plane1 is fully defined.

633) Click **OK** ✔ from the Plane PropertyManager. Plane1 is displayed in the FeatureManager.

Rename Plane1.

634) **Rename** Plane1 to LongRibPlane.

Fit to the Graphics window. Save the model.

635) Press the **f** key.

636) Click **Save** .

Create the second Rib.

637) Right-click **LongRibPlane** for the FeatureManager.

Create the sketch.

638) Click **Sketch** from the Context toolbar.

639) Click **Right view** from the Heads-up View toolbar.

640) Click the **Line** Sketch tool.

641) Sketch a **horizontal line**. Do not select the edges of the Shell1 feature.

Deselect the Line Sketch tool.

642) Right-click **Select**.

Add a Coincident relation.

643) Click **BatteryLoftPlane** from the FeatureManager.

644) Hold the **Ctrl k**ey down.

645) Click the **left end point** of the horizontal sketch line.

646) Release the **Ctrl** key.

647) Click **Coincident**.

648) Click **OK** from the Properties PropertyManager.

Add a dimension.

649) Click the **Smart Dimension** Sketch tool.

650) Click the **horizontal** line.

651) Click the **Origin**.

652) Click a **position** for the vertical linear dimension text.

653) Enter **1.300**in, [**33.02**].

💡 When the sketch and reference geometry become complex, create dimensions by selecting Reference planes and the Origin in the FeatureManager.

💡 Dimension the Rib from the Origin, not from an edge or surface for design flexibility. The Origin remains constant. Modify edges and surfaces with the Fillet feature.

Sketch an arc.

654) **Zoom to Area** 🔍 on the horizontal Sketch line.

655) Click the **Tangent Arc** ⌒ Sketch tool.

656) Click the **left end** point of the horizontal line.

657) Click the **intersection** of the Shell1 and Boss Stop features. The sketch is displayed in black and is fully defined. If needed add an Intersection relation between the endpoint of the Tangent Arc, the left vertical Boss-Stop edge and the Shell1 Silhouette edge of the lower horizontal inside wall.

PAGE 6 - 59

Insert the Rib feature.

658) Click the **Rib** feature tool. The Rib PropertyManager is displayed.

659) Click the **Both Sides** box.

660) Click the **Draft On/Off** box.

661) Enter **1**deg for Angle.

662) Enter **.075**in, [**1.91**] for Rib Thickness.

663) Click the **Draft outward** box.

664) Click the **Flip material side** box if required. The direction arrow points towards the bottom.

665) Rotate the model and click the **inside body**.

666) Click **OK** from the Rib PropertyMana ger. Rib2 is displayed in the FeatureManager.

667) **Hide** LongRibPlane from the FeatureManager.

Display a Trimetric view. Shaded With Edge.

668) Click **Trimetric view** from the Heads-up View toolbar.

669) Click **Shaded With Edges** from the Heads-up View toolbar. View the created Rib feature.

HOUSING Part - Mirror Feature

An additional Rib is required to support the BATTERY. Reuse features with the Mirror feature to create a Rib symmetric about the Right Plane.

The Mirror feature requires:

- Mirror Face or Plane reference

- Features or Faces to Mirror

Utilize the Mirror feature. Select the Right Plane for the Mirror Plane. Select the second Rib for the Features to Mirror.

Activity: HOUSING Part - Mirror Feature.

Insert the Mirror feature.

670) Click the **Mirror** feature tool. The Mirror PropertyManager is displayed.

671) Click inside the **Mirror Face/Plane** box.

672) Click **Right Plane** from the fly-out FeatureManager.

673) Click **Rib2** for Features to Mirror from the fly-out FeatureManager.

674) Click **OK** from the Mirror PropertyManager. Mirror1 is displayed in the FeatureManager.

Display a Trimetric view. Save the model.

675) Click **Trimetric view** from the Heads-Up View toolbar.

676) Click **Save** .

Close all parts.

677) Click **Window**, **Close All** from the Menu bar.

The parts for the FLASHLIGHT are complete. Review the HOUSING before moving on to the FLASHLIGHT assembly.

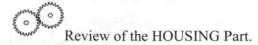 Review of the HOUSING Part.

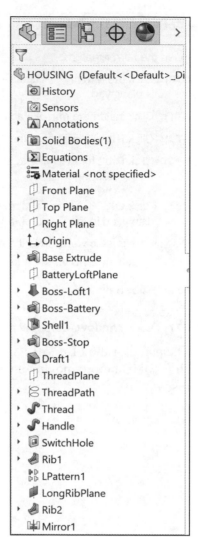

The HOUSING utilized the Extruded Boss/Base feature with the Draft Angle option. The Lofted Boss feature was created to blend the circular face of the LENS with the rectangular face of the BATTERY. The Shell feature removed material with a constant wall thickness. The Draft feature utilized the front face as the Neutral plane.

You created a Thread similar to the LENSCAP Thread. The Thread profile was copied from the LENSCAP and inserted into the Top Plane of the HOUSING. The Extruded Cut feature was utilized to create a hole for the Switch. The Rib features were utilized in a Linear Pattern and Mirror feature.

Each feature has additional options that are applied to create different geometry. The Offset From Surface option creates an Extruded Cut on the curved surface of the HOUSING and LENSCAP. The Reverse offset and Translate surface options produce a cut depth constant throughout the curved surface.

Utilize Tools, Sketch Entities, Text to create the text profile on an Offset Plane.

FLASHLIGHT Assembly

Plan the Sub-assembly component layout diagram.

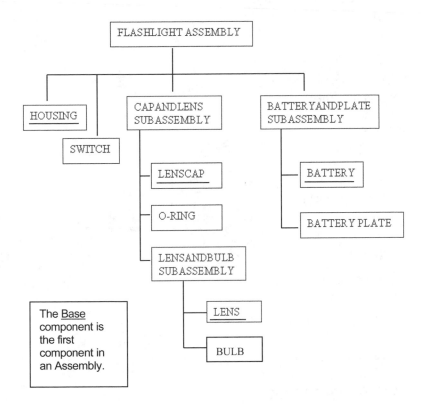

The Base component is the first component in an Assembly.

Assembly Layout Structure

The FLASHLIGHT assembly steps are as follows:

- Create the LENSANDBULB sub-assembly from the LENS and BULB components. The LENS is the Base component.

- Create the BATTERYANDPLATE sub-assembly from the BATTERY and BATTERYPLATE.

- Create the CAPANDLENS sub-assembly from the LENSCAP, O-RING and LENSANDBULB sub-assembly. The LENSCAP is the Base component.

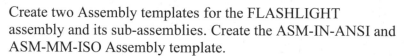

- Create the FLASHLIGHT assembly. The HOUSING is the Base component. Insert the SWITCH, CAPANDLENS and BATTERYANDPLATE.

- Modify the dimensions to complete the FLASHLIGHT assembly.

Assembly Template

An Assembly document template is the foundation of the assembly.

Create two Assembly templates for the FLASHLIGHT assembly and its sub-assemblies. Create the ASM-IN-ANSI and ASM-MM-ISO Assembly template.

LENSANDBULB Sub-assembly

Create the LENSANDBULB sub-assembly. Utilize three Coincident mates to assemble the BULB component to the LENS component. When geometry is complex, select the planes in the Mate Selection box.

 Suppress the Lens Shield feature to view all inside surfaces during the mate process.

Activity: Create two Assembly Templates.

Create the ASM-IN-ANSI Assembly Template.

678) **Close** all models.

679) Click **New** from the Menu bar.

680) Double-click the **Assembly** icon from the default Templates tab. The Begin Assembly PropertyManager and the Window Open dialog box is displayed.

681) Click **Cancel** from the Open Window dialog box.

682) Click **Cancel** ✖ from the Begin Assembly PropertyManager. The Assem1 FeatureManager is displayed.

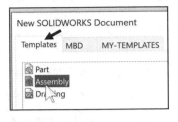

Set Document Properties. Set drafting standard, units and precision.

683) Click **Options** ⚙, **Document Properties** tab from the Menu bar.

Select drafting standard, units, and precision.

684) Select **ANSI** for Overall drafting standard. Click **Units**. Select **IPS** for Unit system.

685) Select **.123** for basic units length decimal places. Select **None** for basic units angle decimal places. Click **OK** from the dialog box.

Save the Assembly template. Enter name.

686) Click **Save As** 📄 .

687) Click **Assembly Templates** (*asmdot) from the Save As dialog box.

688) Select **ENGDESIGN-W-SOLIDWORKS\MY-TEMPLATES** for the Save in folder.

689) Enter **ASM-IN-ANSI** in the File name. Click **Save**.

Create the ASM-MM-ISO Assembly template.

690) Click **New** from the Menu bar.

691) Double-click **Assembly** from the default Templates tab. The Begin Assembly PropertyManager and the Window Open dialog box is displayed.

692) Click **Cancel** from the Open Window dialog box. Click **Cancel** ✖ from the Begin Assembly PropertyManager.

Select drafting standard, units and precision.

693) Click **Options** ⚙, **Document Properties** tab from the Menu bar.

694) Select **ISO** for Overall drafting standard.

695) Click **Units**. Click **MMGS** for Unit system.

696) Select **.12** for basic units length decimal places.

697) Select **None** for basic units angle decimal places. Click **OK**.

Save the Assembly template. Enter name.

698) Click **Save As** .

699) Select **Assembly Templates (*asmdot)** from the Save As dialog box.

700) Select **ENGDESIGN-W-SOLIDWORKS\MY-TEMPLATES** for the Save in folder.

701) Enter **ASM-MM-ISO** for File name.

702) Click **Save**.

Activity: LENSANDBULB Sub-assembly.

Close all documents.

703) Click **Windows**, **Close All** from the Menu bar.

Create the LENSANDBULB sub-assembly.

704) Click **New** from the Menu bar.

705) Click the **MY-TEMPLATES** tab.

706) Double-click **ASM-IN-ANSI**. **Note**: Additional templates are displayed.

Insert the LENS. Save the assembly. Enter name and description.

707) Double-click **LENS** from the ENGDESIGN-W-SOLIDWORKS\PROJECTS folder.

708) Click **OK** from the Begin Assembly PropertyManager. The LENS is fixed to the Origin.

709) Click **Save As** . Click **Save** in the Save All dialog box. Note the location of the assembly.

710) Click **Save As** .

711) Enter **LENSANDBULB** for File name.

712) Enter **LENS AND BULB ASSEMBLY** for Description.

713) Click **Save**.

Insert the BULB.

714) Click **Insert Components** from the Assembly toolbar.

715) Double-click **BULB** from the PROJECTS folder.

716) Click a **position** in front of the LENS as illustrated.

Fit the model to the Graphics window.

717) Press the **f** key.

Move the BULB.

718) Click and drag the **BULB** in the Graphics window.

Save the LENSANDBULB. View the FeatureManager.

719) Click **Save**. View the Assembly FeatureManager.

Pin the Mate PropertyManager. Insert a Coincident mate.

720) Click the **Mate** ^Mate Assembly tool. The Mate PropertyManager is displayed.

721) Click the **Standard** tab. Click the **Keep Visible** 📌 icon.

722) Click **Right Plane** of the LENS from the fly-out FeatureManager.

723) **Expand** BULB in the fly-out FeatureManager.

724) Click **Right Plane** of the BULB. Coincident is selected by default.

725) Click **OK** ✔ from the Coincident PropertyManager. Coincident1 is created.

Insert the second Coincident mate.

726) Click **Top Plane** of the LENS from the fly-out FeatureManager.

727) Click **Top Plane** of the BULB from the fly-out FeatureManager. Coincident is selected by default.

728) Click **OK** ✔ from the Coincident PropertyManager. Coincident2 is created.

729) Click **OK** ✔ from the Mate PropertyManager.

💡 Select face geometry efficiently. Position the mouse pointer in the middle of the face. Do not position the mouse pointer near the edge of the face. Zoom in on geometry. Utilize the Face Selection Filter for narrow faces.

Activate the Face Selection Filter.

730) Click **Right-click** in the Graphics window.

731) Click the **Selection Filters drop-down arrow**. The Selection Filter toolbar is displayed.

732) Click **Filter Faces**. The Filter icon is displayed 🔽.

Only faces are selected until the Face Selection Filter is deactivated. Select Clear All Filters ⬚ from the Selection Filter toolbar to deactivate all filters.

Insert the third Coincident mate.

733) **Zoom to Area** and **Rotate** on the CBORE.

734) Click the **BulbHole front face** of the LENS in the Graphics window.

735) **Rotate** to view the bottom back flat face of the BULB.

736) Hold the **Ctrl** key down.

737) Click the **bottom back flat face**, Revolve1 of the BULB.

738) Release the **Ctrl** key. The Mate pop-up menu is displayed.

739) Click **Coincident** from the Mate pop-up menu. Coincident3 is created.

Clear the Face filter.

740) **Right-click** in the Graphics window.

741) Click the **Selection Filters** drop-down arrow icon.

742) Click **Clear All Filters** from the Selection Filter toolbar.

Display the Mate types.

743) **Expand** the Mates folder in the FeatureManager. View the inserted mates.

Display a Right view in the Wireframe mode.

744) Click **Right view** .

745) Click **Wireframe** from the Heads-up View toolbar.

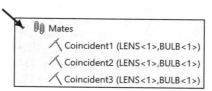

Display an Isometric view with Shaded With Edges.

746) Click **Isometric view** from the Heads-up View toolbar.

747) Click **Shaded With Edges** from the Heads-up View toolbar.

Save the LENSANDBULB.

748) Click **Save** .

749) **View** the results in the Graphics window.

💡 If the wrong face or edge is selected, click the face or edge again to remove it from the Mate Selections text box. Right-click Clear Selections to remove all geometry from the Mate Selections text box. To delete a mate from the FeatureManager, right-click on the mate, click Delete.

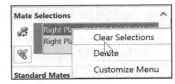

BATTERYANDPLATE Sub-assembly

Create the BATTERYANDPLATE sub-assembly. Utilize two Coincident mates and one Concentric mate to assemble the BATTERYPLATE component to the BATTERY component.

💡 Use the Pack and Go option to save an assembly or drawing with references, decals, studies, etc. The Pack and Go tool saves either to a folder or creates a zip file to e-mail. View SOLIDWORKS help for additional information.

Activity: BATTERYANDPLATE Sub-assembly.

Create the BATTERYANDPLATE sub-assembly.

750) Click **New** ⬜ from the Menu bar.

751) Click the **MY-TEMPLATES** tab.

752) Double-click **ASM-IN-ANSI**.

Insert the BATTERY part.

753) Click **Browse**.

754) Double-click **BATTERY** from the ENGDESIGN-W-SOLIDWORKS\PROJECTS folder.

755) Click **OK** ✔ from the Begin Assembly PropertyManager. The BATTERY is fixed to the Origin.

Save the BATTERYANDPLATE sub-assembly. Enter name and description. Note: Tangent edges are displayed in this example.

756) Click **Save As** 🖫 .

757) Select the **PROJECTS** folder.

758) Enter **BATTERYANDPLATE** for File name.

759) Enter **BATTERY AND PLATE FOR 6-VOLT FLASHLIGHT** for Description.

760) Click **Save**. The BATTERYANDPLATE FeatureManager is displayed.

Insert the BATTERYPLATE part.

761) Click **Insert Components** from the Assembly toolbar.

762) Click **Browse**. Double-click **BATTERYPLATE** from the PROJECTS folder.

763) Click a **position** above the BATTERY as illustrated.

Insert a Coincident mate.

764) Click the **outside bottom face** of the BATTERYPLATE as illustrated.

765) **Rotate** the model to view the top narrow flat face of the BATTERY Base Extrude feature as illustrated.

766) Hold the **Ctrl** key down.

767) Click the **top narrow flat face** of the BATTERY Base Extrude feature.

768) Release the **Ctrl** key. The Mate pop-up menu is displayed.

769) Click **Coincident** from the Mate pop-up menu. Coincident1 is created.

Insert a Coincident mate.

770) Click **Right Plane** of the BATTERY from the FeatureManager.

771) Hold the **Ctrl** key down.

772) Click **Right Plane** of the BATTERYPLATE from the FeatureManager.

773) Release the **Ctrl** key. The Mate pop-up menu is displayed. Click **Coincident** from the Mate pop-up menu. Coincident2 is created.

Insert a Concentric mate.

774) Click the center Terminal feature **cylindrical face** of the BATTERY as illustrated.

775) Hold the **Ctrl** key down.

776) Click the Holder feature **cylindrical face** of the BATTERYPLATE.

777) Release the **Ctrl** key. The Mate pop-up menu is displayed.

778) Click **Concentric** from the Mate pop-up menu. Concentric1 is created.

779) **Expand** the Mates folder. View the created mates.

Display an Isometric view. Save the BATTERYANDPLATE.

780) Click **Isometric view** 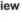.

781) Click **Save** .

CAPANDLENS Sub-assembly

Create the CAPANDLENS sub-assembly. Utilize two Coincident mates and one Distance mate to assemble the O-RING to the LENSCAP. Utilize three Coincident mates to assemble the LENSANDBULB sub-assembly to the LENSCAP component.

Caution: Select the correct reference. Expand the LENSCAP and O-RING. Click the Right Plane within the LENSCAP. Click the Right plane within the O-RING.

Activity: CAPANDLENS Sub-assembly.

Create the CAPANDLENS sub-assembly.

782) Click **New** from the Menu bar.

783) Click the **MY-TEMPLATES** tab.

784) Double-click **ASM-IN-ANSI**.

Insert the LENSCAP.

785) Click **Browse**.

786) Double-click **LENSCAP** from the ENGDESIGN-W-SOLIDWORKS\PROJECTS folder.

787) Click **OK** from the Begin Assembly PropertyManager. The LENSCAP is fixed to the Origin.

Save the CAPANDLENS assembly. Enter name and description.

788) Click **Save As** .

789) Select the **ENGDESIGN-W-SOLIDWORKS\PROJECTS** folder.

790) Enter **CAPANDLENS** for File name.

791) Enter **LENSCAP AND LENS** for Description.

792) Click **Save**. The CAPANDLENS FeatureManager is displayed.

Insert the O-RING part.

793) Click **Insert Components** from the Assembly toolbar.

794) Click **Browse**.

795) Double-click **O-RING** from the ENGDESIGN-W-SOLIDWORKS\PROJECTS folder.

796) Click a **position** behind the LENSCAP as illustrated.

Insert the LENSANDBULB assembly.

797) Click **Insert Components** from the Assembly toolbar.

798) Click **Browse**.

799) Double-click **LENSANDBULB** from the ENGDESIGN-W-SOLIDWORKS\PROJECTS folder.

800) Click a **position** behind the O-RING as illustrated.

801) Click **Isometric view** .

Move and hide components.

802) Click and drag the **O-RING** and **LENSANDBULB** as illustrated in the Graphics window.

803) Right-click **LENSANDBULB** in the FeatureManager.

804) Click **Hide Components** from the Context toolbar.

Insert three mates between the LENSCAP and O-RING.

805) Click **Right Plane** of the LENSCAP from the FeatureManager.

806) Hold the **Ctrl** key down.

807) Click **Right Plane** of the O-RING from the FeatureManager.

808) Release the **Ctrl** key.

809) Click **Coincident** from the Mate pop-up menu. Coincident1 is created.

Insert a second Coincident mate.

810) Click **Top Plane** of the LENSCAP from the FeatureManager.

811) Hold the **Ctrl** key down.

812) Click **Top Plane** of the O-RING from the FeatureManager.

813) Release the **Ctrl** key. The Mate pop-up menu is displayed.

814) Click **Coincident** from the Mate pop-up menu. Coincident2 is created.

Insert a Distance mate. Pin the Mate PropertyManager.

815) Click the **Mate** Mate Assembly tool. The Mate PropertyManager is displayed.

816) Click the **Standard** tab.

817) If needed, click the **Keep Visible** 📌 icon.

818) Click the **Shell1 back inside face** of the LENSCAP as illustrated.

819) Click **Front Plane** of the O-RING in the fly-out FeatureManager.

820) Click **Distance**.

821) Enter **.125/2**in, [**3.175/2mm**].

822) Click **OK** ✔ from the Distance Mate PropertyManager.

823) Click **OK** ✔ from the Mate PropertyManager.

Display an Isometric view. View the created mates.

824) Click **Isometric view** 🔲 from the Heads-up View toolbar.

825) Expand the Mates folder. View the created mates.

Save the model.

826) Click **Save** 💾.

How is the Distance mate, .0625in, [1.588] calculated? Answer:

O-RING Radius (.1250in/2) = .0625in.

O-RING Radius [3.175mm/2] = [1.588mm].

🔆 Utilize the Section view 📖 tool from the Heads-up View toolbar to locate internal geometry for mating and verify position of components.

🔆 Build flexibility into the mate. A Distance mate offers additional flexibility over a Coincident mate. You can modify the value of a Distance mate.

Show the LENSANDBULB.

827) Right-click **LENSANDBULB** in the FeatureManager.

828) Click **Show Components** from the Contexts toolbar.

Fit the model to the Graphics window. Create the first mate.

829) Press the **f** key.

Insert a Coincident mate.

830) Click **Right Plane** of the LENSCAP from the FeatureManager.

831) Hold the **Ctrl** key down.

832) Click **Right Plane** of the LENSANDBULB from the FeatureManager.

833) Release the **Ctrl** key. The Mate pop-up menu is displayed.

834) Click **Coincident** from the Mate pop-up menu.

Insert a Coincident Mate.

835) Click **Top Plane** of the LENSCAP from the FeatureManager.

836) Hold the **Ctrl** key down.

837) Click **Top Plane** of the LENSANDBULB from the FeatureManager.

838) Release the **Ctrl** key.

839) Click **Coincident** from the Mate pop-up menu.

Insert a Coincident Mate.

840) Click the flat inside **narrow back face** of the LENSCAP as illustrated.

841) **Rotate** the model to view the front flat face of the LENSANDBULB.

842) Hold the **Ctrl** key down.

843) Click the **front flat face** of the LENSANDBULB.

844) Release the **Ctrl** key.

845) Click **Coincident** from the Mate pop-up menu.

View the created mates.

846) **Expand** the Mates folder. View the created mates.

Confirm the location of the O-RING.

847) Click **Right Plane** of the CAPANDLENS from the FeatureManager.

848) Click **Section view** 📖 from the Heads-up View toolbar.

849) Click **Isometric view** 📦 from the Heads-up View toolbar.

850) **Expand** the Section 2 box.

851) Click **inside** the Reference Section Plane box.

852) Click **Top Plane** of the CAPANDLENS from the fly-out FeatureManager.

853) Click **OK** ✔ from the Section View PropertyManager.

Return to a full view. Save the CAPANDLENS sub-assembly.

854) Click **Section view** 📖 from the Heads-up View toolbar.

855) Click **Save** 💾 .

The LENSANDBULB, BATTERYANDPLATE and CAPANDLENS sub-assemblies are complete. The components in each assembly are fully defined. Insert the sub-assemblies into the final FLASHLIGHT assembly.

FLASHLIGHT Assembly

Create the FLASHLIGHT assembly. The HOUSING is the Base component. The FLASHLIGHT assembly mates the HOUSING to the SWITCH component. The FLASHLIGHT assembly mates the CAPANDLENS and BATTERYANDPLATE.

> **Activity: FLASHLIGHT Assembly**.

Create the FLASHLIGHT assembly.

856) Click **New** 🗋 from the Menu bar.

857) Click the **MY-TEMPLATES** tab.

858) Double-click **ASM-IN-ANSI**.

Insert the HOUSING and SWITCH.

859) Click **Browse**.

860) Double-click **HOUSING** from the ENGDESIGN-W-SOLIDWORKS\PROJECTS folder.

861) Click **OK** ✔ from the Begin Assembly PropertyManager. The HOUSING is fixed to the Origin.

862) Click **Insert Components** from the Assembly toolbar.

863) Click **Browse**. Double-click **SWITCH** from the PROJECTS folder.

864) Click a **position** in front of the HOUSING as illustrated.

Save the FLASHLIGHT assembly. Enter name and description.

865) Click **Save As** .

866) Select the **ENGDESIGN-W-SOLIDWORKS\PROJECTS** folder.

867) Enter **FLASHLIGHT** for File name.

868) Enter **FLASHLIGHT ASSEMBLY** for Description.

869) Click **Save**. The FLASHLIGHT FeatureManager is displayed.

Pin the Mate PropertyManager. Insert a Coincident mate.

870) Click the **Mate** Mate Assembly tool.

871) Click the **Standard** tab. Click the **Keep Visible** icon.

872) Click **Right Plane** of the HOUSING from the fly-out FeatureManager.

873) Click **Right Plane** of the SWITCH from the fly-out FeatureManager. Coincident is selected by default.

874) Click **OK** from the Coincident1 Mate PropertyManager.

Insert a second Coincident mate.

875) Click **View**, **Hide/Show,** check **Temporary Axes** from the Menu bar.

876) Click the **Temporary axis** inside the Switch Hole of the HOUSING.

877) Click **Front Plane** of the SWITCH from the fly-out FeatureManager. Coincident is selected by default.

878) Click **OK** from the Coincident2 Mate PropertyManager.

Insert a Distance mate.

879) Click the **Vertex** on the Loft top face of the SWITCH.

880) Click the **top face** of the Handle.

881) Click **Distance**.

882) Enter **.100**in, [**2.54**].

883) Check the **Flip Direction** box if needed.

884) Click **OK** ✔ from the Distance Mate PropertyManager.

885) Un-pin the Mate PropertyManager. Click the **Keep Visible** 📌 icon.

886) Click **OK** ✔ from the Mate PropertyManager.

Insert the CAPANDLENS assembly.

887) Click **View**, **Hide/Show**, un-check **Temporary Axis** from the Menu bar.

888) Click **Insert Components** 📂 from the Assembly toolbar.

889) Click **Browse**.

890) Double-click **CAPANDLENS** from the PROJECTS folder.

Place the sub-assembly.

891) Click a **position** in front of the HOUSING as illustrated.

Insert mates between the HOUSING component and the CAPANDLENS sub-assembly. Insert a Coincident mate.

892) Click **Right Plane** of the HOUSING from the FeatureManager.

893) Hold the **Ctrl** key down.

894) Click **Right Plane** of the CAPANDLENS from the FeatureManager.

895) Release the **Ctrl** key. The Mate pop-up menu is displayed.

896) Click **Coincident** from the Mate pop-up menu.

Insert a second Coincident mate.

897) Click **Top Plane** of the HOUSING from the FeatureManager.

898) Hold the **Ctrl** key down.

899) Click **Top Plane** of the CAPANDLENS from the FeatureManager.

900) Release the **Ctrl** key.

901) Click **Coincident** from the Mate pop-up menu.

Insert a Coincident mate.
902) Click the **front face** of the Boss-Stop on the HOUSING.

Rotate the view.
903) **Rotate** the model to view the back face.

904) Hold the **Ctrl** key down.

905) Click the **back face** of the CAPANDLENS.

906) Release the **Ctrl** key.

907) Click **Coincident** from the Mate pop-up menu.

Display an Isometric view. Save the FLASHLIGHT assembly.

908) Click **Isometric view** .

909) Click **Save** .

Insert the BATTERYANDPLATE sub-assembly.

910) Click **Insert Components** from the Assembly toolbar.

911) Click **Browse**.

912) Double-click **BATTERYANDPLATE** from the PROJECTS folder.

913) Click a **position** as illustrated.

Rotate the part.
914) Click **BATTERYANDPLATE** in the FeatureManager.

915) Click **Rotate Component** from the Assembly toolbar.

916) Rotate the **BATTERYANDPLATE** until it is approximately parallel with the HOUSING.

917) Click **OK** from the Rotate Component PropertyManager.

Insert a Coincident mate.
918) Click **Right Plane** of the HOUSING from the FeatureManager.

919) Hold the **Ctrl** key down.

920) Click **Front Plane** of the BATTERYANDPLATE from the FeatureManager.

921) Release the **Ctrl** key.

922) Click **Coincident** from the Mate pop-up menu.

923) Move the **BATTERYANDPLATE** in front of the HOUSING.

Insert a Coincident mate.

924) Click **Top Plane** of the HOUSING from the FeatureManager.

925) Hold the **Ctrl** key down.

926) Click **Right Plane** of the BATTERYANDPLATE from the FeatureManager.

927) Release the **Ctrl** key. The Mate pop-up menu is displayed.

928) Click **Coincident** from the Mate pop-up menu.

Display the Section view.

929) Click **Right Plane** in the FLASHLIGHT Assembly FeatureManager.

930) Click **Section view** 🗐 from the Heads-up View toolbar.

931) Click **OK** ✔ from the Section View PropertyManager.

Move the BATTERYANDPLATE in front of the HOUSING.

932) Click and drag the **BATTERYANDPLATE** in front of the HOUSING.

Insert a Coincident mate.

933) Click the **back center Rib1 face** of the HOUSING.

934) **Rotate** the model to view the bottom face of the BATTERYANDPLATE.

935) Hold the **Ctrl** key down.

936) Click the **bottom face** of the BATTERYANDPLATE.

937) Release the **Ctrl** key.

938) Click **Coincident** from the Mate pop-up menu.

939) Click **Isometric view** 🔲 from the Heads-up View toolbar.

Display the Full view.

940) Click **Section view** from the Heads-up View toolbar.

Save the FLASHLIGHT assembly.

941) Click **Save** .

 Review of the FLASHLIGHT Assembly.

The FLASHLIGHT assembly consisted of the HOUSING part, SWITCH part, CAPANDLENS sub-assembly and the BATTERYANDPLATE sub-assembly.

The CAPANDLENS sub-assembly contained the BULBANDLENS sub-assembly, the O-RING and the LENSCAP part. The BATTERYANDPLATE sub-assembly contained the BATTERY and BATTERYPLATE part.

You inserted eight Coincident mates and a Distance mate. Through the Assembly Layout illustration, you simplified the number of components into a series of smaller assemblies. You also enhanced your modeling techniques and skills.

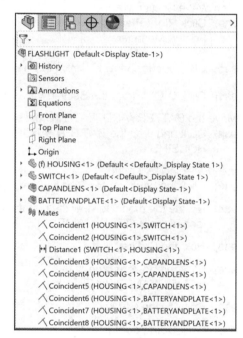

You still have a few more areas to address. One of the biggest design issues in assembly modeling is interference. Let's investigate the FLASHLIGHT assembly.

Clearance Verification checks the minimum distance between components and reports any value that fails to meet your input value of the minimum clearance. View Help for additional information.

Addressing Interference Issues

There is an interference issue between the FLASHLIGHT components. Address the design issue. Adjust Rib2 on the HOUSING. Test with the Interference Check command. The FLASHLIGHT assembly is illustrated in inches.

Activity: Addressing Interference Issues.

Check for interference.

942) Click the **Interference Detection** 🔳 tool from the Evaluate tab in the CommandManager.

943) Delete **FLASHLIGHT.SLDASM** from the Selected Components box.

944) Click **BATTERYANDPLATE** from the fly-out FeatureManager.

945) Click **HOUSING** from the fly-out FeatureManager.

946) Click the **Calculate** button. The interference is displayed in red in the Graphics window.

947) Click each **Interference** in the Results box to view the interference in red with Rib2 of the HOUSING.

948) Click **OK** ✔ from the Interference Detection PropertyManager.

Modify the Rib2 dimension to address the interference issue.
949) **Open** the HOUSING in the FeatureManager.

950) Double-click on the **Rib2** feature.

951) Double click **1.300**in, [**33.02**].

952) Enter **1.350**in, [**34.29**].

953) **Rebuild** 🔳 the model.

954) **Return** to the FLASHLIGHT assembly.

Recheck for Interference.

955) Click **Interference Detection** 🔳 from the Evaluate tab. The Interference dialog box is displayed.

956) Delete **FLASHLIGHT.SLDASM** from the Selected Components box.

957) Click **BATTERYANDPLATE** from the FeatureManager.

958) Click **HOUSING** from the FeatureManager.

959) Click the **Calculate** button. No Interference is displayed in the Results box.

960) Click **OK** ✔ from the Interference Detection PropertyManager.

Save the model. The FLASHLIGHT assembly is finished.

961) Click **Save** .

962) Click **Save All**.

Export Files and eDrawings

You receive a call from the sales department. They inform you that the customer increased the initial order by 200,000 units. However, the customer requires a prototype to verify the design in six days. What do you do? Answer: Contact a Rapid Prototype supplier. You export three SOLIDWORKS files:

- HOUSING

- LENSCAP

- BATTERYPLATE

Use the Stereo Lithography (STL) format. Email the three files to a Rapid Prototype supplier. A Stereolithography (SLA) supplier provides physical models from 3D drawings. 2D drawings are not required. Export the HOUSING. SOLIDWORKS eDrawings provides a facility for you to animate, view and create compressed documents to send to colleagues, customers and vendors. Publish an eDrawing of the FLASHLIGHT assembly.

Activity: Export Files and eDrawings.

Open and Export the HOUSING.

963) Right-click **HOUSING** from the FeatureManager.

964) Click **Open Part** from the Context toolbar.

965) Click **Save As** .

966) Click **Save as copy and open**.

967) Select **STL (*.stl)** from the Save as type drop-down menu.

968) Click the **Options** button.

969) Click the **Binary** box from the Output as format box.

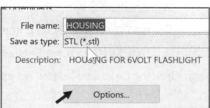

File name:	HOUSING
Save as type:	STL (*.stl)
Description:	HOUSING FOR 6VOLT FLASHLIGHT

Options...

970) Click the **Fine** box for Resolution.

Create the binary STL file.
971) Click **OK** from the Export Options dialog box.

972) Click **Save** from the Save dialog box.

973) Click **Yes**.

Publish an eDrawing and email the document to a colleague.

Create the eDrawing and animation.
974) Click **File**, **Publish to eDrawings** from the Menu bar.

975) Click **OK**.

976) Click the **Run Animation** button. Click **Play**. View the results.

Stop the animation.

977) Click the **Stop** button.

978) Click the **Reset** button to return to the original position.

Save the eDrawing.

979) Click **File**, **Save As** from the eDrawing Main menu.

980) Select the **ENGDESIGN-W-SOLIDWORKS\PROJECTS** folder.

981) Enter **HOUSING** for File name.

982) Click **Save**.

983) **Close** ✖ the eDrawing dialog box.

984) **Close** all models in the session.

It is time to go home. The telephone rings. The customer is ready to place the order. Tomorrow you will receive the purchase order.

The customer also discusses a new purchase order that requires a major design change to the handle. You work with your industrial designer and discuss the options. One option is to utilize Guide Curves on a Sweep feature.

Guide Curves   Deform

You contact your mold maker and send an eDrawing of the LENSCAP.

The mold maker recommends placing the parting line at the edge of the Revolved Cut surface and reversing the Draft Angle direction.

The mold maker also recommends a snap fit versus a thread to reduce cost. The Core-Cavity mold tooling is explored in the project exercises.

Parting Line

Draft Angle

Parting Line

Project Summary

The FLASHLIGHT assembly contains over 100 features, reference planes, sketches and components. You organized the features in each part. You developed an assembly layout structure to organize your components.

The O-RING utilized a Swept Base feature. The SWITCH utilized a Loft Base feature. The simple Swept feature requires two sketches: a path and a profile. A complex Swept requires multiple sketches and guide curves. The Loft feature requires two or more sketches created on different planes.

The LENSCAP and HOUSING utilized a variety of features. You applied design intent to reuse geometry through Geometric relationships, symmetry and patterns.

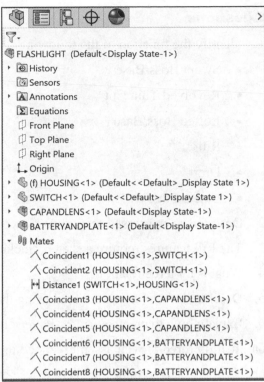

The assembly required an Assembly Template. You utilized the ASM-IN-ANSI Template to create the LENSANDBULB, CAPANDLENS, BATTERYANDPLATE and FLASHLIGHT assemblies.

You created an STL file of the Housing and an eDrawing of the Housing to communicate with your vendor, mold maker and customer. Review the project exercises before moving on to the next project.

Questions

1. Identify the function of the following features:

    - Swept Boss/Base

    - Revolved Thin Cut

    - Lofted Boss/Base

    - Rib

    - Draft

    - Linear Pattern

2. Describe a Suppressed feature. Why would you suppress a feature?

3. The Rib features require a sketch, thickness and a _____ direction.

4. What is a Pierce relation?

5. Describe how to create a thread using the Swept feature. Provide an example.

6. Explain how to create a Linear Pattern feature. Provide an example.

6a. Identify three proven assembly modeling techniques.

6b. How do you determine the Interference between components in an assembly?

7. Identify two advantages of utilizing Convert Entities in a sketch to obtain the profile.

8. How is symmetry built into a:

    - Sketch. Provide an example.

    - Feature. Provide an example.

9. Define a Guide Curve. Identify the features that utilize Guide Curves.

10. Identify the differences between a Draft feature and the Draft Angle option in the Extruded Boss/Base feature.

11. Describe the differences between a Circular Pattern feature and a Linear Pattern feature.

12. Identify the advantages of the Convert Entities tool.

13. True or False. A Lofted feature can only be inserted as the first feature in a part. Explain your answer.

14. A Swept feature adds and removes material. Identify the location on the Main Pull down menu that contains the Swept Cut feature. Hint: SOLIDWORKS Help Topics.

15. Describe the difference between a Distance Mate and a Coincident Mate.

Exercises

Exercise 6.1:

Create a Rib feature. Add a rib to increase structural integrity. Apply the Parallel to Sketch option with a draft.

1. Open **Rib 6-1** from the Chapter 6 Homework folder.

2. Right-click **Top Plane** from the FeatureManager. This is your Sketch plane.

3. Click **Sketch** from the Context toolbar.

4. Click the **Line** Sketch tool.

5. Sketch a **horizontal line** as illustrated on the model. The endpoints are located on either side of the Handle as illustrated.

Add a dimension.

6. Click the **Smart Dimension** Sketch tool. The Dimension PropertyManager is displayed.

7. **Dimension** the model as illustrated.

Insert a Rib feature.

8. Click the **Rib** Features tool. The Rib PropertyManager is displayed.

9. Click the **Both Sides** button.

10. Enter **.100**in for Rib Thickness.

11. Click the **Parallel to Sketch** button. The Rib direction arrow points to the back. If needed, click the **Flip material side** check box.

12. Click the **Draft On/Off** button.

13. Enter **1**deg for Draft Angle.

14. Display a **Front view**.

15. Click the **back inside face** of the HOUSING for the Body. The selected entity is displayed.

16. Click **OK** ✔ from the Rib PropertyManager.

17. Display an **Isometric view**. Rib1 is created and is displayed in the FeatureManager.

18. **View** the Rib in the Graphics window.

19. **Close** the model.

Exercise 6.2:

Create a Rib feature. Apply the Parallel to Sketch option with a draft.

1. Open **Rib 6-2** from the Chapter 6 Homework folder.

2. **Edit** Sketch9 from the FeatureManager.

Activate the Rib feature.

3. Click the **Rib** 🍃 Features tool. The Rib PropertyManager is displayed.

4. Click the **Both Sides** button.

5. Enter **.100**in for Rib Thickness.

6. Click the **Parallel to Sketch** button. The Rib direction arrow points downwards.

7. Check the **Flip material side** box if required. The direction arrow points downwards.

8. Click the **Draft On/Off** button.

9. Enter **2**deg for Draft Angle.

10. Click **OK** ✔ from the Rib PropertyManager. Rib1 is displayed.

11. **Close** the model.

Exercise 6.3:

Create a Rib feature with the Next Reference option.

1. Open **Rib 6-3** from the Chapter 6 Homework folder.

2. **Edit** Sketch9 from the FeatureManager.

Activate the Rib feature.

3. Click the **Rib** Features tool. The Rib PropertyManager is displayed.

4. Click the **Both Sides** button.

5. Enter **.150**in for Rib Thickness.

6. Click the **Parallel to Sketch** button. The Rib direction arrow points to the back.

7. Click the **Flip material side** box. The direction arrow points downwards.

8. Click the **Draft On/Off** button.

9. Enter **3**deg for Draft Angle.

10. Click the **Next Reference** button until you select the contour as illustrated. You can only select one reference contour on which to base your draft.

11. Click **OK** from the Rib PropertyManager. Rib1 is created and is displayed.

12. **Close** the model.

Exercise 6.4:

Create a Linear Pattern feature using the Pattern seed only option and the Geometry Pattern option.

1. Open **Linear Pattern 6-4** from the Chapter 6 Homework folder.

2. Click the **Linear Pattern** Features tool.

3. Expand the **Features and Faces** option box.

4. Click inside the **Pattern Direction** box for Features and Faces.

5. Click **Rib1** from the fly-out FeatureManager. Rib1 is displayed in the Features to Pattern box.

6. Click inside the **Pattern Direction** box for Direction 1.

7. Click the **hidden upper back vertical edge** of Shell1. Edge<1> is displayed in the Pattern Direction box. The direction arrow points upward. Click the Reverse direction button if required.

8. Click **Spacing and instances**.

9. Enter **.500**in for Spacing.

10. Enter **3** for Number of Instances.

11. Expand the **Direction 2** option box.

12. Click the **hidden lower back vertical edge** of Shell1. The direction arrow points downward. Click the Reverse direction button if required.

13. Enter **.500**in for Spacing. Enter **3** for Number of Instances.

14. Check the **Pattern seed only** box.

15. Check the **Geometry Pattern** box in the Options box.

Utilize the Geometry Pattern option to efficiently create and rebuild patterns.

16. Click **OK** from the Linear Pattern PropertyManager. LPattern1 is created and is displayed in the FeatureManager.

17. **Close** the model.

Exercise 6-5:

Create a Linear Pattern feature using the Instances to skip option.

1. Open **Linear Pattern 6-5** from the Chapter 6 Homework folder.

2. Click the **Linear Pattern** ⊞⊞ Features tool.

3. Expand the **Features and Faces** option box.

4. Click inside the **Pattern Direction** box for Features and Faces.

5. Click **Cut-Extrude1** from the fly-out FeatureManager.

6. Click inside the **Pattern Direction** box (Direction 1).

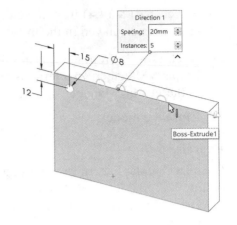

7. Click the **top horizontal edge** of Boss-Extrude1. The direction arrow points to the right. Click the Reverse direction button if required.

8. Click **Spacing and instances**.

9. Enter **20**mm for Spacing.

10. Enter **5** for Number of Instances.

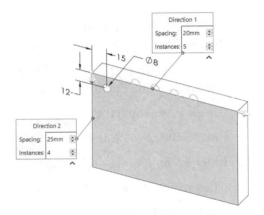

11. Expand the **Direction 2** option box.

12. Click inside the **Pattern Direction** box for Direction 2.

13. Click the **left vertical edge** of Boss-Extrude1. Edge<2> is displayed in the Pattern Direction box.

14. Click **Spacing and instances**.

15. Enter **25**mm for Spacing.

16. Enter **4** for Number of Instances.

17. Check the **Geometry pattern** box in the Options dialog box.

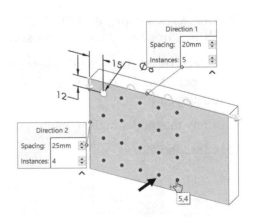

18. Expand the **Instances to Skip** box.

19. Select (**5,4**) & (**4,4**) from the Graphics window. (5,4) & (4,4) is displayed in the Instances to Skip box.

20. Click **OK** ✔ from the Linear Pattern PropertyManager. LPattern1 is created and is displayed in the FeatureManager.

21. **Close** the model.

Exercise 6-6:

Create a Circular Pattern feature using the Faces to Pattern option.

1. Open **Circular Pattern 6-6** from the Chapter 6 Homework folder.

2. Click the **Circular Pattern** 🔧 Features tool. The Circular Pattern PropertyManager is displayed.

3. Click **Axis1** from the fly-out FeatureManager. Axis1 is displayed in the Pattern Axis box.

4. Check **Equal spacing**.

5. Enter **360**deg for Angle.

6. Enter **32** for Number of Instances.

7. Click inside the **Faces to Pattern** box.

8. Click the **9 faces** of the Gear tooth in the Graphics window as illustrated.

9. Check the **Geometry pattern** box in the Options section.

10. Check the **Propagate visual Properties** box in the Options section.

11. Click **OK** ✔ from the Circular Pattern PropertyManager. CirPattern1 is displayed.

12. **View** the model in the Graphics window.

13. **Close** the model.

Exercise 6-7:

Create an O-ring part. Create the part with the Swept Base feature. The Swept Base feature requires a sketch path and a circular profile diameter.

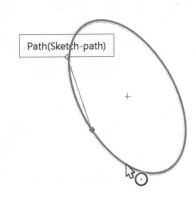

Path(Sketch-path)

1. Open **Swept Base 6-7** from the Chapter 6 Homework folder. The FeatureManager is displayed.

2. Click the **Swept Boss/Base** Features tool. The Sweep PropertyManager is displayed.

3. Click the **Circular Profile** box.

4. Enter .**125**mm for Diameter.

5. Click the **Circular Profile** in the Graphics window. Sketch-path is displayed in the path (Sketch-path) box.

6. Click **OK** ✔ from the Sweep PropertyManager. Sweep1 is created and is displayed in the FeatureManager.

7. **Expand** Sweep1. View the results.

8. **Close** the model.

🔆 For non-circular sketch profiles, create the sketch profile on a perpendicular plane to the path and use the pierce relation to locate the profile on the path.

Use the Circular Profile option to create a solid rod or hollow tube along a sketch line, edge, or curve directly on a model without having to sketch. This enhancement is available for Swept Boss/Base, Swept Cut, and Swept Surface features.

🔆 The PropertyManager provides the ability to select a configuration after the Sweep feature is created.

Exercise 6-8:

Create a Swept Base feature. Create a simple spring (six turns).

1. Open **Swept Base 6-8** from the Chapter 6 Homework folder. The FeatureManager is displayed. View two simple sketches on the Front Plane.

2. Click the **circumference** of the circle in the Graphics window, Sketch2.

3. Click the **Swept Boss/Base** Features tool. The Sweep PropertyManager is displayed. Sketch2 is displayed in the Profile box.

4. Click inside the **Path** box.

5. Click **Sketch1** from the fly-out FeatureManager. Sketch1 is displayed in the Path box.

6. Click the **drop-down arrow** to expand the Options box.

7. Select **Follow Path** for Profile orientation. View your options.

8. Select **Specify Twist Value** for Profile Twist. View your options.

9. Select **Degrees** for Twist control.

10. Enter **2160** for Twist angle. View your options.

11. Click **OK** ✔ from the Sweep PropertyManager. Sweep1 is created and is displayed in the FeatureManager.

12. Display an **Isometric** view.

13. **Close** the model.

Exercise 6-9:

Create a Swept Boss feature using the Helix/Spiral PropertyManager. Create a variable pitch helix for a spring.

1. Open **Swept Boss 6-9** from the Chapter 6 Homework folder. The FeatureManager is displayed.

2. Edit **Sketch1**.

3. Click **Insert, Curve, Helix/Spiral** from the Menu bar menu. The Helix/Spiral PropertyManager is displayed. The default direction of rotation is clockwise. Pitch and Revolution Type is selected by default.

4. Check the **Variable pitch** box.

5. Create a **Region parameters table**. Create a spring. Enter the following information as illustrated. Coils 1, 2, 5, 6 & 7 are the closed ends of the spring. The pitch needs to be slightly larger than the wire.

6. Enter **.021**in for the Pitch.

7. Enter Start angle of **0**deg.

8. Click **Clockwise**.

9. Click **OK** ✔ from the Helix/Spiral PropertyManager.

10. Display an **Isometric** view. View the results.

Create the Swept Boss feature.

11. Click the **Swept Boss/Base** 🐎 Features tool. The Sweep PropertyManager is displayed.

12. Click **Circular Profile**.

13. Click **Helix/Spiral1** from the Graphics window as illustrated.

14. Enter **.015in** for Depth.

15. Click **OK** ✔ from the PropertyManager. View the results.

16. **Close** the model.

Notes:

Project 7

Top-Down Assembly Modeling and Sheet Metal Parts

Below are the desired outcomes and usage competencies based on the completion of Project 7.

Project Desired Outcomes:	Usage Competencies:

• Three BOX configurations:
 	
   ○ Small
 	
   ○ Medium
 	
   ○ Large
	• Understand three key methods for the Top-down assembly modeling. Develop components In-Context with InPlace Mates.

	• Ability to import parts using the Top-down assembly method.
	• Familiarity to develop a Sheet Metal Design Table and equations.

• CABINET drawing with two configurations:
 	
   ○ Default - 3D formed state
 	
   ○ Flat - 2D flatten state
	• Competence to convert a solid part into a Sheet Metal part and insert Sheet Metal features.

	• Knowledge of Sheet Metal configurations in the drawing: 3D Formed and 2D Flat Pattern states.
• BRACKET part	• Skill to insert Sheet Metal features: Base, Edge, Miter Flange, Hem and Flat Pattern.

Notes:

Project 7 - Top-Down Assembly Modeling and Sheet Metal Parts

Project Objective

Create three BOX configurations utilizing the Top-down assembly approach. The Top-down approach is a conceptual approach used to develop products from within the assembly. Major design requirements are translated into sub-assemblies or individual components and key relationships.

Model Sheet metal components in their 3D formed state. Manufacture the Sheet metal components in a flatten state. Control the formed state and flatten state through configurations.

Create the parameters and equations utilized to control the configurations in a Design Table.

On the completion of this project, you will be able to:

- Set Document Properties and System Options.

- Create a new Top-down assembly and Layout sketch.

- Choose the proper Sketch plane.

- Insert Sheet Metal bends to transform a solid part into a sheet metal part.

- Use the Linear Pattern feature to create holes.

- Insert IGES format PEM® self-clinching fasteners.

- Use the Component Pattern feature to maintain a relationship between the holes and the fasteners.

- Create Global Variables.

- Create and Edit Equations.

- Insert a Sheet Metal Library feature and Forming tool feature.

- Address and understand general sheet metal manufacturing considerations.

- Insert Geometric relations.

- Insert Sheet Metal assembly features.

- Replace Components in an assembly.

- Insert a Design Table.

- Create a sheet metal part by starting with a solid part and inserting a Shell feature, Rip feature and Sheet Metal Bends feature.

- Create a Sheet Metal part by starting with a Base Flange feature.

- Use the following SOLIDWORKS features: Extruded Boss/Base, Extruded Cut, Shell, Hole Wizard, Linear Pattern, Sketch Driven Pattern, Pattern Driven Component Pattern, Rip, Insert Bends, Flatten, Hem, Edge Flange, Jog, Flat Pattern, Break Corner/Corner Trim, Miter Flange, Base Flange/Tab, Unfold/Fold, Mirror Components and Replace Component.

Project Situation

You now work for a different company. Life is full with opportunities. You are part of a global project design team that is required to create a family of electrical boxes for general industrial use. You are the project manager.

You receive a customer request from the Sales department for three different size electrical boxes.

Delivery time to the customer is a concern. You work in a concurrent engineering environment. Your company is expecting a sales order for 5,000 units in each requested BOX configuration.

The BOX contains the following key components:

- Power supply.

- Motherboard.

The size of the power supply is the same for all three boxes. The size of the power supply is the constant. The size of the BOX is a variable. The available space for the motherboard is dependent on the size of the BOX. The depth of the three boxes is 100mm.

You contact the customer to discuss and obtain design options and product specifications. Key customer requirements:

Three different BOX sizes:

- 300mm x 400mm x 100mm   Small.

- 400mm x 500mm x 100mm   Medium.

- 550mm x 600mm x100mm   Large.

- Adequate spacing between the power supply, motherboard and internal walls.

- Field serviceable.

You are responsible to produce a sketched layout from the provided critical dimensions. You are also required to design the three boxes. The BOX is used in an outside environment.

Top-Down Assembly Modeling

Top-down assembly is a conceptual method used to develop products from within the assembly. In Top-down assembly modeling, one or more features of the part are defined by something in the assembly. Example: A Layout sketch or the geometry of another part. The design intent of the model, for example the size of the features, location of the components in the assembly, etc. take place from the top level of the assembly and translate downward from the assembly to the component.

A few advantages of the Top-down assembly modeling are that design details of all components are not required and much less rework is required when a design change is needed. The model requires individual relationships between components. The parts know how to update themselves based on the way you created them.

Designers usually use Top-down assembly modeling to lay their assemblies out and to capture key design aspects of custom parts specific in the assemblies. There are three key methods to use for the Top-down assembly modeling approach.

- *Individual features method.* The Individual features method provides the ability to reference the various components and sub-components in an existing assembly. Example: Creating a structural brace in a box by using the Extruded Boss/Base feature tool. You might use the Up to Surface option for the End Condition and select the bottom of the box, which is a different part. The Individual features method maintains the correct support brace length, even if you modify the box in the future. The length of the structural brace is defined in the assembly. The length is not defined by a static dimension in the part. The Individual features method is useful for parts that are typically static but have various features which interface with other assembly components in the model.

- *Entire assembly method.* The Entire assembly method provides the ability to create an assembly from a layout sketch. The layout sketch defines the component locations, key dimensions, etc. A major advantage of designing an assembly using a layout sketch is that if you modify the layout sketch, the assembly and its related parts are automatically updated. The entire assembly method is useful when you create changes quickly, and in a single location.

 You can create an assembly and its components from a layout of sketch blocks.

- *Complete parts method.* The Complete parts method provides the ability to build your model by creating new components In-Context of the assembly. The component you build is actually mated to another existing component in the assembly. The geometry for the component you build is based upon the existing component.

This method is useful for parts like brackets and fixtures, which are mostly or completely dependent on other parts to define their shape and size.

 Whenever you create a part or feature using the Top-Down approach, external references are created to the geometry you referenced.

 The Top-down assembly approach is also referred to as "In-Context design."

Conceptual Top-down design approach

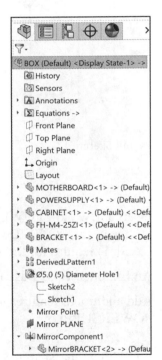

Use a combination of the three assembly methods to create the BOX assembly. Consider the following in a preliminary design product specification:

- What are the major design components?

- The motherboard and power supply are the major design components.

- What are the key design constraints?

The customer specified three different BOX sizes. How does each part relate to the other? From experience and discussions with the electrical engineering department, a 25mm physical gap is required between the power supply and the motherboard. A 20mm physical gap is required between the internal components and the side of the BOX.

How will the customer use the product? The customer does not disclose the specific usage of the BOX. The customer is in a very competitive market.

What is the most cost-effective material for the product? Aluminum is cost-effective, strong, relatively easy to fabricate, corrosion resistant, and non-magnetic.

Incorporate the design specifications into the BOX assembly. Use a sketch, solid parts and sheet metal parts. Obtain additional parts from your vendors.

BOX Assembly Overview

Create the sketch "layout" in the BOX assembly. Insert a new part, MOTHERBOARD. Convert edges from the sketch to develop the outline of the MOTHERBOARD. Use the outline sketch as the Extruded Base feature for the MOTHERBOARD component. An Extruded Boss locates a key electrical connector for a wire harness. Design the location of major electrical connections early in the assembly process.

POWERSUPPLY

Power Cable location

External Connector. Location for Wire Harness.

Layout Sketch

MOTHERBOARD

To PCB - Manufacturing

Insert a new part, POWERSUPPLY. Convert edges from the sketch to develop the outline of the POWERSUPPLY. Use the outline as the Extruded Boss/Base (Boss-Extude1) feature for the POWERSUPPLY component.

An Extruded Boss/Base feature represents the location of the power cable.

Add additional features to complete the POWERSUPPLY.

☀ Utilize the Search feature in the Systems Options dialog box to quickly locate information for System Options or Document Properties.

POWER SUPPLY:
Send the part to a colleague to add additional features.

Insert a new part, CABINET. Convert edges from the sketch.

Use the outline sketch as the Extruded Base feature for the CABINET component.

Shell the Extruded feature. Add a Rip feature to the four CABINET corners.

Insert sheet metal bends to transform a solid part into a sheet metal part.

Sheet metal components utilize features to create flanges, cuts and forms.

Save Model sheet metal components in their 3D formed state.

Manufacture sheet metal components in their 2D flatten state.

Work with your sheet metal manufacturer. Discuss cost-effective options.

Create a sketched pattern to add square cuts instead of formed louvers.

Create a Linear Pattern feature of holes for the CABINET.

2D flatten state

Formed State

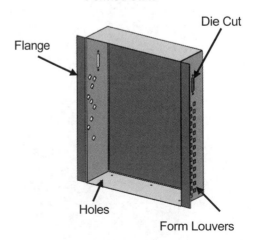

Flange

Die Cut

Holes

Form Louvers

Insert IGES format PEM® self-clinching fasteners. Use a Pattern Driven Component Pattern to maintain a relationship between the holes and the fasteners.

Utilize the Replace Component tool to modify the Ø5.0mm fastener to an Ø 4.0mm fastener.

Create equations to control the sketch and Linear Pattern feature variables in the BOX assembly.

Insert a Design Table to create BOX assembly configurations. Insert a sheet metal BRACKET In-Context of the BOX assembly.

Utilize the Mirror Components feature to create a right- and left-hand version of the BRACKET.

PEM Fastener

Linear step & repeat

InPlace Mates and In-Context Features

Top-down assembly modeling techniques develop InPlace mates. An InPlace mate is a Coincident mate between the Front Plane of the new component and the selected plane or selected face of the assembly.

The new component is fully positioned with an InPlace mate. No additional mates are required. The component is now dependent on the assembly.

An InPlace ✎ mate creates an External reference. The component references geometry in the assembly.

Mates
InPlace1 (MOTHERBOARD<1>,Fron
InPlace2 (POWERSUPPLY<1>,Front
InPlace3 (CABINET<2>,Front Plane)
Concentric1 (CABINET<2>,FH-M4-2
Coincident1 (CABINET<2>,FH-M4-2
Parallel1 (CABINET<2>,FH-M4-25Z
InPlace4 (BRACKET<1>,Front Plane
Concentric3 (FH-M4-25ZI<5>,Mirrc
Concentric4 (FH-M4-25ZI<4>,Mirrc
Coincident4 (CABINET<2>,MirrorBF

If the referenced document changes, the dependent document changes. The InPlace mates are listed under the Mates folder in the FeatureManager. Example:

In a Top-down assembly, you create In-Context features. An In-Context feature is a feature that references geometry from another component. The In-Context feature has an External reference to that component. If you change geometry on the referenced component, the associated In-Context feature also changes.

An In-Context feature is indicated in the FeatureManager with the ">" symbol. The External references exist with the feature, Sketch plane or sketch geometry. The update path to the referenced component is contained in the assembly. When the update path is not valid, or the referenced component is not found the "?" symbol is displayed in the FeatureManager.

In the BOX assembly, develop InPlace Mates for the MOTHERBOARD, POWERSUPPLY and CABINET components.

Create External references with the Convert Entities Sketch tool and extracting geometry from the sketch of the BOX assembly.

Working with In-Context features requires practice and time. Planning and selecting the correct reference and understanding how to incorporate changes are important. Explore various techniques using InPlace mates and External references developed In-Context of another component.

Assembly Modeling Techniques with InPlace Mates:
Plan the Top-down assembly method. Start from a sketch or with a component in the assembly.
Prepare the references. Utilize descriptive feature names for referenced features and sketches.
Utilize InPlace Mates sparingly. Load all related components into memory to propagate changes. Do not use InPlace Mates for purchased parts or hardware.
Group references. Select references from one component at a time.
Ask questions. Will the part be used again in a different assembly? If the answer is yes, do not use InPlace Mates. If the answer is no, use InPlace Mates.
Will the part be used in physical dynamics or multiple configurations? If the answer is yes, do not use InPlace mates.
Work in the Edit Component tool to obtain the required External references in the assembly. Create all non-referenced features in the part, not in the assembly.
Obtain knowledge of your company's policy on InPlace mates or develop one as part of an engineering standard.

Part Template and Assembly Template

The parts and assemblies in this project require Templates created in Project 2 and Project 3. The MY-TEMPLATES tab is located in the New SOLIDWORKS dialog box. If your MY-TEMPLATES folder contains the ASM-MM-ANSI and PART-MM-ANSI Templates, skip the next few steps.

Activity: Create an Assembly and Part Template.

Create the ASM-MM-ANSI Assembly Template for the BOX.

1) Click **New** from the Menu bar.

2) Double-click the **Assembly** icon from the defaults Templates tab. Note: The MY-TEMPLATES tab was created in Project 2.

3) Click **Cancel** from the Window open dialog box.

4) Click **Cancel** ✖ from the Begin Assembly PropertyManager. The Assembly FeatureManager is displayed.

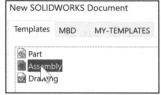

Set the Assembly Document Template options.

5) Click **Options** ⚙, **Document Properties** tab from the Menu bar. Select **ANSI** for the Overall drafting standard.

6) Click **Units**. Click **MMGS** for Units system.

7) Select **.12** (two decimal places) for Length Basic Units.

8) Select **None** for Angular units Decimal places. Click **OK** from the Document Properties - Units dialog box.

Save the Assembly template. Enter name.

9) Click **Save As** .

10) Click the **Assembly Templates (*.asmdot)** from the Save As type box.

11) Select **ENGDESIGN-W-SOLIDWORKS\MY-TEMPLATES** for Save in folder.

12) Enter **ASM-MM-ANSI** for File name. Click **Save**. The ASM-MM-ANSI FeatureManager is displayed.

Create the PART-MM-ANSI Part Template for the BOX.

13) Click **New** from the Menu bar. Double-click the **Part** icon from the default Templates tab. The Part FeatureManager is displayed.

14) Click **Options** ⚙, **Document Properties** tab from the Menu bar.

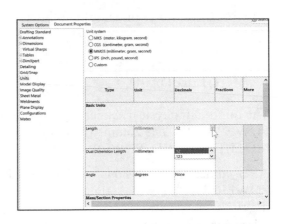

15) Select **ANSI** for the Overall drafting standard.

Set units and precision.
16) Click **Units**.

17) Click **MMGS** for Unit system.

18) Select **.12** (two decimal places) for Length Basic Units. Select **None** for Angular units Decimal places.

19) Click **OK** from the Document Properties - Units dialog box.

Save the Part Template. Enter name.

20) Click **Save As** 💾 .

21) Click **Part Templates (*.prtdot)** from the Save As type box. Select **ENGDESIGN-W-SOLIDWORKS\MY-TEMPLATES** for Save in folder.

22) Enter **PART-MM-ANSI** in the File name text box.

23) Click **Save**.

Close all active documents.
24) Click **Windows**, **Close All** from the Menu bar.

BOX Assembly and Sketch

The BOX assembly utilizes both the Top-down assembly design approach and the Bottom-up assembly design approach. Begin the BOX assembly with a sketch.

Create a New assembly named BOX. Insert a sketch to develop component space allocations and relations in the BOX assembly.

Add dimensions and relations to the sketch. Components and assemblies reference the sketch.

💡 In Layout-based assembly design, you can switch back and forth between Top-down and Bottom-up design methods.

The BOX assembly contains the following key components:

- POWERSUPPLY.

- MOTHERBOARD.

The minimum physical spatial gap between the MOTHERBOARD and the POWERSUPPLY is 25mm. The minimum physical spatial gap between the MOTHERBOARD, POWERSUPPLY and the internal sheet metal BOX wall is 20mm.

After numerous discussions with the electrical engineer, you standardize on a POWERSUPPLY size: 150mm x 75mm x 50mm. You know the overall dimensions for the BOX, POWERSUPPLY and MOTHERBOARD. You also know the dimensional relationship between these components.

Now you must build these parameters into the design intent of the sketch. There is no symmetry between the major components. Locate the sketch with respect to the BOX assembly Origin.

💡 FeatureManager tree folders will vary depending on the SOLIDWORKS system setup and Add-ins.

Activity: Box Assembly and Sketch Layout.

Create the BOX assembly.

25) Click **New** ⬜ from the Menu bar.

26) Click the **MY-TEMPLATES** tab.

27) Double-click **ASM-MM-ANSI**.

28) Click **Cancel** from the Window open dialog box.

29) Click **Cancel** ✖ from the Begin Assembly PropertyManager. The Assembly FeatureManager is displayed.

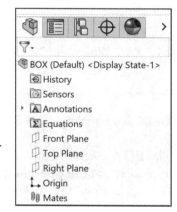

Save the BOX assembly. Enter name.

30) Click **Save As** 🖫 .

31) Select **ENGDESIGN-W-SOLIDWORKS\PROJECTS** for folder.

32) Enter **BOX** for File name. Click **Save**. The BOX Assembly FeatureManager is displayed.

Each part in the BOX assembly requires a template. Utilize the PART-MM-ANSI and ASM-MM-ANSI templates. Under the Option, System Options, Default Templates section prompts you to select a different document template. Otherwise, SOLIDWORKS utilizes the default templates.

Set System Options.

33) Click **Options** ⚙, **Default Templates** from the Menu bar.

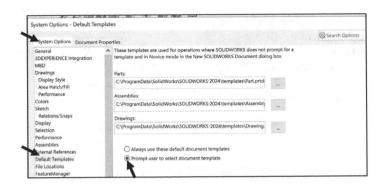

34) Click the **Prompt user to select document template** button.

35) Click **OK**.

The 2D sketch contains the 2D relationships between all major components in the BOX assembly.

Select the Sketch plane for the sketch.
36) Right-click **Front Plane** from the FeatureManager.

Sketch the profile of the BOX.
37) Click **Sketch** 🖊 from the Context toolbar. The Sketch toolbar is displayed. Click **Corner Rectangle** ⬜ from the Sketch toolbar.

38) Click the **Origin**. Sketch a **rectangle** as illustrated. The first point is coincident with the Origin.

Add dimensions.
39) Click **Smart Dimension** 📐 from the Sketch toolbar.

40) Click the **left vertical line**. Click a **position** to the left of the profile. Enter **400**mm.

41) Click the **bottom horizontal line**. Click a **position** below the profile. Enter **300**mm.

Fit the model to the Graphics window.
42) Press the **f** key.

Sketch the 2D profile for the POWER SUPPLY.
43) Click **Corner Rectangle** ⬜ from the Sketch toolbar.

44) Sketch a **small rectangle** inside the BOX as illustrated.

Add dimensions.
45) Click **Smart Dimension** 📐 from the Sketch toolbar.

46) Create a vertical dimension. Enter **150**mm.

47) Create a horizontal dimension. Enter **75**mm.

Dimension the POWERSUPPLY 20mm from the left and top edges of the BOX.

48) Create a horizontal dimension. Click the **BOX left vertical** edge. Click the **POWERSUPPLY left vertical** edge. Click a position **above** the BOX. Enter **20mm**.

49) Create a vertical dimension. Click the **BOX top** horizontal edge.

50) Click the **POWERSUPPLY top horizontal** edge.

51) Click a **position above** and to the right of the BOX. Enter **20**mm.

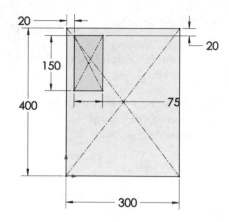

Sketch the profile for the MOTHERBOARD.

52) Click **Corner Rectangle** from the Sketch toolbar.

53) Sketch a **rectangle** to the right of the POWERSUPPLY as illustrated.

Add dimensions.

54) Click **Smart Dimension** from the Sketch toolbar. Click the **left vertical edge** of the MOTHERBOARD.

55) Click the **right vertical edge** of the POWERSUPPLY.

56) Click a **position** above the BOX. Enter **25**mm.

57) Click the horizontal **top edge** of the MOTHERBOARD. Click the horizontal **top edge** of the BOX. Click a position off the profile. Enter **20**mm.

58) Click the vertical **right edge** of the MOTHERBOARD.

59) Click the vertical **right edge** of the BOX. Click a **position below** the Box.

60) Enter **20**mm. Click the horizontal **bottom edge** of the MOTHERBOARD.

61) Click the horizontal **bottom edge** of the BOX.

62) Click a **position** to the right of the BOX. Enter **20**mm.

💡 Sketch1 is fully defined and is displayed in black. Insert a Collinear relation between the top lines of the two inside rectangles if required.

Close and rename the sketch.

63) Right-click **Exit Sketch** ⎘ .

64) Rename **Sketch1** to **Layout**.

Save the BOX assembly.

65) Click **Save** 💾 .

What happens when the size of the MOTHERBOARD changes? How do you ensure that the BOX maintains the required 20mm spatial gap between the internal components and the BOX boundary? How do you design for future revisions? Answer: Through Global Variables and equations.

Global Variables and Equations

Use Global Variables as linked values. Global Variables are much easier to find, change and manage than linked values.

To use a Global Variable to link dimensions: 1.) Create a Global Variable in the Equations dialog box or the Modify dialog box for dimensions, and 2.) Set two or more dimensions equal to the Global Variable.

The project goal is to create three boxes of different sizes. Ensure that the models remain valid when dimensions change for various internal components. This is key for proper design intent. Apply Global Variables and equations.

Activity: Create Global Variables and Equations.

If needed, display the Equations folder in the FeatureManager and display the dimensions in the Graphics window.

66) Click **Options** ⚙, **System Options** tab from the Main menu.

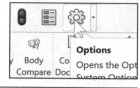

67) Click the **FeatureManager** folder from the System Options column.

68) Select **Show** from the Equations drop-down menu.

69) Click **OK** from the Systems Option - FeatureManager. View the Equations folder in the FeatureManager.

70) Right-click the **Annotations** folder in the FeatureManager.

71) Click **Show Feature Dimensions**.

72) Double-click **Layout** from the FeatureManager.

Insert a Global Variable.

73) Right-click the **Equations** folder from the FeatureManager.

74) Click **Manage Equations**. The Equations, Global Variables, and Dimensions dialog box is displayed.

75) Click the **Equation view** Σₒ button icon.

76) Click a **position** in the first cell under Global Variables.

77) Enter the Global Variable name **Gap**.

78) Press the **Tab** key.

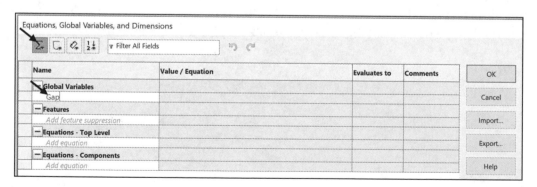

79) Click the **lower right horizontal dimension 20** in the Graphics window. The dimension name is displayed in the Value/Equation column. Note: Your dimension name may be different. A green check mark indicates the value can be calculated.

80) Press the **Tab** key. The Evaluates to column displays 20. The first Global Variable Gap is defined. Next define equations to equate the other variables to Gap.

81) Click **inside** the first cell below Equations-Top Level.

82) Click the **lower right vertical dimension 20** in the Graphics window.

83) Click inside the **Value / Equation** box.

84) Slide the **mouse pointer** to the right of Global Variables to display Gap (20mm).

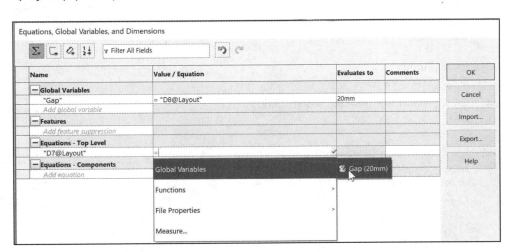

85) Click **Gap (20mm).**

86) Press the **Enter** key to create a new row under Equations-Top Level.

87) Repeat the **above procedure** to enter the rest of the 20mm dimensions on the Layout sketch.

88) Click **OK** from the Equations, Global Variables, and Dimensions dialog box.

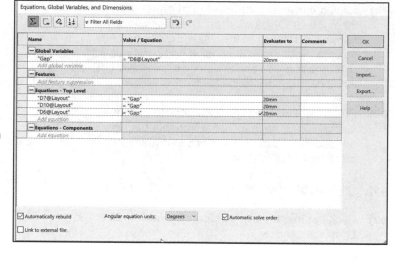

Test the Global Variables.
89) Double-click the lower right **20mm** dimension as illustrated.

90) Enter **10**mm.

91) Click **Rebuild** from the Modify dialog box. The four Global variables change and display the equation symbol Σ.

Return to the original value.
92) Double-click the lower right **10**mm dimension.

93) Enter **20**mm.

94) Click **Rebuild** from the Modify dialog box.

95) Click the **Green Check mark** from the Modify dialog box. All Global Variables are equal to 20mm.

Additional dimensions are required for Equations. Each dimension has a unique variable name. The names are used as Equation variables. The default names are based on the Sketch, Feature or Part. Feature names do not have to be changed. Rename variables for clarity when creating numerous Equations.

Rename the BOX assembly width and height. The full variable name is "box-width@ Layout." The system automatically appends Layout. If features are created or deleted in a different order, your variable names will be different.

Rename for overall box width.
96) Click the horizontal dimension **300** in the Graphics window. The PropertyManager is displayed.

97) Delete **D2@Layout**.

98) Enter **box_width**.

99) Click **OK** ✔ from the PropertyManager.

Rename the overall box height.
100) Click the vertical dimension **400** in the Graphics window. The PropertyManager is displayed.

101) Delete **D1@Layout**.

102) Enter **box height**.

103) Click **OK** ✔ from the PropertyManager.

Display an Isometric view.
104) Click **Isometric view** from the Heads-up View toolbar.

Hide all dimensions.
105) Right-click **the Annotations** folder.

106) Uncheck **Show Feature Dimensions**.

Save the BOX assembly.
107) Click **Save** 💾.

The sketch is complete. Develop the MOTHERBOARD part and the POWERSUPPLY part In-context of the BOX assembly Layout sketch.

Select the BOX assembly Front Plane to create InPlace Mates. An InPlace Mate is a Coincident Mate developed between the Front Plane of the component and the selected plane in the assembly. Utilize Convert Entities and extract geometry from the Layout Sketch to create external references for both the MOTHERBOARD part and POWERSUPPLY part. Utilize Ctrl-Tab to switch between part and assembly windows.

MOTHERBOARD-Insert Component

The MOTHERBOARD requires the greatest amount of lead-time to design and manufacture. The outline of the MOTHERBOARD is created. Create the MOTHERBOARD component from the Layout Sketch. An electrical engineer develops the Logic Diagram and Schematic required for the Printed Circuit Board (PCB). The MOTHERBOARD rectangular sketch represents the special constraints of a blank Printed Circuit Board (PCB).

Logic Diagram (partial)

Schematic

PCB

Courtesy of Electronic Workbench
(www.electronicworkbench.com)

A rough design of a critical connector is located on the MOTHERBOARD in the upper right corner.

CircuitWorks Add-in is a fully integrated data interface between SOLIDWORKS and PCB Design systems.

As the project manager, your job is to create the board outline with the corresponding dimensions from the Layout Sketch. Export the MOTHERBOARD data in an industry-standard Intermediate Data Format (IDF) from SOLIDWORKS.

The IDF file is sent to the PCB designer to populate the board with the correct 2D electronic components.

Board Outline

2D PCB

CircuitWorks utilizes industry-standard IDF files or PADS files, and produces the 3D SOLIDWORKS assembly of the MOTHERBOARD. IDF and PADS are common file formats utilized in the PCB industry.

The MOTHERBOARD is fully populated with the components at the correct height. Your colleagues use the MOTHERBOARD assembly to develop other areas of the BOX assembly. An engineer develops the wire harness from the MOTHERBOARD to electrical components in the BOX.

SOLIDWORKS assembly developed with CircuitWorks

A second engineer uses the 3D geometry from each electrical component on the MOTHERBOARD to create a heat sink. As the project manager, you move components that interfere with other mechanical parts, cables and or wire harness. You distribute the updated information to your colleagues and manufacturing partners.

To save a virtual component to its own external file, right-click the component and select **Save Part (in External File)** or **Save Assembly (in External file)**.

To rename the virtual component in the assembly, right-click the **virtual component** and click **Rename Part**.

Remember, the **Prompt user to select document template** option is selected in System Options, Default Templates.

If you do not want to create any External references in your part, click **Options, External References** and uncheck the **Allow creation of references external to the Model**.

Activity: MOTHERBOARD-Insert Component.

Create the MOTHERBOARD component.

108) The BOX assembly is the open document. Click **New Part** from the Consolidated Insert Components toolbar as illustrated.

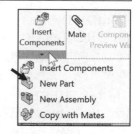

Select the Part Template.

109) Double-click **PART-MM-ANSI** from the MY-TEMPLATES tab in the New SOLIDWORKS Document dialog box. The new part is displayed in the FeatureManager design tree with a name in the form [Partn^assembly_name]. The square brackets indicate that the part is a virtual component. The new Component pointer ✔ icon is displayed.

The default component is empty and requires a Sketch plane. The Front Plane of the default component is mated with the Front Plane of the BOX.

Select the Sketch plane for the default component.

110) Click **Front Plane** in the BOX assembly FeatureManager.

111) **Expand** the new inserted component [Part1^BOX] in the FeatureManager. The FeatureManager is displayed in light blue.

Components added In-Context of the assembly automatically received an InPlace Mate within the Mates entry in the FeatureManager and under the component.

The InPlace mate is a Coincident mate between the Front Plane of the [Part1^BOX] component and the Front Plane of the BOX.

The [Part1^BOX] entry is added to the FeatureManager. The system automatically selects the Edit Component 🐷 mode.

The [Part1^BOX] text is displayed in blue to indicate that the component is actively being edited. The current Sketch plane is the Front Plane.

The current sketch name is ⬜ (-) Sketch1.

Create the sketch.

112) Click **Convert Entities** 🗇 from the Sketch toolbar.

113) Click the **right vertical edge** of the right inside rectangle as illustrated.

114) Click the **three other sketch** lines of the inside rectangle. Check the **Select chain** box.

115) Click **OK** ✔ from the Convert Entities PropertyManager. The outside perimeter of the Layout Sketch is the current sketch.

Insert an Extruded Boss/Base feature for the [Part1^BOX] component.

116) Click **Extruded Boss/Base** 🗐 from the Features toolbar. The Boss-Extrude PropertyManager is displayed.

117) Enter **10**mm for Depth in Direction 1. Accept the default settings.

118) Click **OK** ✔ from the Boss-Extrude PropertyManager. Boss-Extrude1 is displayed in the FeatureManager.

Rename Boss-Extrude1.

119) Rename **Boss-Extrude1** to **Base Extrude**.

The default component, "[Part1^BOX]<#> ->", Base Extrude -> and Sketch1 -> all contain the "->" symbol indicating External references to the BOX assembly.

The Edit Component feature acts as a switch between the assembly and the component edited In-Context.

🔆 If you do not want to create any External references in your model, click **Options**, **System Options**, **External References** and un-check the Allow creation of references external to the **model** box.

Return to the BOX assembly.

120) Click **Edit Component** 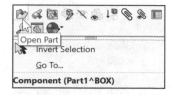 from the Assembly toolbar. The FeatureManager is displayed in black. Rebuild the model.

Fit the BOX to the Graphics window. Save the BOX.

121) Press the **f** key. Click **Save** .

122) Click **Save All** to save all modified documents.

123) Click **OK** to Save internally (inside the assembly).

Save and rename the default virtual component [Part1^BOX]<#> -> to MOTHERBOARD.
124) Right-click **[Part1^BOX]<#> ->**. Click **Open Part**.

125) Click **Save As** from the Menu bar. Click **OK** from the dialog box.

126) Enter **MOTHERBOARD** in the File name box. Note: Save in the VENDOR-COMPONENTS folder. Click **Save as**. Click **Save**. **Close** the MOTHERBOARD part. Return to the **BOX** assembly. View the updated FeatureManager.

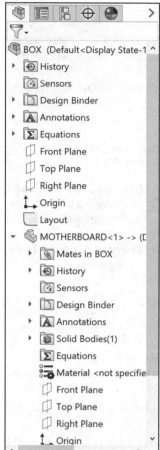

The Reference models are the MOTHERBOARD and the BOX assembly. Additional features are required that do not reference the Layout Sketch or other components in the BOX assembly. An Extruded Boss feature indicates the approximate position of an electrical connector. The actual measurement of the connector or type of connector has not been determined.

Perform the following steps to avoid unwanted assembly references:

1. Open the part from the assembly. **2.** Add features. **3.** Save the part. **4.** Return to the assembly. **5.** Save the assembly.

Insert an Extruded Boss/Base feature for the MOTHERBOARD.
127) Right-click **MOTHERBOARD** in the FeatureManager.

128) Click **Open Part**. The MOTHERBOARD FeatureManager is displayed. Press the **f** key.

Insert the Sketch.
129) Right-click the **front face** of Base Extrude. This is your Sketch

plane. Click **Sketch** from the Context toolbar. The Sketch toolbar is displayed. Click the **outside right vertical edge**. Hold the **Ctrl** key down. Click the **top horizontal edge**. Release the **Ctrl** key.

Front face

130) Click **Convert Entities** from the Sketch toolbar.

131) Drag the **bottom end point** of the converted line three quarters upward.

132) Drag the **left end point** of the converted line three quarters of the way to the right. Click **Front view** from the Heads-up View toolbar.

133) Click **Line** from the Sketch toolbar.

134) Sketch a **vertical line** as illustrated.

135) Sketch a **horizontal line** to complete the rectangle.

Add dimensions.

136) Click **Smart Dimension** from the Sketch toolbar.

137) Enter **30**mm for the horizontal dimension. Enter **50**mm for the vertical dimension. The sketch is fully defined.

Extrude the sketch.

138) Click **Extruded Boss/Base** from the Features toolbar. The Boss-Extrude PropertyManager is displayed.

139) Enter **10**mm for Depth in Direction 1.

140) Click **OK** from the Boss-Extrude PropertyManager.

Rename Boss-Extrude2.
141) Rename **Boss-Extrude** to **Connector1**.

Display an Isometric view. Save the model.
142) Click **Isometric view**.

143) Click **Save** .

Use color to indicate electrical connectors. Color the face of the Extruded Boss feature.

Color the Front face of Connector1.
144) Right-click the **Front face** of Connector1.

145) Click the **Appearances** drop-down arrow.

146) Click **Connector1** as illustrated.

147) Select **yellow** as the color.

148) Click **OK** ✔ from the Color PropertyManager. The selected face is displayed in yellow.

Save the MOTHERBOARD.

149) Click **Save** 💾.

Return to the BOX assembly.
150) Press **Ctrl Tab**. Click **Yes** to rebuild the assembly.

The MOTHERBOARD contains the new Extruded Boss feature. For now, the MOTHERBOARD is complete. The POWERSUPPLY is the second key component to be defined In-Context of the BOX assembly.

🔆 A learning goal of this book is to expose the new user to various tools, techniques and procedures. It may not always use the most direct tool or process.

🔆 Locate External references. Open the BOX assembly before opening individual components referenced by the assembly when you start a new session of SOLIDWORKS. The components load and locate their external references to the original assembly.

POWERSUPPLY-Insert Component

Create a simplified POWERSUPPLY part In-Context of the BOX assembly. Utilize the Layout Sketch in the BOX assembly. Open the POWERSUPPLY part and insert an Extruded Boss to represent the connection to a cable.

Activity: POWERSUPPLY-Insert Component.

Insert the POWERSUPPLY component.
151) Click **New Part** from the Consolidated Insert Components toolbar.

Select the Part Template.
152) Double-click the **PART-MM-ANSI** template from the MY-TEMPLATES tab. A new default virtual component is added to the FeatureManager. Note: Save the new component and name it POWERSUPPLY later in the procedure.

The Component Pointer 🔧✔ is displayed. The new default component is empty and requires a Sketch plane.

Select the Sketch plane.

153) Click the **Front Plane** of the BOX in the FeatureManager. An InPlace mate is added to the BOX FeatureManager.

The Front Plane of the default component [Part#^BOX]<#> is mated to the Front Plane of the BOX. Sketch1 of the [Part#^BOX]<1> is the active sketch.

The system automatically selects the Edit Component tool. The default component [Part#^BOX]<#> entry in the FeatureManager is displayed in blue. The current Sketch plane is the Front Plane. The current sketch name is Sketch1. The name is indicated on the current document window title.

[Part#^BOX]<#> is the default name of the component created In-Context of the BOX assembly. Create the first Extruded Base feature In-Context of the Layout Sketch.

Create the sketch.

154) Click **Convert Entities** from the Sketch toolbar.

155) Click the **Select chain** box.

156) Click the **right vertical edge** of [Part#^BOX]<1> as illustrated. The **three other sides** of the inside small rectangle are selected.

157) Click **OK** from the Convert Entities PropertyManager.

Extrude the sketch.

158) Click **Extruded Boss/Base** from the Features toolbar. Blind is the default End Condition in Direction 1.

159) Enter **50**mm for Depth in Direction 1. Accept the default settings.

160) Click **OK** from the Boss-Extrude PropertyManager. Boss-Extrude1 is displayed in the FeatureManager under [Part#^BOX]<#>.

The [Part#^BOX]<#> default name is displayed in blue. The component is being edited In-Context of the BOX assembly.

Return to the BOX assembly.

161) Click **Edit Component** from the Assembly toolbar. The [Part#^BOX]<#> displayed in black.

Save the BOX assembly.

162) Click **Save** . Click **Save All**. Click **OK** to save internally.

Rename the default component [Part2^BOX]<1> to POWERSUPPLY.
163) Right-click **[Part#^BOX]<1> ->**. Click **Open Part**. The [Part#^BOX]<1> -> FeatureManager is displayed.

164) Click **Save As** from the Menu bar. Enter **POWERSUPPLY** in the File name box. Save in the VENDOR-COMPONENTS folder. Click **Save as**.

165) Click **Save**. **Close** the part. **Return** to the BOX assembly. View the updated FeatureManager.

The Extruded Boss/Base feature represents the location of the cable that connects to the POWERSUPPLY. Think about where the cables and wire harness connect to key components. You do not have all of the required details for the cables.

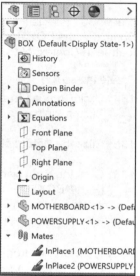

In a concurrent engineering environment, create a simplified version early in the design process. No other information is required from the BOX assembly to create additional features for the POWERSUPPLY.

The design intent for the POWERSUPPLY is for the cable connection to be centered on the top face of the POWERSUPPLY. Utilize a centerline and Midpoint relation to construct the Extruded Boss/Base feature centered on the top face. Open the POWERSUPPLY from the BOX assembly.

Open the POWERSUPPLY. Fit the model to the Graphics window.
166) Right-click **POWERSUPPLY** in the Graphics window. Click **Open Part**. The POWERSUPPLY FeatureManager is displayed. Press the f key. Click **Top view** from the Heads-up View toolbar.

Create the sketch.

167) Right-click the **top face** of Boss-Extrude1. Click **Sketch** from the Context toolbar. Click **Centerline** from the Sketch toolbar. The Insert Line PropertyManager is displayed.

168) Sketch a **diagonal centerline** from the top left to the bottom right as illustrated. Align the endpoints with the corners of the POWERSUPPLY.

Sketch a circle.

169) Click **Circle** from the Sketch toolbar. Click the **Midpoint** of the centerline. Sketch a circle as illustrated. The centerline midpoint is the center of the circle.

Add a dimension.

170) Click **Smart Dimension** from the Sketch toolbar.

171) Click the **circumference** of the circle. Click a **position off** the profile. Enter **15**mm for diameter. The circle is displayed in black.

Extrude the sketch.

172) Click **Extruded Boss/Base** from the Features toolbar. The Boss-Extrude PropertyManager is displayed. Blind is the default End Condition in   Direction 1.

173) Enter **10**mm for End Condition in Direction 1.

174) Click **OK** ✔ from the Boss-Extrude PropertyManager. The feature is displayed in the FeatureManager.

Rename the Boss-Extrude2 feature. Display an Isometric view.
175) Rename the **Boss-Extrude2** to **Cable1**.

176) Click **Isometric view**.

Save the POWERSUPPLY.

177) Click **Save** .

Return to the BOX assembly.
178) Press **Ctrl Tab**. Click **Yes** to update the assembly.

The POWERSUPPLY contains the new Extruded Boss feature, Cable1. Recall the initial design parameters. The requirement calls for three different size boxes. Test the Layout Sketch dimensions for the three configurations: *Small, Medium and Large*.

Do the Layout Sketch, MOTHERBOARD and POWERSUPPLY reflect the design intent of the BOX assembly? Review the next steps to confirm the original design intent.

Display all dimensions.
179) Click **Front view** from the Heads-up View toolbar.

180) Right-click the **Annotations** folder.

181) Click **Show Feature Dimensions**.

182) Double-click **300**. Enter **550** for the horizontal dimension.

183) Double-click **400**. Enter **600** for the vertical dimension.

Return to the original dimensions.
184) Double-click **550**. Enter **300** for width dimension.

185) Double-click **600**. Enter **400** for height dimension.

Display an Isometric view. Save the assembly.
186) Click **Isometric view** from the Heads-up View toolbar.

187) Click **Save** .

Hide all dimensions.
188) Right-click the **Annotations** folder.

189) Un-check **Show Feature Dimensions**.

Suppress the MOTHERBOARD and POWERSUPPLY.
190) Click **MOTHERBOARD** from the FeatureManager.

191) Hold the **Ctrl** key down.

192) Click **POWERSUPPLY** from the FeatureManager.

193) Release the **Ctrl** key.

194) Right-click **Suppress**. Both components and their mates are Suppressed. They are displayed in light gray in the FeatureManager.

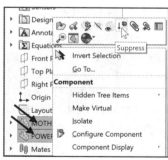

Electro-mechanical assemblies contain hundreds of parts. Suppressing components saves rebuild time and simplifies the model. Only display the components and geometry required to create a new part or to mate a component.

Improve rebuild time and display performance for large assemblies. Review the System Options listed in Large Assembly Mode. Adjust the large assembly threshold to match your computer performance.

Check Remove detail during zoom/pan/rotate. Check Hide all planes, axes and sketches.

Review the System Option, Image Quality. Drag the resolution slider towards Low to improve computer performance.

Review the BOX Assembly.

The BOX assembly utilized the Top-down assembly approach by incorporating a Layout Sketch. The Layout Sketch consisted of 2D profiles and dimensions of the BOX, MOTHERBOARD and POWERSUPPLY.

The MOTHERBOARD was developed as a new part In-Context of the BOX assembly. An InPlace mate was created between the MOTHERBOARD Front Plane and the BOX assembly Front Plane. In the Edit Component mode, you extracted the Layout Sketch geometry to create the first feature of the MOTHERBOARD part. The first feature, sketch and Sketch plane contained External references to the BOX assembly.

You opened the MOTHERBOARD from the BOX assembly. Additional features were added to the MOTHERBOARD part. The POWERSUPPLY was developed as a new part In-Context of the BOX assembly. The BOX assembly design intent and requirements were verified by modifying the dimensions to the different sizes of the BOX configurations.

The MOTHERBOARD and POWERSUPPLY are only two of numerous components in the BOX assembly. They are partially complete but represent the overall dimensions of the final component. Electro-Mechanical assemblies contain parts fabricated from sheet metal. Fabricate the CABINET and BRACKET for the BOX assembly from Sheet metal.

Sheet Metal Overview

Sheet metal manufacturers create sheet metal parts from a flat piece of raw material. To produce the final part, the material is cut, formed and folded. In SOLIDWORKS, the material thickness does not change.

Talk to colleagues. Talk to sheet metal manufacturers. Review other sheet metal parts previously designed by your company. You need to understand a few basic sheet metal definitions before starting the next parts, CABINET and BRACKET.

The CABINET begins as a solid part. The Shell feature determines the material thickness. The Rip feature removes material at the edges to prepare for sheet metal bends. The Insert Sheet Metal Bends feature transforms the solid part into a sheet metal part.

The BRACKET begins as a sheet metal part. The Base Flange feature is the first feature in a sheet metal part. Create the Base Flange feature from a sketch. The material thickness and bend radius of the Base Flange are applied to all other sheet metal features in the part.

There are two design states for sheet metal parts:

- Formed.

- Flat.

Formed State   Flat State
BRACKET

Design the CABINET and BRACKET in the formed state. The Flatten feature creates the Flat Pattern for the manufactured flat state.

Sheet metal parts can be created from the flat state. Sketch a line on a face to indicate bend location. Insert the Sketched Bend feature to create the formed state.

Flat State   Sketched Bend Lines   Formed State

Bends

Example: Use a flexible eraser, 50mm or longer. Bend the eraser in a U shape. The eraser displays tension and compression forces.

The area where there is no compression or tension is called the neutral axis or neutral bend line.

Neutral Bend Line

Tension Forces

Compression Forces

Eraser Example

Assume the material has no thickness. The length of the material formed into a 360° circle is the same as its circumference.

The length of a 90° bend would be ¼ of the circumference of a circle.

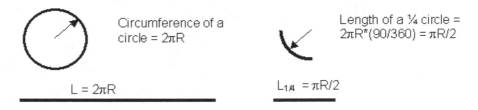

Circumference of a circle = 2πR

L = 2πR

Length of a ¼ circle = 2πR*(90/360) = πR/2

$L_{1/4} = \pi R/2$

Length of 90° bend

In the real world, materials have thickness. Materials develop different lengths when formed in a bend depending on their thickness.

There are three major properties which determine the length of a bend:

- Bend radius.

- Material thickness.

- Bend angle.

The distance from the inside radius of the bend to the neutral bend line is labeled δ.

The symbol 'δ' is the Greek letter delta. The amount of flat material required to create a bend is greater than the inside radius and depends on the neutral bend line.

The true developed flat length, L, is measured from the endpoints of the neutral bend line.

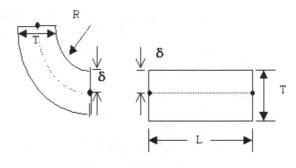

90° Bent Material   Flat Developed Length

True Developed Flat Length

Example:

T is the Material thickness.

Create a 90° bend with an inside radius of R:

$L = \frac{1}{2} \pi R + \delta T$

The ratio between δ and T is called the K-factor. Let K = 0.41 for Aluminum:

$K = \delta / T$

$\delta = KT = 0.41T$

$L = \frac{1}{2} \pi R + 0.41T$

U.S. Sheet metal shops use their own numbers.

Example:

One shop may use K = 0.41 for Aluminum, versus another shop that uses K = 0.45.

Use tables, manufacturer's specifications or experience.

Save design time and manufacturing time. Work with your Sheet metal shop to know their K-factor. Material and their equipment produce different values. Build these values into the initial design.

Where do you obtain the material required for the Bend? The answer comes from the Sheet metal Flange position option. The four options are Material Inside, Material Outside, Bend Outside and Bend from Virtual Sharp.

Material Inside: The outside edge of the Flange coincides with the fixed edge of the sheet metal feature.

Material Outside: The inside edge of the Flange coincides with the fixed edge of the sheet metal feature.

Bend Outside: The Flange is offset by the bend radius.

Material Inside   Material Outside   Bend Outside

Bend from Virtual Sharp: The Flange maintains the dimension to the original edge and varies the bend material condition to automatically match with the flange's end condition.

Bend from Virtual Sharp

Relief

Sheet metal corners are subject to stress. Excess stress will tear material. Remove material to relieve stress with Auto Relief. The three options are Rectangular, Tear and Obround.

- Rectangular: Removes material at the bend with a rectangular shaped cut.

- Tear: Creates a rip at the bend, a cut with no thickness.

- Obround: Removes material at the bend with a rounded shaped cut.

Rectangular   Tear   Obround

CABINET-Insert Component

Create the CABINET component inside the assembly and attach it to the Front Plane. The CABINET component references the Layout sketch.

Create the CABINET component as a solid box. Shell the box to create the constant sheet metal thickness. Utilize the Rip feature to cut the solid edges. Utilize the Insert Bends feature to create a sheet metal part from a solid part.

2D Flat State   3D Formed State

Add additional sheet metal Edge Flange and Hem features. Add dies, louvers, and cuts to complete the CABINET.

Utilize configurations to create the 3D formed state for the BOX assembly and the 2D flat state for the CABINET drawing.

Activity: CABINET-Insert Component.

Insert the CABINET Component.
195) If needed, open the **BOX** assembly.

196) Click **New Part** from the Consolidated Insert Components toolbar as illustrated.

Select the Part Template.
197) Double-click the **PART-MM-ANSI** template from the MY-TEMPLATES tab. A new virtual default component is added to the FeatureManager. Note: Save the new component and name it CABINET later in the procedure.

The Component Pointer is displayed. The default component [Part3^BOX]<1> is empty and requires a Sketch plane. The Front Plane of the [Part3^BOX]<1> component is mated with the Front Plane of the BOX. [Part3^BOX]<1> is the name of the component created In-Context of the BOX assembly.

Select the Sketch plane.
198) Click **Front Plane** from the BOX FeatureManager. The system automatically selects the Sketch toolbar.

Create the sketch.
199) Click the **Convert Entities** Sketch tool.

200) Select the **Select chain** box.

201) Click the **right vertical edge** of the outside BOX as illustrated. The three other sides of the outside BOX are selected.

202) Click **OK** from the Convert Entities PropertyManager. The outside perimeter is the current sketch.

Extrude the sketch.

203) Click **Extruded Boss/Base** from the Features toolbar. The Boss-Extrude PropertyManager is displayed.

204) Enter **100**mm for Depth in Direction 1. Accept the default settings.

205) Click **OK** ✔ from the Boss-Extrude PropertyManager.

Return to the BOX assembly.

206) Click **Edit Component** from the Assembly toolbar.

Rename the default component [Part3^BOX]<1> to CABINET.
207) Right-click **[Part3^BOX]<1> ->**. Click **Open Part**. The [Part3^BOX]<1> -> FeatureManager is displayed.

208) Click **Save As** from the Menu bar. Click **OK**. Enter **CABINET** in the File name box. Note: Save in the VENDOR-COMPONENTS folder.

209) Click **Save**. **Close** the part.

210) Return to the **BOX** assembly. **View** the updated FeatureManager

Caution: Do not create unwanted geometry references. Open the part when creating features that require no references from the assembly. The features of the CABINET require no additional references from the BOX assembly, MOTHERBOARD or POWERSUPPLY.

Open the CABINET part.
211) Right-click **CABINET** from the FeatureManager. Click **Open Part**. The CABINET FeatureManager is displayed.

Fit the model to the Graphics window. Display an Isometric view.
212) Press the **f** key. Click the **Isometric view** from the Heads-up View toolbar.

Create the Shell.
213) Click the **front face** of the Boss-Extrude1 feature.

214) Click **Shell** from the Features toolbar. The Shell1 PropertyManager is displayed. Check the **Shell outward** box. Enter **1.00**mm for Thickness.

215) Click **OK** ✔ from the Shell1 PropertyManager. Shell1 is displayed in the FeatureManager.

Display the Sheet Metal toolbar.

216) Click **View**, **Toolbars** from the Main menu.

217) Check **Sheet Metal**. The Sheet Metal toolbar is displayed.

218) Click and drag to a location inside the Graphics window.

💡 Maintain constant thickness. Sheet metal features only work with constant wall thickness. The Shell feature maintains constant wall thickness. The solid Extruded Base feature represents the overall size of the sheet metal CABINET. The CABINET part is solid. The Rip feature and Insert Sheet Metal Bends feature convert a shelled solid part into a sheet metal part.

CABINET-Rip Feature and Sheet Metal Bends

The Rip feature creates a cut of no thickness along the edges of the Extruded Base feature. Rip the Extruded Base feature along the four edges. The Bend feature creates sheet metal bends.

Specify bend parameters: bend radius, bend allowance and relief. Select the bottom face to remain fixed during bending and unbending. In the next example, you will create the Rip and the Insert Bend features in two steps. The Bends PropertyManager contains the Rip parameter. Utilize the Rip feature to create a simple Rip and Bend in a single step.

> **Activity: CABINET-Rip Feature and Sheet Metal Bends**.

Fit the model to the Graphics window.

219) Press the **f** key.

Insert a Rip feature. A Rip feature creates a gap between two edges in a sheet metal part.

220) Click **Rip** 🠗 from the Sheet Metal toolbar. The Rip PropertyManager is displayed.

221) Rotate the part to view the inside edges.

222) Click the **inside lower left** edge as illustrated.

223) Click the **inside upper left** edge.

224) Click the **inside upper right** edge.

225) Click the **inside lower right** edge. The selected entities are displayed in the Edges to Rip box.

226) Enter **.10mm** for Rip Gap.

227) Click **OK** ✔ from the Rip PropertyManager. Rip1 is displayed in the FeatureManager.

Insert the Sheet Metal Bends.
228) Click the **inside bottom face** to remain fixed.

229) Click **Insert Bends** from the Sheet Metal toolbar. The Bends PropertyManager is displayed. Face<1> is displayed.

230) Enter **2.00**mm for Bend radius.

231) Enter **.45** for K-Factor.

232) Enter **.5** for Rectangular Relief.

233) Click **OK** ✔ from the Bends PropertyManager.

234) Click **OK** to the message, "Auto relief cuts were made for one or more bends."

235) Zoom in on the Rectangular relief in the upper back corner.

Display an Isometric view. Save the assembly.
236) Click **Isometric view** from the Heads-up View toolbar.

237) Click **Save** . View the created features in the FeatureManager.

You just created your first sheet metal part. The .45 K-Factor is based on your machine shop's parameters for Aluminum.

Save manufacturing cost and reduce setup time. A sheet metal manufacturer maintains a turret of standard relief tools for Rectangular and Obround relief. Obtain the dimensions of these tools to utilize in your design.

The CABINET part is in its 3D formed state. Display its 2D flat manufactured state. Test every feature to determine if the part can be manufactured as a sheet metal part.

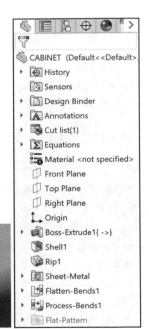

Alternate between 3D formed and 2D flat
for every additional sheet metal feature
you create. The Flatten feature alternates
a sheet metal part between the flat state
and formed state.

Display the Flat State.

238) Click **Flatten** ⬜ from the Sheet Metal
toolbar.

Fit the model to the Graphics window.
239) Press the **f** key.

Display the part in its fully formed state.

240) Click **Flatten** ⬜ from the Sheet Metal toolbar.

Save the CABINET.

241) Click **Save** 💾.

Note: To return to the solid part, utilize the No Bends feature to roll back the model
before the first sheet metal bend.

CABINET-Edge Flange

Create the right flange wall and left flange wall with the Edge
Flange feature. The Edge Flange 🦴 feature adds a wall to a
selected edge of a sheet metal part.

Select the inside edges when creating bends. Create the right
hem and left hem with the Hem feature. The Hem feature folds
back at an edge of a sheet metal part. A Hem can be closed,
open, tear-drop or rolled.

💡 In the preliminary design stage, review Hem options with
your sheet metal manufacturer.

Activity: CABINET-Edge Flange.

Insert the front right Flange.
242) Click the **front vertical right** edge of the CABINET as illustrated.

243) Click **Edge Flange** from the Sheet Metal toolbar. The Edge-Flange PropertyManager is displayed. Select **Blind** for Length Edge Condition.

244) Enter **30**mm for Length. Click the **Reverse Direction** button. The Direction arrow points towards the right. Click **Outer Virtual Sharp**. Accept all other defaults.

245) Click **OK** from the Edge-Flange PropertyManager. Edge-Flange1 is displayed in the FeatureManager.

Insert the right Hem.
246) Click **Front view** from the Heads-up View toolbar.

247) Click the **right edge** of the right flange.

248) Click **Hem** from the Sheet Metal toolbar. The Hem PropertyManager is displayed.

249) Click **Isometric view** from the Heads-up View toolbar.

250) Click the **Reverse Direction** button.

251) Enter **10**mm for Length.

252) Enter **.10**mm for Gap Distance.

253) Click **OK** from the Hem PropertyManager. Hem1 is displayed in the FeatureManager.

Insert the left Edge Flange wall.
254) Click **Front view** from the Heads-up View toolbar.

255) Click the **front vertical left** edge as illustrated.

256) Click **Edge Flange** from the Sheet Metal toolbar. The Edge-Flange PropertyManager is displayed.

257) Select **Blind** for Length End Condition.

258) Enter **30**mm for Length. The feature direction arrow points towards the left. Reverse the arrow direction if required.

259) Click **Outer Virtual Sharp**. Accept the default settings.

260) Click **OK** ✔ from the Edge-Flange PropertyManager. Edge-Flange2 is displayed in the FeatureManager.

Insert the left edge Hem.
261) Select the **left edge** of the left flange as illustrated.

262) Click **Isometric view** from the Heads-up View toolbar.

263) Click **Hem** from the Sheet Metal toolbar. The Hem PropertyManager is displayed.

264) Enter **10**mm for Length. Enter **.10**mm for Gap Distance. Click the **Reverse Direction** button.

265) Click **OK** ✔ from the Hem PropertyManager. Hem2 is displayed in the FeatureManager.

Display the part in its flat manufactured state and formed state. Rename the features.
266) Click **Flatten** from the Sheet Metal toolbar.

267) Click **Top view** from the Heads-up View toolbar.

268) Click **Flatten** to display the part in its fully formed state.

269) Click **Isometric view** from the Heads-up View toolbar.

270) Rename **Edge-Flange1** to **Edge-Flange1-Right**.

271) Rename **Edge-Flange2** to **Edge-Flange2-Left**.

272) Rename **Hem1** to **Hem1-Right**.

273) Rename **Hem2** to **Hem2-Left**.

Save the CABINET.

274) Click **Save** 💾.

Simplify the FeatureManager. Sheet metal parts contain numerous flanges and hems. Rename features with descriptive names.

Save design time. Utilize Mirror Feature to mirror sheet metal features about a plane. Example: In a second design iteration, the Edge Flange1-Right and Hem1-Right are mirrored about Plane1.

The Mirror feature was not utilized in the CABINET. A Mirror Plane and Equations are required to determine the flange and hem locations.

☀ Maintain the design intent. In the Top-down design process return to the Layout Sketch after inserting a new component to verify the assembly design intent.

Increase and decrease dimension values in the Layout Sketch. To avoid problems in a top level assembly, select the same face to remain fixed for the Flat Pattern and Bend/Unbend features. Save sheet metal parts in their 3D formed state.

The CABINET requires additional solid and sheet metal features. These features require no references in the BOX assembly. Work in the CABINET part.

CABINET - Hole Wizard and Linear Pattern features

Sheet metal holes are created through a punch or drill process. Each process has advantages and disadvantages: cost, time and accuracy.

Investigate a Linear Pattern feature of holes. Select a self-clinching threaded fastener that is inserted into the sheet metal during the manufacturing process. The fastener requires a thru hole in the sheet metal.

Holes should be of equal size and utilize common fasteners. Why? You need to ensure a cost-effective design that is price competitive. Your company must be profitable with their designs to ensure financial stability and future growth.

Another important reason for fastener commonality and simplicity is the customer. The customer or service engineer does not want to supply a variety of tools for different fasteners.

A designer needs to be prepared for changes. You proposed two fasteners. Ask Purchasing to verify availability of each fastener. Select a Ø5.0mm hole and wait for a return phone call from the Purchasing department to confirm. Design flexibility is key!

Dimension the hole position. Do not dimension to sheet metal bends. If the sheet metal manufacturer modifies the bend radius, the hole position does not maintain the design intent. Reference the hole dimension by selecting the Origin in the FeatureManager.

Utilize the Hole Wizard feature to create a Simple Through All hole. The hole is positioned on the inside bottom face based on the selection point.

Activity: CABINET - Hole Wizard and Linear Pattern feature.

Insert the first hole. Open the Hole Wizard.

275) Click **Hole Wizard** from the Features toolbar. The Hole Specification PropertyManager is displayed.

276) Click the **Hole** tab.

277) Select **ANSI Metric** for Standard.

278) Select **Drill sizes** for Type.

279) Select **Ø5.0** for Size.

280) Select **Through All** for End Condition.

281) Click the **Positions** tab.

282) Click the **inside bottom face** as illustrated. The Point Sketch tool is active.

283) Click **again to place** the center point of the hole.

284) Right-click **Select** to deselect the Point Sketch tool.

Dimension the hole. Create a horizontal and vertical dimension from the Origin.

285) Click **Smart Dimension** .

286) Click **Top view** from the Heads-up View toolbar.

287) **Expand** CABINET from the fly-out FeatureManager.

288) Click the **Origin** from the FeatureManager.

289) Click the **center point of the hole** in the Graphics window.

290) Enter **25**mm for the horizontal dimension. Click the **Origin** from the FeatureManager.

291) Click the **center point of the hole** in the Graphics window.

292) Enter **20**mm for the vertical dimension.

293) Click **OK** ✔ from the Dimension PropertyManager.

294) Click **OK** ✔ from the Hole Position PropertyManager. The Hole is displayed in the FeatureManager.

Display an Isometric view. Save the model.
295) Click **Isometric view** from the Heads-up View toolbar.

296) Click **Save** 💾.

Insert a Linear Pattern of holes.

297) Click **Linear Pattern** from the Features toolbar. The Linear Pattern PropertyManager is displayed.

298) Click inside the **Features to Pattern** box.

299) Click **(5) Diameter Hole1**.

300) Click inside the **Pattern Direction** box.

301) Click the **back inside edge** for Direction 1. Edge<1> is displayed in the Pattern Direction box. The direction arrow points to the right. Click the Reverse Direction button if required.

302) Enter **125**mm for Spacing.

303) Enter **3** for Number of Instances.

304) Click the **left edge** for Direction 2. Edge<2> is displayed in the Pattern Direction box. The second direction arrow points to the front. Click the Reverse Direction button if required.

305) Enter **50**mm for Spacing.

306) Enter **2** for Number of Instances.

Remove an Instance.
307) Click **inside the Instances to Skip** box.

308) Click the **front middle hole, (2,2)** in the Graphics window as illustrated. (2,2) is displayed in the Instances to Skip box.

309) Check the **Geometry pattern** box. Accept the default settings.

310) Click **OK** ✔ from the Linear Pattern PropertyManager. LPattern1 is displayed in the FeatureManager.

Display an Isometric view.
311) Click **Isometric view** from the Heads-up View toolbar.

Use the Geometry pattern option to improve system performance for sheet metal parts. The Geometry pattern option copies faces and edges of the seed feature. Type options such as Up to Surface are not copied.

The Instances to Skip option removes a selected instance from the pattern. The value (2, 2) represents the instance position in the first direction - second instance, second direction - second instance.

CABINET-Sheetmetal Design Library Feature

Sheet metal manufacturers utilize dies and forms to create specialty cuts and shapes. The Design Library contains information on dies and forms. The Design Library, features folder contains examples of predefined sheet metal shapes. Insert a die cut on the right wall of the CABINET. The die cut is for a data cable. The team will discuss sealing issues at a later date.

Activity: CABINET-Sheetmetal Design Library Feature.

Insert a Sheet metal Library Feature.

312) Click **Design Library** 🗔 from the Task Pane.

313) Expand the Design Library 🗔 folder.

314) Expand the features folder.

315) Click the **Sheetmetal** folder.

316) Click and drag the **d-cutout** feature into the Graphics window.

317) Release the **mouse button on the right flange outside face** of the CABINET as illustrated. The d-cutout PropertyManager is displayed.

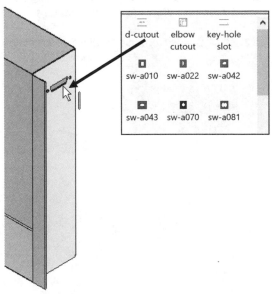

Edit the sketch.

318) Click the **Edit Sketch** button from the d-cutout PropertyManager. The Library Feature Profile dialog box is displayed. DO NOT SELECT THE FINISH BUTTON.

Rotate the d-cutout.

319) **Expand** CABINET in the fly-out FeatureManager.

320) Click **Tools**, **Sketch Tools**, **Modify** ⌐ Modify... from the Menu bar. The mouse pointer displays the modify move/rotate icon . The Modify Sketch dialog box is displayed.

321) Enter **90deg** in the Rotate box. Note: Press the Enter key. View the d-cutout feature.

322) Click **Close** from the Modify Sketch dialog box.

Display the Right view.
323) Click **Right view** from the Heads-up View toolbar.

324) Zoom to Area on the d-cutout.

Dimension the d-cutout.
325) Click **Smart Dimension** from the Sketch toolbar.

Create a vertical dimension.
326) Click the **midpoint** of the d-cutout. The Point PropertyManager is displayed.

327) Expand CABINET from the fly-out FeatureManager.

328) Click the **Origin** of the CABINET.

329) Enter **320**mm.

Create a horizontal dimension.
330) Click the **left vertical line** of the d-cutout.

331) Expand the CABINET from the fly-out FeatureManager.

332) Click the **Origin** of the CABINET.

333) Enter **50**mm.

334) Click **Finish** from the Library Feature Profile dialog box. The d-cutout entry is displayed in the FeatureManager.

Display an Isometric view. Save the model.
335) Click **Isometric view** from the Heads-up View toolbar.

336) Click **Save**.

With thin sheet metal parts, select the dimension references with care. Use the Zoom and Rotate commands to view the correct edge. Use reference planes or the Origin to create dimensions. The Origin and planes do not change during the flat and formed states.

Save time with the Modify Sketch tool. Utilize the left and right mouse point positioned on the large black dots. The left button Pans and the right button Rotates. The center dot reorients the Origin of the sketch and Mirrors the

sketch about y, x or both.

The d-cutout goes through the right and left side. Through All is the current End Condition Type. This is not the design intent. Redefine the end condition through the right flange.

The d-cutout feature is positioned before the Flat Pattern icon in the FeatureManager. The d-cutout is incorporated into the Flat Pattern. The sheet metal manufacturing process is cost effective to perform cuts and holes in the flat state.

CABINET-Louver Forming Tool

Form features such as louvers are added in the formed state. Formed features are normally more expensive than cut features. Louvers are utilized to direct air flow. The louvers are used to dissipate the heat created by the internal electronic components.

The forming tools folder contains numerous sheet metal forming shapes. In SOLIDWORKS, the forming tools are inserted after the Bends are processed. Suppress forming tools in the Flat Pattern.

Activity: CABINET-Louver Forming Tool.

Insert a Louver Forming tool.

337) Rotate the model to display the inside right flange.

338) Expand the forming tools folder in the Design Library.

339) Click the **louvers** folder. A louver is displayed.

340) Click and drag the **louver to the inside right flange** of the CABINET. The Form Tool Feature PropertyManager is displayed.

341) Rotate the Louver as illustrated. Enter **270.00** deg.

Dimension the location of the louver.

342) Click the **Position** tab as illustrated.

343) Click the **Smart Dimension** tool.

344) Click the **center point of the Louver**.

345) Click the **Origin** of the CABINET from the FeatureManager.

346) Enter **100**mm for the vertical dimension.

347) Click the **Origin** of the CABINET from the fly-out FeatureManager.

348) Click the **vertical centerline** of the Louver. Note: If needed, create a vertical centerline from the center point of the Louver. Enter **50**mm for the horizontal dimension.

Display the Louver.
349) Click **OK** from the FeatureManager. Press the **f** key.

Add a Linear Pattern of louvers.

350) Click **Linear Pattern** from the Features toolbar.

351) Click the CABINET **back vertical edge** for Direction 1. Edge<1> is displayed in the Pattern Direction box.

352) Enter **25**mm for Spacing.

353) Enter **6** for Number of Instances. The direction arrow points upward. Click the Reverse Direction button if required.

354) Click inside the **Features and Faces** box. Click **louver1** from the fly-out FeatureManager.

355) Click **OK** from the Linear Pattern PropertyManager. LPattern2 is displayed in the FeatureManager.

Display an Isometric view.
356) Click **Isometric view**.

Save the CABINET.

357) Click **Save**.

Manufacturing Considerations

How do you determine the size and shape of the louver form? Are additional die cuts or forms required in the project? Work with a sheet metal manufacturer. Ask questions. What are the standards? Identify the type of tooling in stock. Inquire about form advantages and disadvantages.

One company that has taken design and form information to the Internet is Lehi Sheetmetal, Westboro, MA (www.lehisheetmetal.com).

Standard dies, punches and manufacturing equipment such as brakes and turrets are listed in the Engineering helpers section.

Dimensions are provided for their standard forms. The form tool you used for this project creates a 32mm x 6mm louver. The tool is commercially available.

Your manufacturer only stocks 3in. (75mm) and 4in. (100mm) louvers. Work with the sheet metal manufacturer to obtain a cost effective alternative. Create a pattern of standard square cuts to dissipate the heat.

If a custom form is required, most custom sheet metal manufacturers can accommodate your requirement. Example: Wilson Tool, Great Bear Lake, MN (www.wilsontool.com).

However, you will be charged for the tool. How do you select material? Consider the following:

- Strength.

- Fit.

- Bend Characteristics.

- Weight.

- Cost.

All of these factors influence material selection. Raw aluminum is a commodity. Large manufacturers such as Alcoa and Reynolds sell to a material supplier, such as *Pierce Aluminum.

**ALUMINUM SHEET
NON HEAT TREATABLE, 1100-0
QQ-A-250/1 ASTM B 209**

All thickness and widths available from coil for custom blanks

SIZE IN INCHES	WGT/SHEET
.032 x 36 x 96	11.05
.040 x 36 x 96	13.82
.040 x 48 x 144	27.65
.050 x 36 x 96	17.28
.050 x 48 x 144	34.56
.063 x 36 x 96	21.77
.063 x 48 x 144	43.55
.080 x 36 x 96	27.65
.080 x 48 x 144	55.30
.090 x 36 x 144	46.66
.090 x 48 x 144	62.26
.100 x 36 x 96	34.56
.125 x 48 x 144	86.40
.125 x 60 x 144	108.00
.190 x 36 x 144	131.33

*Courtesy of Pierce Aluminum Co, Inc.
Canton, MA

Pierce Aluminum, in turn, sells material of different sizes and shapes to distributors and other manufacturers. Material is sold in sheets or cut to size in rolls. U.S. Sheet metal manufacturers work with standard 8ft - 12ft stock sheets. For larger quantities, the material is usually supplied in rolls. For a few cents more per pound, sheet metal manufacturers request the supplier to shear the material to a custom size.

Check Dual Dimensioning from Tools, Options Document Properties to display both Metric and English units. Do not waste raw material. Optimize flat pattern layout by knowing your sheet metal manufacturer's equipment.

Discuss options for large cabinets that require multiple panels and welds. In Project 1, you were required to be cognizant of the manufacturing process for machined parts.

In Project 4 and 5 you created and purchased parts and recognized the Draft feature required to manufacture plastic parts.

Whether a sheet metal part is produced in or out of house, knowledge of the materials, forms and layout provides the best cost-effective design to the customer.

🔆 Save manufacturing cost and time. Obtain a list of standard dies and forms from your sheet metal manufacturer. Utilize their standard dies and forms at the start of your design. Prepare for lead-time if custom dies and forms are required.

You require both a protective and cosmetic finish for the Aluminum BOX. Review options with the sheet metal vendor. Parts are anodized. Black and clear anodized finishes are the most common. In harsh environments, parts are covered with a protective coating such as Teflon® or Halon®.

The finish adds thickness to the material. A few thousandths could cause problems in the assembly. Think about the finish before the design begins.

Your sheet metal manufacturer suggests a pattern of square cuts replace the Louvers for substantial cost savings. Suppress the Louver. The Linear Pattern is a child of the Louver. The Linear Pattern is suppressed. Utilize the Sketch Linear Step and Repeat option.

Activity: Manufacturing Considerations-Sketch Linear Pattern.

Suppress the louver.
358) Right-click **louver1** from the FeatureManager.

359) Click **Suppress**.

Sketch the profile of squares.
360) Click **Right view**.

361) Right-click the **right face** of the CABINET in the Graphics window. Shell1 is highlighted in the FeatureManager.

362) Click **Sketch** ⌐ from the Context toolbar.

Create the first square.
363) Click **Corner Rectangle** ⬜ from the Sketch toolbar.

364) Sketch a **square** in the lower left corner as illustrated.

Add dimensions.
365) Click **Smart Dimension** ↖ from the Sketch toolbar.

366) Create the vertical and horizontal **10**mm dimensions of the square as illustrated.

367) Create the 10mm dimension from the **bottom horizontal line** of the square to the Origin.

Create a horizontal dimension.
368) Click the **Origin**.

369) Click the **right vertical line** of the square.

370) Click a **position below** the profile.

371) Enter **80**mm.

Create a second square.
372) Click **Corner Rectangle** ⬜ from the Sketch toolbar.

373) Sketch a **square diagonal to the right** of the first square as illustrated.

Create a horizontal dimension.

374) Click **Smart Dimension** from the Sketch toolbar.

375) Click the **right vertical line** of the first box.

376) Click the **left vertical line** of the second box. Enter **5**mm.

Add an Equal relation.
377) Right-click **Select**.

378) Click the **top line** of the second square. Hold the **Ctrl** key down.

379) Click the **left line** of the second square. The selected entities are displayed in the Selected Entities box. Release the **Ctrl** key. Click **Equal** from the Add Relations box.

380) Click **OK** from the Properties PropertyManager.

Add an Equal relation.
381) Click the **top horizontal line** of the first square. Hold the **Ctrl** key down. Click the **top horizontal line** of the second square. The selected entities are displayed in the Selected Entities box. Release the **Ctrl** key. Click **Equal** from the Add Relations box.

382) Click **OK** from the Properties PropertyManager.

Add a Collinear relation.
383) Click the **top horizontal line** of the first square. Hold the **Ctrl** key down. Click the **bottom horizontal line** of the second square. Release the **Ctrl** key.

384) Click **Collinear** from the Add Relations box.

385) Click **OK** from the Properties PropertyManager. The sketches are fully defined.

Repeat the sketch.
386) **Window-Select** the two squares.

387) Click **Linear Sketch Pattern** from the Sketch toolbar. The Linear Pattern PropertyManager is displayed.

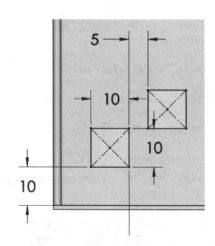

388) Enter **2** for Number in Direction 1.

389) Enter **30**mm for Spacing in Direction 1. The pattern direction arrow points to the right.

390) Enter **13** for Number in Direction 2. The pattern direction arrow points upward.

391) Enter **20**mm for Spacing in Direction 2.

392) Check the **Dimension angle dimension between axes** box.

393) Click **OK** ✔ from the Linear Pattern PropertyManager.

Extrude the sketch.

394) Click **Extruded Cut** from the Features toolbar. The Cut-Extrude PropertyManager is displayed.

395) Check the **Link to thickness** box. Accept the default settings.

396) Click **OK** ✔ from the Cut-Extrude PropertyManager.

Display an Isometric view. Rename the Cut-Extrude2 feature.
397) Click **Isometric view**.

398) Rename the **Cut-Extrude1** to **Vent**.

Save the CABINET in its formed state.

399) Click **Save** .

💡 Before you utilize the Sketch Linear Pattern feature, determine the design intent of the pattern in the part and in a future assembly. If there are mating components, use Pattern features. Pattern features provide additional options compared to a simple sketch tool.

Additional Pattern Options

Three additional pattern options to discuss in this book are Curve Driven Pattern, Sketch Driven Pattern and Table Driven Pattern.

- Curve Driven Pattern creates instances of the seed feature through a sketched curve or edge of a sketch.

- Sketch Driven Pattern creates instances of the seed feature through sketched points.

- Table Driven Pattern creates instances of the seed feature through X-Y coordinates in an existing table file or text file.

Utilize a Sketch Driven Pattern for a random pattern. A Sketch Driven Pattern requires a sketch and a feature. The Library Feature Sheetmetal folder contained the d-cutout and other common profiles utilized in sheet metal parts. Create a sketch with random Points on the CABINET left face. Insert sw-b212 and create a Sketch Pattern. View SOLIDWORKS help for additional information on Pattern features.

Activity: Additional Pattern Options-Sketch Driven Pattern.

Create the sketch.
400) Click **Left view**.

401) Right-click the **left face** of the CABINET for the Sketch plane.

402) Click **Sketch** from the Context toolbar. Click **Point** ▫ from the Sketch toolbar.

403) Sketch a **random series of points on the left face** as illustrated. The Point PropertyManager is displayed. Remember the first point you select. This point is required to constrain the sw-b212 sheet metal feature.

Close the sketch. Save the model.
404) Right-click **Exit Sketch**. Rename Sketch to **Sketch-Random**.

405) **Rotate** the CABINET to view the inside left face.

406) Click **Save** 💾.

Insert the sw-b212 sheet metal tool.
407) **Expand** the Design Library 🗄 folder in the Task Pane.

408) **Expand** the features folder.

409) Click the **Sheetmetal** folder.

410) Drag the **sw-b212 feature** to the inside left face of the CABINET, near the first point. The sw-b212 PropertyManager is displayed.

411) Click the **Edit Sketch** button. DO NOT SELECT THE FINISH BUTTON AT THIS TIME.

Add a Midpoint relation between the center point of the sw-b212 feature and the first random sketch point.
412) Click the **vertical centerline** of sw-b212.

413) Hold the **Ctrl** key down.

414) Click the **Point of the first random sketch point**. The selected entities are displayed in the Selected Entities box. Release the **Ctrl** key.

415) Click **Midpoint** ⬉ from the Add Relations box.

416) Click the **Finish** button.

Edit the End Condition.
417) Right-click **Cut-Extrude1** in the FeatureManager.

418) Click **Edit Feature**. Select **Blind** for End Condition in Direction 1.

419) Check the **Link to thickness** box.

420) Click **OK** ✅ from the Cut-Extrude1 PropertyManager.

Insert the Sketch Driven Pattern feature.
421) Click **Insert**, **Pattern/Mirror**, **Sketch Driven Pattern** from the Menu bar. The Sketch Driven Pattern PropertyManager is displayed.

422) Click inside the **Reference Sketch** box. **Expand** the CABINET fly-out FeatureManager.

423) Click **Sketch-Random**. If required, click inside the Features to Pattern box.

424) Click **Cut-Extrude1** from the fly-out FeatureManager.

425) Click **OK** ✅ from the Sketch Driven Pattern PropertyManager. Sketch-Pattern1 is displayed in the FeatureManager.

Display an Isometric view. Save the model.

426) Click **Isometric view**. Click **Save** 💾.

The Sketch Driven Pattern is complete. Suppress the patterns when not required for future model development.

Utilize the Suppress option with configurations to save the formed and flat state of the CABINET.

CABINET-Formed and Flat States

How do you control the 3D formed state in the assembly and the 2D flat state in the drawing? Answer: Utilize configurations. A configuration is a variation of a part or assembly within a single document. Variations include different dimensions, features, and properties.

The BOX assembly requires the CABINET in the 3D formed state. Utilize the Default Configuration in the assembly.

The CABINET drawing requires the 2D flat state. The drawing requires a view with the Linear Pattern of square cuts and the random sketch pattern suppressed. Create a new configuration named NO-VENT. Suppress the patterns. Create a Derived configuration named NO-VENT-FLAT. Un-suppress the Flat Pattern feature to maintain the CABINET in its flat state. Create two configurations in the 3D formed and 2D flat state.

Activity: CABINET-Formed and Flat States.

Display the ConfigurationManager and add a Configuration.

427) Drag the **Split bar** <img> downward to divide the FeatureManager in half.

428) Click the **ConfigurationManager** <img> tab.

429) Right-click **CABINET**.

430) Click **Add Configuration**. The Add Configuration PropertyManager is displayed.

431) Enter **NO-VENT** for Configuration name.

432) Enter **SUPPRESS VENT CUTS** for Description.

433) Click **OK** ✔ from the Add Configuration PropertyManager.

434) Drag the **Split bar** <img> upward to display a single FeatureManager.

Suppress the Vent Extruded Cut feature.

435) Right-click **Vent** from the FeatureManager.

436) Click **Suppress** ↓⁰ from the Context toolbar.

The NO-VENT Configuration contains the suppressed Vent Extruded Cut feature. A Flat Pattern exists for the Default Configuration. The Flat Pattern Configuration is called a Derived Configuration.

Create a Flat Pattern Derived Configuration for the NO-VENT configuration. Un-suppress the Flat Pattern feature.

Create a Derived Configuration.

437) Right-click **NO-VENT [CABINET]** from the ConfigurationManager.

438) Click **Add Derived Configuration**. The Add Configuration PropertyManager is displayed.

439) Enter **NO-VENT-FLAT-PATTERN** for Configuration name.

440) Click **OK** ✔ from the Add Configuration PropertyManager.

Un-suppress the Flat-Pattern1 feature. Suppress the Random Sketch points.

441) Right-click **Flat-Pattern1** from the CABINET FeatureManager.

442) Click **Unsuppress** ↑⁰ from the Context toolbar to display the Flat State.

443) Right-click **Sketch Random** from the CABINET FeatureManager.

444) Click **Suppress** ↓⁰ from the Context toolbar.

Display the configurations.

445) Double-click **Default [CABINET]** from the ConfigurationManager.

446) UnSuppress ↑[□] **Flat-Pattern1** from the FeatureManager.

447) Double-click **NO-VENT [CABINET]** from the ConfigurationManager. View the results in the Graphics window.

448) Double-click **NO-VENT-FLAT-PATTERN [CABINET]** from the ConfigurationManager.

449) Double-click **Default [CABINET]** from the ConfigurationManager to return to the Default Configuration.

450) Suppress ↓[□] **Flat-Pattern1** from the CABINET FeatureManager.

Return to a single CABINET FeatureManager.

451) Click the **FeatureManager** tab.

Save the CABINET.

452) Click **Save**.

CABINET-Sheet Metal Drawing with Configurations

Sheet metal drawings are produced in the flat state. Create a new C-size [A2] drawing size for the CABINET. Utilize the default SOLIDWORKS Drawing Template. Insert the Top view into the drawing. Modify the Drawing View Properties from the Default configuration to the Flat configuration.

To create a multi-configuration drawing, insert the Default configuration for Model View. Modify the View Properties in the drawing to change the configuration.

Activity: CABINET-Sheet Metal Drawing with Configurations.

Create a new drawing.

453) Click **Make Drawing from Part/Assembly** from the Menu bar.

454) Double-click **Drawing** from the Templates tab.

455) Select **C (ANSI) Landscape** for Sheet Format/Size.

456) Click **OK**. The View Palette is displayed on the right side of the Graphics window.

Insert an Isometric view using the View Palette.
457) Click and drag the **Isometric view** from the View Palette into the drawing sheet as illustrated.

458) Click **OK** from the Drawing View1 PropertyManager.

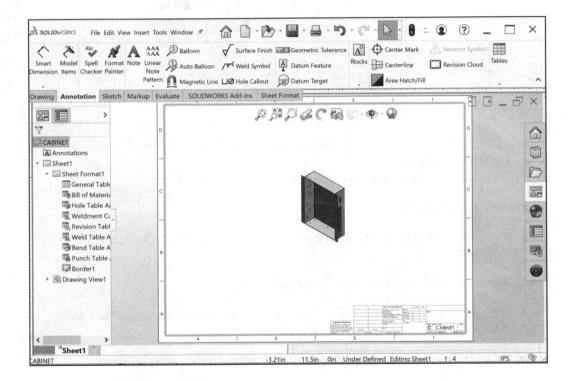

Insert the Default Flat pattern view.
459) Click and drag the **Flat pattern view** on the left side of the drawing.

460) Click **OK** ✅ from the Projected View PropertyManager.

Copy and Paste the two views.
461) Click inside the **Isometric view** boundary.

462) Hold the **Ctrl** key down.

463) Click inside the **Flat pattern view boundary**. The Multiple View PropertyManager is displayed.

464) Release the **Ctrl** key. Press **Ctrl C** to copy.

465) Click a **position below** the views. Click **Ctrl V** to paste. Four Drawing Views are displayed in the CABINET Drawing FeatureManager.

466) Click and drag the **views under each** other. Address the drawing view scale if needed.

Hide Sketches, Points, and Origins in the drawing if required.
467) Click **View**, **Hide/Show**, un-check **Sketches** from the Menu bar.

468) Click **View**, **Hide/Show**, un-check **Points** from the Menu bar.

469) Click **View**, **Show/Hide**, un-check **Origins** from the Menu bar.

Modify the configurations.

470) Click **inside the lower left Flat pattern view** in Sheet1.

471) Right-click **Properties**.

472) Select **NO-VENT-FLAT-PATTERN** from the Configuration information drop-down menu.

473) Click **OK** ✔ from the Drawing View Properties dialog box.

474) Click **inside the lower right Isometric** view.

475) Right-click **Properties**.

476) Select **NO-VENT "SUPPRESS VENT CUTS"** from the Configuration information drop down menu.

477) Click **OK** ✔ from the Drawing View Properties dialog box.

Hide Origins, Planes, and Sketches in drawing views. Insert centerlines to represent sheet metal bend lines.

Save and close the CABINET drawing.

478) Click **Save As** .

479) Enter **CABINET** For File name.

480) Click **Save**. Click **File**, **Close** from the Menu bar.

The CABINET is in the Default configuration.

Return to the BOX assembly. Review the BOX
FeatureManager. The MOTHERBOARD and
POWERSUPPLY are suppressed. Restore the
MOTHERBOARD and POWERSUPPLY. Collapse
entries in the FeatureManager.

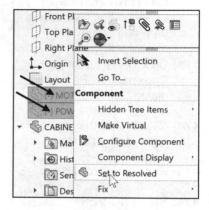

Return to the BOX assembly.
481) Open the **BOX** assembly.

Display the components.
482) Click **MOTHERBOARD** in the BOX FeatureManager.

483) Hold the **Ctrl** key down.

484) Click **POWERSUPPLY** in the FeatureManager.

485) Release the **Ctrl** key.

486) Right-click **Set to Resolved**. The selected entities are
displayed in the Graphics window.

487) Click **OK**.

Save the BOX assembly.

488) Click **Save** .

Work in SOLIDWORKS from the design through the manufacturing process to avoid data conversion issues and decreased shop time. The SOLIDWORKS Manufacturing Partner Network (www.SOLIDWORKS.com) lists Sheet Metal manufacturers that utilize SOLIDWORKS.

Additional details on Sheet Metal, Rip, Insert Bends, Flat Pattern, Forming Tools, Edge Flange, Hem, Library Feature, Linear Step and Repeat and Configuration Properties are available in SOLIDWORKS Help.

 Review the CABINET.

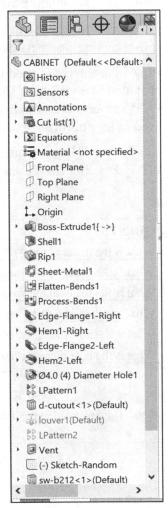

You created the CABINET component inside the BOX assembly. The InPlace Mate inserted a Coincident Mate between the Front Plane of the CABINET and the Front Plane of the BOX. The CABINET component referenced the outside profile of the Layout Sketch.

The first feature of the CABINET was the Extruded Boss/Base (Boss-Extrude1) feature that created a solid box. You utilized the Shell feature to create the constant wall thickness.

The Rip feature cut removed the solid edges. The Insert Bends feature created a sheet metal part from a solid part. The Flat Pattern feature displayed the sheet metal part in the manufactured 2D flat state.

The Edge Flange and Hem features were added to the right and left side of the CABINET. Sheet metal dies and form louvers were inserted from the Library Features.

The Sketch Linear Pattern was utilized to replace the form louvers with more cost-effective cuts. You created a configuration to represent the 2D flat state for the CABINET drawing. The 3D formed state was utilized in the BOX assembly.

The Vent feature located in the Sheet Metal toolbar allows you to create a vent by selecting a closed profile as the outer vent boundary.

The Fill Pattern feature allows you to fill a closed boundary to create a grid for a sheet-metal perforation - style pattern.

PEM® Fasteners and IGES Components

The BOX assembly contains an additional support BRACKET that is fastened to the CABINET. To accommodate the new BRACKET, the Linear Pattern of five holes is added to the CABINET.

How do you fasten the BRACKET to the CABINET?

Answer: Utilize efficient and cost effective self-clinching fasteners. Example: PEM® Fasteners.

PEM® Fastening Systems, Danboro, PA, USA manufactures self-clinching fasteners called PEM® Fasteners.

PEM® Fasteners are used in numerous applications and are provided in a variety of shapes and sizes. IGES standard PEM® models are available at (www.pennfast.com) and are used in the following example.

Answer the following questions in order to select the correct PEM® Fastener for the assembly:

QUESTION:	ANSWER:
What is the material type?	Aluminum
What is the material thickness of the CABINET?	1mm
What is the material thickness of the BRACKET?	19mm
What standard size PEM® Fastener is available?	FHA-M5-30ZI
What is the hole diameter required for the CABINET?	5mm ± 0.08
What is the hole diameter required for the BRACKET?	5.6mm maximum
Total Thickness = CABINET Thickness + BRACKET Thickness + NUT Thickness Total Thickness = 1mm + 19mm + 5mm = 25mm (minimum)	25mm (minimum)

The selected PEM® Fastener for the project is the FHA-M5-30ZI.

- **FHA** designates the stud type and material, Flush-head Aluminum.

- **M5** designates the thread size.

- **30** designates length code: 30mm.

- **ZI** designates finish code.

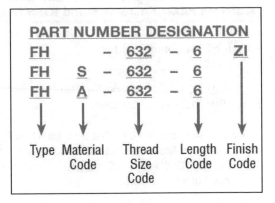

Part Number General Designation for PEM®

The overall dimensions of the FH PEM® Fasteners series are the same for various materials. Do not utilize PEM® Fasteners as a feature in the Flat State of a component. In the sheet metal manufacturing process, PEM® Fasteners are added after the material pattern has been fabricated.

Proper PEM® Fasteners material selection is critical. There are numerous grades of aluminum and stainless steel. Test the installation of the material fastener before final specification.

Note: When the fastener is pressed into ductile host metal, it displaces the host material around the mounting hole, causing metal to cold flow into a designed annular recess in the shank.

Ribs prevent the fastener from rotating in the host material once it has been properly inserted. The fasteners become a permanent part of the panel in which they are installed.

During the manufacturing process, a stud is installed by placing the fastener in a punched or drilled hole and squeezing the stud into place with a press. The squeezing action embeds the head of the stud flush into the sheet metal.

Obtain additional manufacturing information that directly affects the design.

Manufacturing Information:	FH-M4ZI:	FH-M5ZI:
Hole Size in sheet metal (CABINET).	4.0mm ± 0.8	5.0mm ± 0.8
Maximum hole in attached parts (BRACKET).	4.6mm	5.6mm
Centerline to edge minimum distance (from center point of hole to edge of CABINET).	7.2mm	7.2mm

Activity: PEM® Fasteners and IGES Components.

Use the PEM Fasteners for this project from the
Vendor Component folder.

🔅 Part geometry for Aluminum and Stainless
Steel fasteners are the same.

Copy and open the needed files from the COMPONENTS folder.

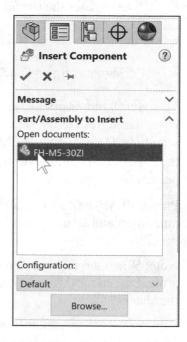

489) Copy **FH-M5-30ZI** and **FH-M4-25ZI** from the ENGDESIGN-
W-SOLIDWORKS 2024\Vendor Component folder on your
hard drive.

490) Open the **FH-M5-30ZI** part.

491) Click **No**. The model has imported geometry.

Insert the FH-M5-30ZI fastener into the BOX assembly.

492) Open the **BOX** assembly.

493) Click **OK**.

494) Click **Insert Components** 📂 from the Menu bar. The
Insert Component PropertyManager is displayed.

495) Double-click **FH-M5-30ZI** in the Open documents box.

496) Click inside the **Graphics window** in the inside bottom
left corner near the seed feature of the Linear Pattern of
holes.

497) Zoom to Area 🔍 on the hole and fastener.

Rotate the FH-M5-30ZI component.

498) Click **Rotate Component** 🔄 from the Assembly toolbar.
The Rotate Component PropertyManager is displayed.

499) Rotate the **FH-M5-30ZI** component until the head faces
downward.

500) Click **OK** ✔ from the Rotate Component
PropertyManager.

Insert a Concentric mate.

501) Click the **cylindrical shaft face** of the FH-M5-30ZI fastener.

502) Hold the **Ctrl** key down.

503) Click the **cylindrical face** of the back left hole.

504) Release the **Ctrl** key. The Mate Pop-up menu is displayed.

505) Click **Concentric** from the Pop-up menu. A Concentric mate is applied.

Insert a Coincident mate.

506) Click the flat **head face** of the FH-M5-30ZI fastener.

507) Rotate **CABINET** to view the outside bottom face.

508) Hold the **Ctrl** key down.

509) Click the **outside bottom face** of the CABINET.

510) Release the **Ctrl** key. The Mate Pop-up menu is displayed.

511) Click **Coincident** ⟨ from the Pop-up menu. A Coincident mate is applied.

Insert a Parallel mate.

512) Click **Mate** 🔗 from the Assembly toolbar.

513) Click the **Standard** tab.

514) **Expand** BOX from the fly-out FeatureManager.

515) Click the **Front Plane** of the FH-M5-30ZI fastener.

516) Click the **Right Plane** of the CABINET. The selected planes are displayed in the Mate Selections box.

517) Click **Parallel**.

518) Click **OK** ✔ from the Mate PropertyManager.

Fit to the Graphics window. Display an Isometric view.
Save the model.
519) Press the **f** key.

520) Click **Isometric view**.

521) Click **Save** .

Three mates fully define the FH-M5-30ZI fastener. The PEM® fastener creates an
interference fit with hole1.

Pattern Driven Component Pattern

A Pattern Driven Component Pattern is a pattern that utilizes an existing component in an
assembly and a driving feature from another component in the same assembly. Utilize the
Pattern Driven Component Pattern tool to create multiple copies of a component in an
assembly. The FH-M5-30ZI references the hole seed feature. The Pattern Driven
Component Pattern displays the five FH-M5-30ZI fasteners based on the CABINET
LPattern1 feature.

Activity: Pattern Driven Component Pattern.

Insert a Pattern Driven Component Pattern.
522) Click the **Pattern Driven Component Pattern** tool.

523) Click inside the **Components to Pattern** box.

524) Click **FH-M5-30ZI** from the fly-out FeatureManager.

525) Click **inside** the Driving Feature or Component box.

526) Click **CABINET\LPattern1** from the BOX fly-out
FeatureManager.

527) Click **OK** ✔ from the Pattern Driven PropertyManager.
DerivedLPattern1 is displayed in the FeatureManager.

Expand DerivedLPattern1.
528) **Expand** DerivedLPattern1 in the FeatureManager. View the
results.

Display an Isometric view. Save the BOX.
529) Click **Isometric view** from the Heads-up View toolbar.

530) Click **Save** .

The additional FH-M5-30ZI instances are located under DerivedLPattern1 in the FeatureManager. The purchasing manager for your company determines that the FH-M5-30ZI fastener has a longer lead-time than the FH-M4-25ZI fastener.

The FH-M4-25 fastener is in stock. Time is critical. You ask the engineer creating the corresponding BRACKET if the FH-M4-25ZI fastener is a reliable substitute.

You place a phone call. You get voice mail. You send email. You wait for the engineer's response and create the next assembly feature.

How do you locate a hole for a fastener that has to go through multiple components in an assembly? Answer: Utilize an Assembly Hole feature.

MOTHERBOARD-Assembly Hole Feature

Assembly features are holes and cuts created in the assembly. Utilize the Assembly Feature option from the Menu bar or from the Assembly tab after components have been assembled. The Assembly Feature option includes Hole Series, Hole Wizard, Simple Hole, Revolved Cut, Extruded Cut, Swept Cut, Fillet, Chamfer, Weld Bead, Linear Pattern, Circular Pattern, Table Driven Pattern, Sketch Driven Pattern and Belt/Chain. Create a Cut or Hole Assembly feature to activate the Assembly Feature Pattern tool.

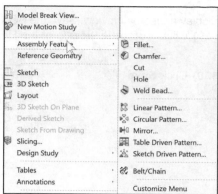

Insert an Assembly feature into the BOX assembly. When creating an Assembly feature, determine the components to be affected by the feature. Assembly features are listed at the bottom of the FeatureManager. They are not displayed in the components affected by the feature.

Additional holes are created through multiple components in the BOX assembly. A through hole is required from the front of the MOTHERBOARD to the back of the CABINET. Since the hole is an Assembly feature, it is not displayed in the CABINET part.

Activity: MOTHERBOARD-Assembly Hole Feature.

Insert an Assembly feature with the Hole Wizard.

531) Click **Hole Wizard** 🖱 from the Assembly toolbar. The Hole Specification PropertyManager is displayed.

532) Click the **Hole** icon type.

533) Select **ANSI Metric** for Standard.

534) Select **Drill sizes** for Type. Select **Ø5.0** for Size.

535) Click **Through All** for End Condition.

536) Click the **Positions** tab. Click the **front face** of the MOTHERBOARD. The Point Sketch tool is displayed. Click **again** to position the center point of the hole.

537) Right-click **Select** to deselect the Sketch Point tool in the Graphics window.

538) Click **Front view**.

Dimension the hole on the MOTHERBOARD.
539) Click **Smart Dimension** from the Sketch toolbar.

Add a horizontal dimension.
540) Click the **left edge** of the MOTHERBOARD. Click the **centerpoint** of the hole. Click a **position** below the profile. Enter **25**mm.

Add a vertical dimension.
541) Click the **Origin** of the CABINET from the fly-out FeatureManager.

542) Click the **centerpoint** of the hole.

543) Click a **position** to the left of the profile.

544) Enter **50**mm.

545) Click **OK** from the Dimension PropertyManager.

546) Click **OK** from the Hole Position PropertyManager. The hole is added to the FeatureManager.

Display an Isometric view. Save the BOX.
547) Click **Isometric view**.

548) Click **Save**.

Close all parts and assemblies.
549) Click **Window**, **Close All** from the Menu bar.

Assembly FeatureManager and External References

The FeatureManager contains numerous entries in an assembly. Understanding the organization of the components and their Mates is critical in creating an assembly without errors. Errors occur within features such as a radius that is too large to create a Fillet feature. When an error occurs in the assembly, the feature or component is labeled in red.

Mate errors occur when you select conflicting geometry such as a Coincident Mate and a Distance Mate with the same faces.

In the initial mating, undo a mate that causes an error. Mate errors occur later on in the design process when you suppress required components and features. Plan ahead to avoid a problem.

Use reference planes for Mates that will not be suppressed.

When a component displays a '->' symbol, an External reference exists between geometry from the assembly or from another component.

When the system does not locate referenced geometry, the part name displays a '->?' symbol to the right of the component name in the FeatureManager.

Explore External references in the next activity. Open the CABINET without opening the BOX assembly. The Open External Referenced documents dialog box is displayed. Click OK (Open all referenced documents).

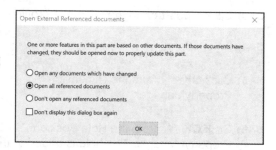

Activity: Assembly FeatureManager and External References.

Open the CABINET part.

550) Click **Open** from the Menu bar.

551) Double-click **CABINET**. The CABINET FeatureManager is displayed.

552) Click **OK** from the Open External Referenced documents dialog box. The CABINET FeatureManager is displayed.

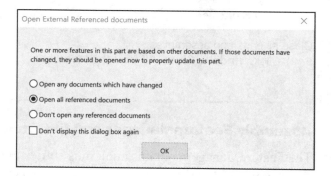

Save the CABINET.

553) Click **Save** .

The CABINET references the BOX assembly. The External references are resolved and the question marks are removed.

Replace Components

The Replace Components option replaces a component in the current assembly with another component. Replace a part with a new part or assembly or an assembly with a part. If multiple instances of a component exist, select the instances to replace.

The BRACKET engineer contacts you. You proceed to replace the FH-M5-30ZI fastener with the FH-M4-25ZI fastener. Use the Replace Components option to exchange the FH-M5-30ZI fastener with FH-M4-25ZI fastener in the BOX assembly.

Edit the Mates and select new Mate references. The Derived Component Pattern updates with the new fastener. Edit the CABINET and change the hole size of the seed feature from Ø5.0mm to Ø4.0mm. The CABINET Linear Pattern updates with the new hole.

Before you replace Component-A in an assembly with Component-B, close Component-B. The Replace Component option produces an error message with open components. Example: Close the FH-M4-25ZI part before replacing the FH-M5-30ZI part in the current assembly.

Activity: Replace Components.

Replace the FH-M5-30ZI component with the FH-M4-25ZI component.

554) Open the **BOX** assembly.

555) Right-click **FH-M5-30ZI**

▸ 🔩 FH-M5-30ZI<1> from the FeatureManager.

Replace five FH-M5-30ZI with FH-M4-25ZI by replacing the seed feature.

556) Click the **More arrow** ⌄ at the bottom of the Pop-up menu.

557) Click **Replace Components** . The Replace PropertyManager is displayed.

558) Click the **Browse** button.

559) Double-click **FH-M4-25ZI** from the folder on your hard drive. ENGDESIGN-W-SOLIDWORKS\Vendor Components folder.

560) Click **OK** ✔ from the Replace PropertyManager. The What's Wrong box is displayed.

561) Click **Close** from the What's Wrong dialog box.

The five FH-M5-30ZI fasteners are replaced with the five FH-M4-25ZI fasteners. Mate Entity errors are displayed.

A red X is displayed on the Mated Entities PropertyManager. Redefine the Concentric Mate and Coincident Mate. Select faces from the FH-M4-25ZI component.

Edit the Concentric Mate.

562) **Expand** the first Mate entity. ? Face of FH-M4-25ZI-1 .

563) Double-click **Concentric1**.

564) Click the **FH-M4-25ZI cylindrical face** as illustrated in the Graphics window. View the results in the Mate Entities box.

Edit the Coincident mate.
565) Expand the second Mate entity.

566) Click **Bottom view**.

567) Double-click **Coincident1**.

568) Click the **FH-M4-25 bottom face**. View the results in the Mate Entities box.

569) Click **OK** ✔ from the Mate Entities PropertyManager.

Edit the Ø5.0mm hole to a Ø4.0mm hole.
570) Click **CABINET** from the FeatureManager.

571) Click **Edit Component** 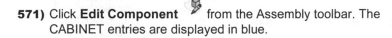 from the Assembly toolbar. The CABINET entries are displayed in blue.

572) Expand CABINET in the FeatureManager.

573) Right-click **(5) Diameter Hole1** from the FeatureManager.

574) Click **Edit Feature**. The Hole Specification PropertyManager is displayed. Select **Ø4.0** for Size.

575) Click **OK** ✔ from the Hole Specification PropertyManager. The (4) Diameter Hole1 is displayed in the FeatureManager.

576) Click **OK**.

577) Click **Edit Component** from the Assembly toolbar to return to the BOX assembly. The CABINET entries are displayed in black.

Save the BOX.
578) Click **Save** 💾. Click **Save All**.

💡 The hole of the CABINET requires an Ø4mm ± 0.8. Add a limit dimension in the drawing and a revision note to document the change from an Ø5.0mm to an Ø4.0mm.

💡 IGES imported geometry requires you to select new mate references during Replace Components. Utilize reference planes for Mate entities to save time. The Replace Components option attempts to replace all Mate entities. Reference planes provide the best security in replacing the appropriate Mate entity during the Replace Components option.

Equations

How do you ensure that the fasteners remain symmetrical with the bottom of the BOX when the box-width dimension varies?

Answer: With an Equation. Equations use shared names to control dimensions. Use Equations to connect values from sketches, features, patterns and various parts in an assembly.

Create an Equation. Each dimension has a unique variable name. The names are used as Equation variables. The default names are based on the Sketch, Feature or Part. Rename default names for clarity.

Increase BOX width. Fasteners are not symmetrical with the bottom of the BOX.

Resolve all Mate Errors before creating Equations. Mate Errors produce unpredictable results with parameters driven by equations.

Activity: Create an Equation.

Display Feature dimensions.
579) Right-click the CABINET **Annotations** folder.

580) Check the **Show Feature Dimensions** box.

581) Click **Front view**.

Edit the LPattern1 dimension name.
582) Click the **125**mm dimension in the Graphics window. The Dimension PropertyManager is displayed.

Rename the D3@LPattern1 dimension.
583) Click inside the **D3@LPattern1** box.

584) Enter **lpattern-bottom** for Name.

585) Click **OK** from the Dimension PropertyManager.

Edit the 4.0mm hole dimension name.
586) Rotate to display the back hole.

587) Click the **25**mm dimension. 25 is the Ø4.0mm hole horizontal dimension in the lower left corner. The Dimension PropertyManager is displayed.

Rename the Sketch dimension name.
588) Enter **hole-bottom** for Name.

589) Click **OK** ✔ from the Dimension PropertyManager.

Display an Isometric view.
590) Click **Isometric view**.

Display the BOX feature dimensions.
591) Right-click the BOX **Annotations** folder.

592) Check the **Show Feature Dimensions** box.

Create an Equation.
593) Click **Front view**.

594) Click **Tools**, **Equations** Σ Equations... from the Menu bar. The Equations, Global Variables, and Dimensions dialog box is displayed.

Create the first half of equation1.
595) Click a **position** in the fifth cell under Equations - Top Level.

596) Click the lpattern-bottom @Pattern1 horizontal dimension, **125**mm in the Graphics window. The variable "lpattern-bottom @LPattern1@ CABINET" is added.

D1@Sketch18 of CABINET<2>

Name	Value / Equation	Evaluates to	Comments
Global Variables			
"Gap"	= "D9@Layout"	20.00mm	Gap from cabinet
Add global variable			
Features			
Add feature suppression			
Equations - Top Level			
"D8@Layout"	= "Gap"	20.00mm	
"D10@Layout"	= "Gap"	20.00mm	
"D5@Layout"	= "Gap"	20.00mm	
"D6@Layout"	= "Gap"	20.00mm	
Equations - Components			
Add equation			

Create the second half of the Equation.

597) Enter **0.5*(** from the keypad.

598) Click the box_width horizontal dimension, **300** from the Graphics window. The variable "box_width@layout" is added to the equation text box.

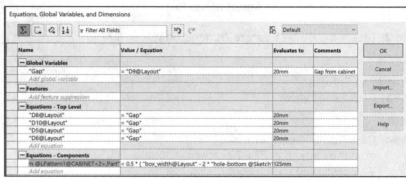

599) Enter **–2*** from the keypad.

600) Click the M4 hole-bottom dimension, **25**. The variable "hole-bottom@Sketch12@CABINET.Part" is added to the equation text box.

601) Enter **)** from the keypad.

602) Press the **Tab** key. View the results.

Click **OK** from the dialog box.

The Equations dialog box contains the complete equation. A green check mark indicates that the Equation is solved. The Equation evaluates to 125mm.

The Equation Σ icon placed in front of the 125 dimension indicates that an equation drives the value.

Remember the initial design parameters. There are three different size boxes. Remember the design intent. You build a Layout Sketch to control the different sizes.

Test the equation by modifying the Layout Sketch dimensions.

Verify the Equation. Modify the dimensions.
603) Click **Front view**.

604) Double-click the horizontal dimension box-width, **300**mm.

605) Enter **400**mm.

606) Click **Rebuild** . The Equation dimension displays 175.

607) Double-click the dimension box-height, **400**mm.

608) Enter **500**mm.

609) Click **Rebuild** .

Return to the original dimensions.
610) Double-click on the horizontal dimension, **400**mm.

611) Enter **300**mm.

612) Double-click the vertical dimension, **500**mm.

613) Enter **400**mm.

614) Click **Rebuild** .

Hide the dimensions.
615) Right-click the CABINET **Annotations** folder.

616) Un-check the **Show Feature Dimension** box.

617) Right-click the BOX **Annotations** folder.

618) Uncheck the **Show Feature Dimension** box. View the Equations folder.

Display an Isometric view. Save the BOX.
619) Click **Isometric view**.

620) Click **Save** .

The BOX assembly requires three configurations. Recall the CABINET drawing. You had two configurations that controlled the Flat Pattern feature state.

Now control dimensions, features and parts in an assembly. A Design Table is an efficient tool to control variations in assemblies and parts.

Design Tables

A Design Table is an Excel spreadsheet used to create multiple configurations in a part or assembly. Utilize the Design Table to create three configurations of the BOX assembly:

- Small.

- Medium.

- Large.

Activity: Design Tables.

Create a Design Table.
621) Click **Insert**, **Tables**, **Excel Design Table** from the Main menu. The Excel Design Table PropertyManager is displayed. Accept the Auto-create default setting.

622) Click **OK** ✔ from the Excel Design Table PropertyManager.

Select the input dimension.
623) Hold the **Ctrl** key down.

624) Click the **box_width@Layout** dimension.

625) Click the **box_height@Layout** dimension.

626) Release the **Ctrl** key.

627) Click **OK** from the Dimension dialog box.

The input dimension names and default values are automatically entered into the Design Table.

The value Default is entered in Cell A3. The value 400 is entered in Cell B3 for box_height and the value 300 is entered in Cell C3 for box_width.

Enter the three configuration names.
628) Double click Cell **A4**. Enter **Small**.

629) Double click Cell **A5**. Enter **Medium**.

630) Double click Cell **A6**. Enter **Large**.

Enter the dimension values for the Small configuration.
631) Double click Cell **B4**. Enter **400** for box_height.

632) Double click Cell **C4**. Enter **300** for box_width.

Enter the dimension values for the Medium configuration.
633) Double click Cell **B5**. Enter **600**. Double click Cell **C5**.
Enter **400**.

Enter the dimension values for the Large configuration.
634) Double click Cell **B6**. Enter **650**. Double click Cell **C6**.
Enter **500**.

Build the three configurations.
635) Click a **position** outside the Excel Design Table in the
Graphics window.

636) Click **OK** to generate the three configurations.

Display the configurations.
637) Click the **ConfigurationManager** tab. The Design
Table icon is displayed.

638) Double-click **Small**. Double-click **Medium**.

639) Double-click **Large**.

Return to the Default configuration.
640) Double-click **Default**.

Fit the model to the Graphics window.
641) Press the **f** key.

Edit the Design Table.
642) Right-click **Design Table** from the BOX
ConfigurationManager.

643) Click **Edit Table**.

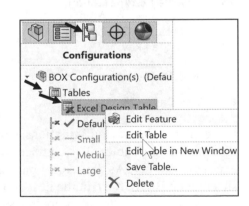

644) Click **OK** from the Add Rows and Columns dialog box.

645) Double click Cell **D2**. Enter **$State@POWERSUPPLY<1>**.

646) Double click Cell **D3**. Enter **R** for Resolved.

647) Copy Cell **D3**. Select Cell **D4** and Cell **D5**. Click **Paste**. Click Cell **D6**. Enter **S** for Suppressed. Note: If needed remove any hyperlinks.

648) Update the configurations. Click a **position** outside the EXCEL Design Table.

	A	B	C	D
1	Design Table for: BOX			
2		box height@Layout	box_width@Layout	$State@POWERSUPPLY<1>
3	Default	400	300	R
4	Small	400	300	R
5	Medium	600	400	S
6	Large	650	500	S
7				
8				
9				
10				

Sheet1 ⊕

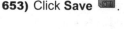 The Design Table component variables must match the FeatureManager names in order to create configurations. The Design Table variable $State@POWERSUPPLY<1> is exactly as displayed in the BOX FeatureManager. If the name in the BOX FeatureManager was P-SUPPLY<2> then the Design Table variable is $State@ P-SUPPLY<2>.

Display the Large Configuration.
649) Double-click **Large** from the ConfigurationManager. The POWERSUPPLY is suppressed in the Large configuration.

Return to the Default configuration and FeatureManager.
650) Double-click **Default**.

651) Click the **FeatureManager** tab.

Fit the model to the Graphics window.
652) Press the **f** key.

Save the BOX assembly.

653) Click **Save**.

Dimensions, Color, Configurations, Custom Properties, and other variables are controlled through Design Tables. The engineer working on the sheet metal bracket had to visit another customer. As the project manager, it is up to you to get the job done. Create the sheet metal BRACKET in the next activity. The sheet metal BRACKET is created In-Context to illustrate the Top-down Assembly Modeling process. Sheet metal parts are created in a Bottom-up approach and then inserted and mated in the assembly.

BRACKET Part-Sheet Metal Features

The CABINET was converted from a solid part to a sheet metal part by using the Rip and Insert Bends feature. The sheet metal BRACKET part starts with the Base Flange feature. The Base Flange feature utilizes a U-shaped sketch profile. The material thickness and bend radius are set in the Base Flange.

Create the BRACKET In-Context of the BOX assembly. The BRACKET references the inside bottom edge and left corner of the CABINET.

☼ Orient the Flat Pattern. The first sketched edge remains fixed with the Flatten tool.

Sketch the first line from left to right Coincident with the inside bottom edge. The BRACKET is not symmetrical and requires a right hand and left hand version. Utilize Mirror Component feature to create a right-hand copy of the BRACKET.

Activity: BRACKET Part-Sheet Metal Features.

Insert the BRACKET component.
654) If required, open the **BOX** assembly.

655) Click **New Part** for the Consolidated Insert Components drop-down menu. A new virtual default component is added to the FeatureManager. Note: Save the new component and name it BRACKET later in the procedure.

The Component Pointer ✓ is displayed on the mouse pointer.

656) Click **Front Plane** from the BOX FeatureManager.

657) **Expand** the new inserted component [Part1^BOX]<1> in the FeatureManager. The FeatureManager is displayed in light blue.

Sketch the profile for the Base Flange.
658) Click **Front view**.

659) Check **View**, **Hide/Show**, **Temporary Axis** from the Menu bar.

660) Click **Hidden Lines Visible** from the Heads-up View toolbar.

661) **Zoom to Area** on the front lower inside edge of the BOX as illustrated.

662) Click **Line** from the Sketch toolbar.

663) Sketch a **horizontal line** Collinear with the inside edge of the CABINET. Note: Sketch the horizontal line from left to right.

664) Sketch a **vertical line** and a **horizontal line** to complete the U-shaped profile.

Add an Equal relation.
665) Right-click **Select**. Click the **top horizontal** line. Hold the **Ctrl** key down.

666) Click the **Bottom horizontal** line.

667) Release the **Ctrl** key.

Inside edge of the BOX

668) Click **Equal** $=$ from the Add Relations box.

669) Click **OK** ✔ from the PropertyManager.

The bottom horizontal line is displayed in black. The bottom line is Collinear with the inside bottom edge. Add a Collinear relation between the bottom line and the inside bottom edge if required.

Add dimensions.
670) Click **Smart Dimension** ✎ from the Sketch toolbar.

671) Click the **top horizontal line**. Click the **Top** of the fastener. Click a **position** to the right of the fastener. Enter **2**mm.

672) Click the **centerline** of the fastener. Click the **vertical line**. Click a **position** below the fastener.

673) Enter **10**mm. Click the **left edge** of the CABINET.

674) Click the **left end point** of the top horizontal line.

675) Click a **position** above the profile.

676) Enter **5**mm.

Insert the Base Flange feature.
677) Click **Isometric view**.

678) Click **Base Flange/Tab** ⋃ from the Sheet Metal toolbar. The Base Flange PropertyManager is displayed.

679) Select **Up to Vertex** for End Condition in Direction 1.

680) Click the CABINET **front left vertex** as illustrated.

681) Enter **1**mm for Thickness. If required, click the Reverse direction box.

682) Enter **2**mm for Bend Radius.

683) Enter **.45** for K factor. If needed, check the Reverse direction box. Accept the default settings.

684) Click **OK** ✅ from the Base Flange PropertyManager.

685) Click **Front view**. Note the location of the BRACKET.

The PEM® fasteners protrude through the BRACKET in the BOX assembly.

Save the BOX assembly.

686) Click **Edit Component** ✏️ from the Assembly toolbar. Click **Save** 💾. Click **Save All**. Click **OK**.

Rename the default component [Part1^BOX]<1> to BRACKET.
687) Right-click **[Part1^BOX]<1>** from the FeatureManager. Click **Open Part**. Click **Save As** from the Menu bar. Enter **BRACKET**. Click **Save**. The BRACKET FeatureManager is displayed. **Close** the BRACKET part.

▸	🧊 MOTHERBOARD<1> ->
▸	🧊 POWERSUPPLY<1> -> (
▸	🧊 CABINET<2> -> (Defaul
▸	🧊 FH-M4-25ZI<1> (Defau
▾	🧊 BRACKET<1> -> (Defau

688) Return to **BOX** assembly.

The BRACKET part contains no holes. The BRACKET requires two holes that reference the location of the fasteners. Utilize In-Context Extruded Cut feature to locate the holes. Display Wireframe mode to view the fastener position. Allow for a 0.8mm clearance.

Activity: BRACKET Part-In-context Features.

Insert the Extruded Cut feature In-Context of the BOX assembly.
689) Click **BRACKET** from the FeatureManager.

690) Right-click **Edit Part**. The Edit Component icon is activated and the BRACKET entry in the FeatureManager is displayed in blue.

691) Press the **f** key. **Rotate** to view the top face of the BRACKET.

Base-Flange1 of BRACKET<1>

692) Right-click the **top face** of the BRACKET. Click **Sketch** 📝 from the Context toolbar.

693) Click **Top view**.

694) Click **Wireframe** from the Heads-up View toolbar.

Sketch two circular profiles referencing the PEM center points.
695) Click **Circle** from the Sketch toolbar.

696) Drag the **mouse pointer** over the centerpoint of the top
PEM circular edge to "wake up" the PEM center point.

697) Click the **PEM centerpoint**. Click a **position** to the right of
the center point.

698) Drag the **mouse pointer** over the centerpoint of the
bottom PEM circular edge to "wake up" the PEM
centerpoint.

699) Click the **PEM center point**.

700) Click a **position** to the right of the centerpoint.

Deselect the Circle Sketch tool and add an Equal relation.
701) Right-click **Select**. Click the **circumference** of the top
circle.

702) Hold the **Ctrl** key down. Click the **circumference** of the
bottom circle.

703) Release the **Ctrl** key. Click **Equal** from the Add Relations
box.

704) Click **OK** ✔ from the Properties PropertyManager.

Add dimensions.
705) Click **Smart Dimension** ⟲ from the Sketch toolbar.

706) Click the **circumference** of the top circle.

707) Click a **position** off the profile.

708) Enter **4.80**mm.

709) Click **Extruded Cut** 🔲 from the Features toolbar. The Cut-Extrude PropertyManager is
displayed.

710) Select **Through All** for End Condition in Direction1. Accept the default settings.

711) Click **OK** ✔ from the Cut-Extrude PropertyManager.

$\emptyset$ 4.80

Return to the BOX assembly.

712) Click **Edit Component** from the Assembly toolbar.

Display an Isometric view - Shaded With Edges.
713) Click **Isometric view**.

714) Click **Shaded With Edges** from the Heads-up View toolbar.

Save the Box assembly and its referenced components.

715) Click **Save** . Click **Save All**.

No additional references from the BOX components are required. The Edge Flange/Tab feature, Break Corner feature, and Miter Flange feature are developed in the BRACKET part.

The Edge Flange feature adds a flange on the selected edge. Edit the Flange Profile to create a 30mm tab, 20mm from the front face. The tab requires fillet corners. The Break Corner feature adds fillets or chamfers to sheet metal edges.

> **Activity: BRACKET Part-Edge Flange Tab, Break Corner, and Miter Flange Features**.

Open the BRACKET.
716) Right-click **BRACKET** in the Graphics window.

717) Click **Open Part** from the Context toolbar.

Display the 2D flat state.
718) Right-click **Flat-Pattern1** from the FeatureManager.

719) Click **Unsuppress** from the Context toolbar.

Display the 3D formed state.
720) Right-click **Flat-Pattern1** from the FeatureManager.

721) Click **Suppress** from the Context toolbar.

Insert a tab using the Edge Flange feature.
722) Click the **top left edge** of the BRACKET as illustrated.

723) Click **Edge Flange** from the Sheet Metal toolbar. The Edge Flange PropertyManager is displayed.

724) Click a **position** above the BRACKET for direction.

725) Enter **20**mm for Flange Length.

726) Click the **Edit Flange Profile** button. Do not click the Finish button at this time. Move the Profile Sketch dialog box off to the side of the Graphics window.

Profile Sketch

The sketch is valid.

< Back | Finish | Cancel | Help

727) Drag the **front edge** of the tab towards the front hole as illustrated.

728) Drag the **back edge** of the tab towards the front hole as illustrated.

Add dimensions.

729) Click **Smart Dimension** from the Sketch toolbar.

730) Click the **front left vertex** of Base-Flange1.

731) Click the **front vertical** line as illustrated.

732) Click a **position** above the profile.

733) Enter **20**mm.

734) Click the **top horizontal** line.

735) Click a **position** above the profile.

736) Enter **30**mm.

737) Click **Finish** from the Profile Sketch box.

738) Click **OK** from the PropertyManager. Edge-Flange1 is displayed in the FeatureManager.

Insert the Break Corner/Corner Trim Feature.

739) **Zoom in** on the tab.

740) Click **Break Corner/Corner Trim** from the Sheet Metal toolbar. The Break Corner PropertyManager is displayed.

741) Click the small **front edge** of the flange as illustrated. Edge<1> is displayed in the Corner Edges box.

742) Click the small **back edge** of the flange. Edge<2> is displayed in the Corner Edges box. Select **Fillet** for Break type.

743) Enter **10**mm for Radius.

744) Click **OK** from the Break Corner PropertyManager. Break Corner1 is displayed in the FeatureManager.

Save the model.

745) Click **Save** .

The Miter Flange feature inserts a series of flanges to one or more edges of a sheet metal part. A Miter Flange requires a sketch. The sketch can contain lines or arcs and can contain more than one continuous line segment.

The Sketch plane is normal (perpendicular) to the first edge closest to the selection point. The geometry you create must be physically possible. New flanges cannot cross or deform. The current 2mm bend radius of the BRACKET is too small to insert a Miter Flange feature. Leave the area open for stress relief and utilize a Gap distance of 0.5mm.

Gap for stress relief

Insert the Miter Flange Feature.

746) Click **Miter Flange** from the Sheet Metal toolbar.

747) Click the BRACKET **front inside edge** as illustrated. The new sketch Origin is displayed in the left corner.

748) **Zoom to Area** on the left corner.

Base-Flange1

749) Click **Line** ✎ from the Sketch toolbar.

750) Sketch a **small vertical line** at the front inside left corner. The first point is coincident with the new sketch Origin as illustrated.

Add dimensions.

751) Click **Smart Dimension** ⤢ from the Sketch toolbar.

752) Click the **vertical line**.

753) Click a **position** to the front of the BRACKET.

754) Enter **2**mm.

755) Click **Exit Sketch**.

756) Click the **Propagate** 🗗 icon to select inside tangent edges.

757) Uncheck the **Use default radius** box.

758) Enter **.5**mm for Bend Radius.

759) Check the **Trim side bends** box.

760) Enter **.5**mm for Rip Gap distance.

761) Click **OK** ✅ from the Miter Flange PropertyManager. Miter Flange1 is displayed in the FeatureManager.

Display the flat and formed states.

762) Click **Isometric view**.

763) Click **Flatten** from the Sheet Metal toolbar to display the 2D flat state.

764) Click **Flatten** from the Sheet Metal toolbar to display the 3D formed state.

Save the BRACKET.

765) If required, click **View**, **Hide/Show**, un-check **Planes** from the Menu bar.

766) Click **Save** .

The Trim side bends option creates a Chamfer at the inside corners of the Miter Flange feature. Miter profiles produce various results. Combine lines and arcs for more complex miters.

Covers are very common sheet metal parts. Covers protect equipment and operators from harm.

Utilize the Base Flange, Miter Flange, Close Corner, and Edge Flange to create simple covers. Add holes for fasteners. The Base Flange is a single sketched line that represents the area to be covered.

Insert the Miter Flange feature and select the 4 edges on the top face. Utilize the Material Outside option for Flange position to add the walls to the outside of the Base Flange.

Base Flange
Sketch Line on Right Plane

Miter Flange
Select 4 edges around the top face.

The Miter Flange creates a gap between the perpendicular walls. The Closed Corner extends the faces to eliminate the gap.

Create the Edge Flanges after the corners are closed.

A Sketched Bend inserts a bend on a flat face or plane. The Sketch Bend requires a profile.

Holes and cuts are added to the cover. A hole or cut can cross a bend line.

The Base Flange utilized the MidPlane option. Utilize the same plane to mirror the sets of holes.

There is plenty of time to develop additional sheet metal parts in the provided exercises at the end of the chapter.

Return to the BOX assembly.

A second BRACKET is required for the right side of the CABINET. Design for the three configurations and reuse geometry.

BRACKET-Mirror Component

Insert a reference plane through the center PEM® fastener. The right-side BRACKET requires a right hand version of the left side BRACKET. Utilize the Mirror Component option to create the mirrored right-side BRACKET.

Activity: BRACKET-Mirror Component.

767) Open the **BOX** assembly.

Display Sketches and Temporary Axes.
768) Click **View**, **Hide/Show**, and check **Sketches** from the Menu bar.

769) Click **View**, **Hide/**Show, and check **Temporary Axes** from the Menu bar.

770) Zoom to Area on the top circular face of the middle PEM® fastener.

Insert a Reference Point.
771) Click **Insert**, **Reference Geometry**, **Point** from the Menu bar. The Point PropertyManager is displayed.

772) Click the **top circular edge** of the back middle fastener. The Reference Point is located at the center point of the top circle.

773) Click **OK** ✔ from the Point PropertyManager. Point1 is displayed.

Fit the model to the Graphics window. Rename Point1.
774) Press the **f** key.

775) Rename **Point1** to **Mirror Point**.

Insert a Reference Plane.
776) Click **Insert**, **Reference Geometry, Plane** from the Menu bar. The Plane PropertyManager is displayed.

777) Click **Mirror Point** from the fly-out FeatureManager. Mirror Point is displayed in the First Reference box.

778) Click **Right Plane** in the BOX assembly from the fly-out FeatureManager. Right Plane is displayed in the Second Reference box.

779) Click **OK** ✔ from the Plane PropertyManager. PLANE1 is displayed in the FeatureManager.

Fit the model to the Graphics window.
780) Press the **f** key.

Rename PLANE1. Save the assembly.
781) Rename **PLANE1** to **Mirror PLANE**.

782) Click **Save** 💾.

783) Click **Save All**.

💡 CommandManager and FeatureManager tabs and file folders will vary depending on system setup and Add-ins.

Mirror the Component.

784) Click **Mirror PLANE** from the FeatureManager.

785) Click **Insert**, **Mirror Components** from the Menu bar. The Mirror PLANE is selected.

786) Click **BRACKET** from the BOX fly-out FeatureManager.

787) Click **Next** .

788) Click the **Create opposite hand version** button. Accept the default conditions.

789) Click **OK** ✔ from the Mirror Components PropertyManager.

790) Click **Yes**. MirrorComponent1 is displayed in the FeatureManager.

791) **Expand** MirrorComponent1 in the FeatureManager. View the new name of the mirror component.

The right-hand version of the BRACKET is named MirrorBRACKET<#>. An interference exists between the MOTHERBOARD and MirrorBRACKET<#>. Address the interference in the next section.

MirrorBRACKET Part-Bends, Unfold and Fold and Jog Features

Open the MirrorBRACKET part. The MirrorBRACKET is not a sheet metal part. The MirrorBRACKET references only the BRACKET's geometry. The MirrorBRACKET must be manufactured in a 2D Flat State. Insert Sheet Metal Bends to create the Flat State.

Control the unfold/fold process. Insert an Extruded Cut across a bend radius. Utilize the Unfold and Fold features to flatten and form two bends. Perform cuts across bends after the Unfold. Utilize unfold/fold steps for additional complex sheet metal components.

Sketch the cut profile across the Bend Radius. The Extruded Cut is linked to the Thickness. A Jog feature creates a small, short bend. Insert a Jog feature to offset the right tab and to complete the MirrorBRACKET.

> **Activity: MirrorBRACKET Part-Bends, Unfold and Fold and Jog Features**.

Open MirrorBRACKET and Insert Bends.
792) Right-click **MirrorBRACKET** in the FeatureManager.

793) Click **Open Part** from the Context toolbar. The MirrorBRACKET FeatureManager is displayed.

Fit the Model to the Graphics window.
794) Press the **f** key.

795) Click the **inside bottom face** of MirrorBRACKET.

BRACKET_Mirrored

796) Click **Insert Bends** from the Sheet Metal toolbar. The Bends PropertyManager is displayed. Face<1> is displayed in the Fixed Face box. Accept the default conditions.

797) Click **OK** from the Bends PropertyManager. Process-Bends1 is displayed in the FeatureManager.

Unfold two Bends.

798) Click **Unfold** from the Sheet Metal toolbar. The Unfold PropertyManager is displayed.

799) **Zoom to Area** on the right inside bottom face.

800) Click the **inside bottom face** as illustrated.

801) Rotate the **model** to view the two long bends.

802) Click the two long **bends** to unfold as illustrated. RoundBend5 and RoundBend6 is displayed in the Bends to unfold box.

803) Click **OK** ✔ from the Unfold PropertyManager. Unfold1 is displayed in the PropertyManager.

Display a Top view.
804) Click **Top view** from the Heads-up View toolbar.

Note: Only the selected two bends unfold. The tab remains folded.

Create the Sketch for an Extruded Cut.
805) Right-click the **right bottom face**. This is your Sketch plane.

806) Click **Sketch** from the Context toolbar. The Sketch toolbar is displayed.

807) Click **Corner Rectangle** ▭ from the Sketch toolbar.

808) Click the **lower right corner** of the Mirror BRACKET.

809) Click a **position** coincident with the left bend line as illustrated.

Add a dimension.

810) Click **Smart Dimension** from the Sketch toolbar.

811) Click the **top horizontal edge** of the MirrorBRACKET.

812) Click the **bottom horizontal line** of the rectangle.

813) Enter **15**mm.

Extrude the sketch.

814) Click **Extruded Cut** from the Sheet Metal toolbar. The Cut-Extrude PropertyManager is displayed.

815) Check the **Link to Thickness** box.

816) Click **OK** from the Cut-Extrude PropertyManager. The Cut-Extrude feature is displayed in the FeatureManager.

817) Click **Wireframe** from the Heads-up View toolbar.

Fold the bends.

818) Click **Fold** from the Sheet Metal toolbar. The Fold PropertyManager is displayed.

819) Click the **Collect All Bends** button. All Unfolded bends are displayed in the Bends to fold box.

820) Click **OK** from the Fold PropertyManager. Fold1 is displayed in the FeatureManager. View the Cut-Extrude1 feature in the Graphics window.

Display Shaded With Edges - Isometric view.
821) Click **Shaded With Edges**.

822) Click **Isometric view**.

Insert a Jog feature.
823) Click the **back face** of the tab.

824) Click **Jog** 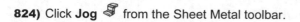 from the Sheet Metal toolbar.

825) Click **Line** from the Sketch toolbar.

826) Sketch a **horizontal line** across the midpoints of the tab.

827) Click **Exit Sketch**. The Jog PropertyManager is displayed.

828) Click inside the **Fixed Face** box.

829) Click the **back face** of the tab below the horizontal line. The direction arrow points to the right.

830) Enter **5**mm for Offset Distance.

831) Click the **Reverse Direction** button. The direction arrow points to the left.

832) Click **OK** from the Jog PropertyManager. Jog1 is displayed in the FeatureManager.

Display the Jog.
833) Use the **arrow keys** to rotate and view the Jog feature.

Display the flat/formed state.
834) Click **Flatten** from the Sheet Metal toolbar to display the flat state.

835) Click **Flatten** from the Sheet Metal toolbar to display the formed state.

Display an Isometric view. Save the model.
836) Click **Isometric view**.

837) Click **Save** .

Return to the BOX assembly.
838) Open the **BOX** assembly.

Save the BOX assembly.

839) Click **Save** 💾.

The BRACKET and
MirrorBRACKET remain
centered around the Mirror
Plane for the BOX assembly
configurations. If the
MirrorBRACKET is not
centered for the Small, Medium and Large
Configurations, Unsuppress the Mirror Point
and Mirror Plane in the BOX FeatureManager.
Un-suppress the Mates created for the
MirrorBRACKET part.

The ConfigurationManager, Add Configuration, Advanced,
Suppress new features and mates option controls the state of
new features and mates in multiple configuration assemblies.

The BOX configurations are complete.

 Fully define all sketches. Rebuild errors occur in a Top
Down design approach if the sketches are not fully defined.

Review the PEM® fasteners, Configurations and
Sheet Metal features.

IGES files are opened in SOLIDWORKS and saved as part
files. The PEM® fasteners were inserted into the BOX
assembly seed hole. You developed a Component Pattern for
the fasteners that referenced the CABINET Linear Pattern of
Holes.

You added an Assembly Cut feature to the BOX assembly.
The Cut feature created a hole through the
MOTHERBOARD and CABINET components.

Due to part availability, you utilized the Replace Components feature to exchange the Ø5.0mm fastener with the Ø4.0mm fastener. The new mates were redefined to update the Ø 4.0mm fastener. The Ø 4.0mm fastener required the CABINET holes to be modified from the Ø5.0mm to the Ø4.0mm.

Equations controlled the location of the CABINET holes based on the Layout Sketch dimensions. Variables in the Layout Sketch were inserted into a Design Table to create the BOX assembly Configurations: Small, Medium and Large.

The sheet metal BRACKET was created In-Context of the BOX assembly and referenced the location of the PEM® fasteners. You inserted the Base Flange, Edge Flange, and Miter Flange features.

The Mirror Components option was utilized to create a right hand version of the BRACKET. The MirrorBRACKET required Insert Bends, Fold/Unfold, and an Extruded Cut across the Bends for manufacturing. The Jog feature provided an offset for the tab by inserting two additional bends from a single sketched line.

Project Summary

You created three different BOX sizes utilizing the Top-down assembly modeling approach. The Top-down design approach is a conceptual approach used to develop products from within the assembly.

The BOX assembly began with a Layout Sketch. The Layout Sketch built in relationships between 2D geometry that represented the MOTHERBOARD, POWERSUPPLY, and CABINET components. These three components were developed In-Context of the Layout Sketch. The components contain InPlace Mates and External references to the BOX assembly.

Additional Extruded features were inserted in the MOTHERBOARD and POWERSUPPLY parts.

The CABINET part developed an Extruded Boss/Base feature In-Context of the BOX assembly. The Shell feature created a constant wall thickness for the solid part.

The Rip feature and Insert Bends feature were added to convert the solid part into a sheet metal part.

Sheet metal components utilized the Edge, Hem, Extruded Cut, Die Cuts, Forms, and Flat Pattern features. Model sheet metal components in their 3D formed state. Manufacture sheet metal components in their 2D flatten state.

The Component Pattern maintained the relationship between the Cabinet holes and the PEM® self-clinching fasteners. You utilized Replace Components to modify the Ø5.0mm fastener to the Ø 4.0mm fastener.

The new mates were redefined to update the Ø4.0mm fastener.

Equations controlled the location of the Cabinet holes. Variables in the Layout Sketch were inserted into a Design Table to create the BOX assembly Configurations: Small, Medium and Large.

The BRACKET was created utilizing sheet metal features: Base Flange, Edge Flange, and Miter Flange. You utilized Mirror Components to create a right hand version: MirrorBRACKET. The MirrorBRACKET required Insert Bends and an Extruded Cut feature through the sheet metal bends.

The Mirror Plane built in the design intent for the BRACKET and MirrorBRACKET to remain centered for each BOX assembly configuration.

Sheet metal parts can be created individually or In-Context of an assembly.

Questions

1. Explain the Top-down assembly approach. When do you create a Layout Sketch?

2. How do you create a new component In-Context of the assembly?

3. What is the difference between a Link Value and an Equation?

4. Name three characteristics unique to Sheet metal parts.

5. Identify the indicator for a part in an Edit Component state.

6. Where should you position the Layout Sketch?

7. For a solid extruded block, what features do you add to create a Sheet metal part?

8. Name the two primary states of a Sheet metal part.

9. Explain the procedure on how to insert a formed Sheet metal feature such as a dimple or louver.

10. Identify the type of information that a Sheet metal manufacturer provides.

11. What is an Assembly Feature?

12. What features are required before you create the Component Pattern in the assembly?

13. True or False. If you utilize Replace Component, you have to modify all the mate references.

14. Define a Design Table. Provide an example.

Exercises

Exercise 7.1: Insert an Extruded Cut feature In-Context of an assembly.

- Open the MOTOR assembly from the Chapter 7 Homework folder. Insert mounting holes through the plate In-Context of the assembly.

Boss-Extrude1 of b5g-plate<1>

- Edit the b5g-plate In-Context of the assembly. Right-click the right face of b5g-plate from the Graphics window as illustrated.

- Click Edit Part.

- Display a Bottom, Wireframe view.

- Right-click the bottom face of b5g-plate for the sketch plane. Create a sketch.

- Ctrl-select the 4 circular edges of the f718b-30-b5-g mounting holes.

- Apply the Convert Entities sketch tool.

- Apply an Extruded Cut feature for the 4 selected circular holes Thru-all thru the plate.

- Return to the assembly. Display an Isometric with Hidden Lines Visible view.

- Rotate the model to view the 4 Thru-all holes in the plate.

Exercise 7.2: Create a New Part In-Context of an assembly.

- Open the PLATE1 assembly from the Chapter 7 Homework folder located in the book.

- Insert a new part into the assembly. Use the default part template. The mouse pointer displays the ![icon] icon. The new virtual component name is displayed in the FeatureManager.

- Click the top face of PLATE1 as illustrated.

- Click the Convert Entities Sketch tool.

- Create a top plate from the bottom plate. Click the top four edges.

Base-Extrude of Support Plate_PCS2B<1>

Base-Extrude of Support Plate_PCS2B<1>

- Create a Boss-Extrude feature. Extrude upwards 50mm.

- Return to the assembly. Name the new virtual component Top-Plate1.

Exercise 7.3: L-BRACKET Part.

Create the L-BRACKET Sheet metal parts.

- Create a family of sheet metal L-BRACKETS. L-BRACKETS are used in the construction industry.

- Create an eight hole L-BRACKET.

- Use Equations to control the hole spacing.

Strong-Tie Reinforcing Brackets
Courtesy of Simpson Strong Tie
Corporation of California

Review Design Tables in On-line Help and the SOLIDWORKS Tutorial.

	A	B	C	D	E
1	Design Table for: lbracket				
2		height@Sketch1	width@Sketch1	depth@Base-Extrude-Thin	hole1dia@Sketch2
3	First Instance	2	1.5	1.375	0.156
4	Small2x1.5x1	2	1.5	1	0.156
5	Medium3x2x2	3	2	2	0.156
6	Large4x3x4	4	3	4	0.25

L-BRACKET DESIGN TABLE

- Create a Design Table for the L-Bracket.

- Rename the dimensions of the L-Bracket.

- Create a drawing that contains the flat state and formed state of the L-BRACKET.

Exercise 7.4: Create a lofted bend feature in a Sheet Metal part.

- Open the Lofted part from the Chapter 7 Homework folder.

- Apply a Lofted-Bend feature on both sketches as illustrated.

- Thickness = .01in. The direction arrow is inward.

- View the results.

Profile(Sketch2-middle)

Exercise 7.5: Create a Vent in a Sheet Metal box.

- Open Vent from the Chapter 7 Homework folder. View the FeatureManager.

- Create the illustrated sketch for the Vent on the Right top side of the box.

- Click the Vent tool.

- Click the large circle circumference as illustrated. The 2D Sketch is a closed profile.

- Click inside the Ribs box.

- Click the horizontal sketch line.

- Click the three remaining sketch lines.

- Click OK to create Vent1. The Area of Vent1 is 1590.67 square mms. The Open area is 51.13%.

- Insert a Mirror Vent on the left side of the Box. Create Mirror1.

- Modify the vent design and obtain an open flow area of 1,700 square mms.

Geometry Properties

Face<1>

1.00deg

☐ Draft inward

0.00mm

☑ Show preview

Flow Area

Area = 1590.67 square mm

Open area = 51.13 %

PROJECT 8: SOLIDWORKS SIMULATION

Project Objective

Provides coverage of the material in the Certified SOLIDWORKS Associate - Simulation (CSWA-S) exam. Knowledge to create and modify models, exam layout, question types and grading procedure.

The project is not designed to educate a new user on SOLIDWORKS Simulation and Finite Element Analysis (FEA).

If your school is an academic certification provider, your instructor can allocate a free CSWA-S exam credit. The instructor will require your .edu email address.

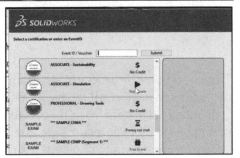

Finite Element Modeling

SOLIDWORKS Simulation uses the displacement formulation of the finite element method to calculate component displacements, strains, and stresses under internal and external loads. The geometry under analysis is discretized using tetrahedral (3D), triangular (2D), and beam elements, and solved by either a direct sparse or iterative solver. SOLIDWORKS Simulation also offers the 2D simplification assumption for plane stress, plane strain, extruded, or axisymmetric options.

Introduction

The Certified SOLIDWORKS Associate - Simulation (CSWA-S) certification indicates a foundation in and apprentice knowledge of demonstrating an understanding in the principles of stress analysis and the Finite Element Method (FEM).

The main requirement for obtaining the CSWA-S certification is to take and pass the on-line **120-minute** exam which consists of **20** questions.

The questions consist of **4 hands-on problems (13 points, 7 points, 5 points and 20 points)** single answer, and the rest are multiple choice questions for a total of 100 points.

Point allocation ranges from **3 points** to **20 points**, depending on the difficulty. A passing grade is 70% (**70 out of 100**) or higher.

You are allowed to answer the questions in any order. Use the Summary Screen during the CSWA-S exam to view the list of all questions you have or have not answered.

The SOLIDWORKS Simulation hands-on problems, **4 hands-on problems (13 points, 7 points, 5 points and 20 points)** require you to create and modify various elements (beam, solid, shell), apply and modify fixtures types, loads types, and results. You need to enter the answer in the correct units with the correct decimal places. There is no partial credit for any questions.

Timing yourself is crucial. One exam strategy is to answer all the hands-on questions first within 90 minutes. Here you will gain the most points. Leave 30 minutes to answer the multiple choice yes/no, and multiple answer questions. Use SOLIDWORKS Simulation Help during the exam.

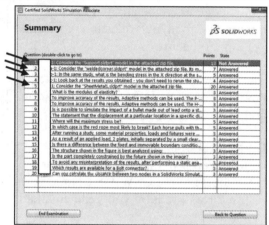

Screen shots from the exam

☼ Download all needed model files from the SDC Publications website.

☼ Table 1 displays the User Interface updates for Contact terms between SOLIDWORKS 2020 and newer versions. The CSWA-S exam uses the older contact terms. The older and newer terms are used in this project.

Table 1: User Interface Updates for Contact Terms

2020	2021
Contacts	Interactions
Contact Visualization Plot	Interaction Viewer
Contact set	Local interaction
Component contact	Component interaction
Global contact	Global interaction
Allow penetration	Free
No penetration	Contact
Incompatible mesh	Independent mesh
Compatible mesh	Continuous mesh
Compatible bonding	Enforce continuous mesh between touching faces

CSWA-S Exam Audience

The intended audience for the exam is anyone with a minimum of 3 - 6 months of SOLIDWORKS and SOLIDWORKS Simulation experience and knowledge in the following areas:

- Engineering Mechanics - Statics.

- Strength of Materials.

- Finite Element Method/Finite Element Analysis Theory.

- Define a Static Analysis Study.

- Knowledge of setting and modifying Unit systems.

- Knowledge of Simulation Options (System Options dialog box).

- Apply material to a part or assembly.

- Define a Beam, Solid or Shell element.

- Define Connector properties such as Contact Sets (Local Interaction), No Penetration (Contact) and Bonded.

- Apply and modify Connections, Contact Set (Local Interactions) and Contact Type (Interactions).

- Define Standard and Advanced Fixtures and External loads.

Force in direction -Y

4 bottom faces are Immovable

Sample CSWA-S model types

- Define Local and Global coordinate systems.

- Apply either the h-Adaptive or p-Adaptive method to the static study.

- Understand axial forces, sheer forces, bending moments and factor of safety.

- Set and modify plots to display in the correct results folders.

- Work with Multi-body parts as different solid bodies.

- Select different solvers as directed to optimize problems.

- Determine if the results are valid.

- Understand the types of problems SOLIDWORKS Simulation (Static Linear) can solve.

- Ability to use SOLIDWORKS Simulation Help.

The CSWA-S **only** covers Static Linear analysis.

🔆 During the exam, you may be asked to apply a German material standard (DIN) to the model.

Note: Simulations using **SW 2024** are more accurate (than previous years) for curved surfaces that come into contact. Geometry corrections factors in the contact detection algorithms improve the representation of curved geometries such as cylindrical and spherical surfaces.

🔆 Illustrations will vary depending on your SOLIDWORKS version and operating system.

Sample CSWA-S model type

Basic FEA Concepts

SOLIDWORKS Simulation uses the Finite Element Method (FEM). FEM is a numerical technique for analyzing engineering designs. FEM is accepted as the standard analysis method due to its generality and suitability for computer implementation.

FEM divides the model into many small pieces of simple shapes called elements effectively replacing a complex problem by many simple problems that need to be solved simultaneously.

Elements share common points called nodes. The process of dividing the model into small pieces is called meshing.

The behavior of each element is well-known under all possible support and load scenarios. The finite element method uses elements with different shapes.

The response at any point in an element is interpolated from the response at the element

Node

Tetrahedral Element

nodes. Each node is fully described by a number of parameters depending on the analysis type and the element used.

For example, the temperature of a node fully describes its response in thermal analysis. For structural analyses, the response of a node is described, in general, by three translations and three rotations. These are called degrees of freedom (DOFs). Analysis using the FEM is called a Finite Element Analysis (FEA).

SOLIDWORKS Simulation formulates the equations governing the behavior of each element taking into consideration its connectivity to other elements. These equations relate the response to known material properties, restraints, and loads.

Next, SOLIDWORKS Simulation organizes the equations into a large set of simultaneous algebraic equations and solves for the unknowns.

In stress analysis, for example, the solver finds the displacements at each node and then the program calculates strains and finally stresses.

Static studies calculate displacements, reaction forces, strains, stresses, and factor of safety distribution. Material fails at locations where stresses exceed a certain level. Factor of safety calculations are based on one of the following failure criterion:

- **Maximum von Mises Stress**.

- **Maximum shear stress (Tresca)**.

- **Mohr-Coulomb stress**.

- **Maximum Normal stress**.

- **Automatic** (Automatically selects the most appropriate failure criterion across all element types).

Static studies can help avoid failure due to high stresses. A factor of safety less than unity indicates material failure. Large factors of safety in a contiguous region indicate low stresses and that you can probably remove some material from this region.

Simulation Advisor

Simulation Advisor is a set of tools that guide you through the analysis process. By answering a series of questions, these tools collect the necessary data to help you perform your analysis. Simulation Advisor includes:

- **Study Advisor**. Recommends study types and outputs to expect. Helps you define sensors and creates studies automatically.

- **Interactions Advisor**. Defines internal interactions between bodies in the model as well as external interactions between the model and the environment. Interactions can include loads, fixtures, connectors, and contacts.

- **Mesh and Run Advisor**. Helps you specify the mesh and run the study.

- **Results Advisor**. Provides tips for interpreting and viewing the output of the simulation. Also, helps determine if frequency or buckling might be areas of concern.

Simulation Advisor works with the SOLIDWORKS
Simulation interface by starting the appropriate
PropertyManagers and linking to online help topics
for additional information. Simulation Advisor leads
you through the analysis workflow from determining
the study type through analyzing the simulation
output. By following the workflow, you use,
depending on your requirements, each of the
individual Advisors.

The purpose of this chapter is not to educate a
new or intermediate user on SOLIDWORKS
Simulation, but to cover and to inform on the types of
questions, layout and what to expect when taking the
CSWA-S exam.

Illustrations and values will vary slightly depending on your SOLIDWORKS
release.

Simulation Help & Tutorials

SOLIDWORKS Simulation is an
Add-in. Use SOLIDWORKS
Simulation during the CSWA-S exam to discover information during the exam. Utilize
the Contents and Search tabs to locate subject matter.

Review the SOLIDWORKS Simulation Tutorials on Static parts and assemblies.
Understand your options and setup parameters. Questions in these areas will be on the
exam.

Review the SOLIDWORKS Simulation Validation, Verification Problems, Static section
and the SOLIDWORKS Simulation, Verification, NAFEMS Benchmarks and Linear
Static sections.

Access SOLIDWORKS Simulation directly from the SOLIDWORKS Add-Ins tab in the
CommandManager.

A model needs to be open to obtain access to SOLIDWORKS Simulation Help or
SOLIDWORKS Simulation Tutorials.

Linear Static Analysis

This section provides the basic theoretical information required for a Static analysis using SOLIDWORKS Simulation. The CSWA-S only covers Static analysis. SOLIDWORKS Simulation and SOLIDWORKS Simulation Professional cover the following topics:

General Simulation:

- *Static Analysis (default)*
- *Use 2D Simplification*
- *Frequency Analysis*

Design Insight:

- *Topology Study*
- *Design Study*

Advanced Simulation:

- *Thermal*
- *Buckling*
- *Fatigue*
- *NonLinear*
- *Linear Dynamic*

Specialized Simulation:

- *Submodeling*
- *Drop Test*
- *Pressure Vessel Design*

Linear Static Analysis

When loads are applied to a body, the body deforms and the effect of loads is transmitted throughout the body. The external loads induce internal forces and reactions to render the body into a state of equilibrium.

Linear Static analysis calculates displacements, strains, stresses, and reaction forces under the effect of applied loads.

Strain ε is the ratio of change, δ L, to the original length, L. Strain, is a dimensionless quantity. Stress σ is defined in terms of Force per unit Area.

Linear Static analysis makes the following assumptions:

- **Static Assumption**. All loads are applied *slowly* and gradually until they reach their full magnitudes. After reaching their full magnitudes, loads *remain constant* (time-invariant). This assumption *neglects inertial and damping forces*.

  Time-variant loads that induce considerable inertial and/or damping forces and require dynamic analysis.

  Dynamic loads change with time and in many cases induce considerable inertial and damping forces that cannot be neglected.

- **Linearity Assumption**. The relationship between loads and induced responses is linear. For example, **if you double the loads, the response of the model (displacements, strains, and stresses), will also double**. Apply the linearity assumption if:

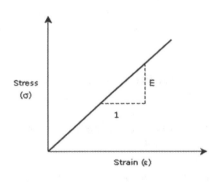

  - All materials in the model comply with Hooke's law; that is, stress is directly proportional to strain.
  - The induced displacements are small enough to ignore the change in stiffness caused by loading.
  - Boundary conditions do not vary during the application of loads. Loads must be constant in magnitude, direction, and distribution. They should not change while the model is deforming.

💡 In SOLIDWORKS Simulation for Static analysis, all displacements **are small relative** to the model geometry (unless the Large Displacements option is activated).

💡 FeatureManager and CommandManager tabs and tree folders will vary depending on system setup, year, and SOLIDWORKS Add-ins.

SOLIDWORKS Simulation assumes that the normals to contact areas do not change direction during loading. Hence, it applies the full load in one step. This approach may lead to inaccurate results or convergence difficulties in cases where these assumptions are not valid.

- Elastic Modulus, E, is the stress required to cause one unit of strain. The material behaves linearly at low stresses.

- Elastic Modulus (Young's Modulus) is the slope defined as stress divided by strain. $E = modulus\ of\ elasticity$ $(Pa\ (N/m^2),\ N/mm^2,\ psi)$.

- Stress σ is proportional to strain in a Linear Elastic Material. Units: $(Pa\ (N/m^2),\ N/mm^2,\ psi)$.

You must be able to work in SI and English units for the CSWA-S exam within the same problem. For example, you apply a Force in Newtons and then you determine displacement in inches.

Different materials have different stress property levels. Mathematical equations derived from Elasticity theory and Strength of Materials are utilized to solve for displacement and stress. These analytical equations solve for displacement and stress for simple cross sections.

General Procedure to Perform a Linear Static Analysis

- Complete a Linear Static study by performing the following steps:

- Open the model (part or assembly).

- Add-in SOLIDWORKS Simulation.

- Select the Simulation tab.

- Create a new Static study.

- Apply material (part or assembly).

- Define and apply various Connections and Fixtures. Standard and Advanced options.

- Define and apply various External Loads. Standard and Advanced options.

- Knowledge in defining coordinates of the reference node location and remote forces and remote translations.

- Define and apply component contacts (Local interactions) and contact sets (interactions) for assemblies and multi-body parts.

- Mesh and Run the study. Select correct mesh type and parameters.

- Create Iso Clipping or Section Clipping plot.

- Define what criteria are best suited to check the failure of ductile or brittle material in a SOLIDWORKS Simulation.

- Assess failure based on a yield criterion (Define Factor of Safety Plot).

Sequence of Calculations in General

Given a meshed model with a set of displacement restraints and loads, the linear static analysis program proceeds as follows:

- The program constructs and solves a system of linear simultaneous finite element equilibrium equations to calculate displacement components at each node.

- The program then uses the displacement results to calculate the strain components.

- The program uses the strain results and the stress-strain relationships to calculate the stresses.

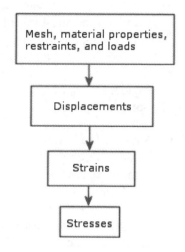

Stress Calculations in General

Stress results are first calculated at special points, called Gaussian points or Quadrature points, located inside each element.

💡 SOLIDWORKS Simulation utilizes a tetrahedral element (High quality mesh for a Solid). Each node contains a series of equations.

These points are selected to give optimal numerical results. The program calculates stresses at the nodes of each element by extrapolating the results available at the Gaussian points.

Node

Tetrahedral Element

After a successful run, nodal stress results at each node of every element are available in the database. Nodes common to two or more elements have multiple results. In general, these results are not identical because the finite element method is an approximate method. For example, if a node is common to three elements, there can be three slightly different values for every stress component at that node.

Overview of the Yield or Inflection Point in a Stress-Strain curve

When viewing stress results, you can ask for element stresses or nodal stresses. To calculate element stresses, the program averages the corresponding nodal stresses for each element.

Stress vs. Strain Plot
Linearly Elastic Material

To calculate nodal stresses, the program averages the corresponding results from all elements sharing that node.

Stress vs. Strain Plot
Linearly Elastic Material

- The material remains in the Elastic Range until it reaches the elastic limit.

- The point E is the elastic limit. The material begins Plastic deformation.

- Yield Stress is the stress level at which the material ceases to behave elastically.

- The point Y is called the Yield Point. The material begins to deform at a faster rate. In the Plastic Range the material behaves non-linearly.

- The point U is called the ultimate tensile strength. Point U is the maximum value of the non-linear curve. Point U represents the maximum tensile stress a material can handle before a fracture or failure.

- Point F represents where the material will fracture.

- Designers utilize maximum and minimum stress calculations to determine if a part is safe. Simulation reports a recommended Factor of Safety during the analysis.

- The Simulation Factor of Safety is a ratio between the material strength and the calculated stress.

Brittle materials do not have a specific yield point and hence it is not recommended to use the yield strength to define the limit stress for the criterion.

Material Properties in General

Before running a study, you must define all material properties required by the associated analysis type and the specified material model. A material model describes the behavior of the material and determines the required material properties. Linear isotropic and orthotropic material models are available for all structural and thermal studies. Other material models are available for nonlinear stress studies. The von Mises plasticity model is available for drop test studies. Material properties can be specified as function of temperature.

- For solid assemblies, each component can have a different material.

- For shell models, each shell can have a different material and thickness.

- For beam models, each beam can have a different material.

- For mixed mesh models, you must define the required material properties for solid and shell separately.

Connections in General

A connection replaces a piece of hardware or fastener by simulating its effect on the rest of the model. Connections include Bolts, Springs, Flexible Support, Bearings, Bonding - Weld/Adhesives, Welds, etc.

The automatic detection tool in SOLIDWORKS Simulation defines Contact Sets **(SW 2021 and newer versions changed to Local interactions)**. Sometimes additional Contact Sets (Local Interactions) and types need to be defined. The SOLIDWORKS Simulation Study Advisor can help.

For example, the behavior of an adhesive depends on its strength and thickness. You can select the Type manually.

Fixtures: Adequate restraints to prevent the body from rigid body motion. If your model is not adequately constrained, check the Use soft springs to stabilize the model option in the Static dialog box.

When importing loads from SOLIDWORKS Motion, make sure that Use inertial relief option is checked. These options are available for the Direct Sparse solver and FFEPlus solver.

See SOLIDWORKS Simulation Help for additional information.

Restraint Types

The Fixture PropertyManager provides the ability to prescribe zero or non-zero displacements on vertices, edges, or faces for use with static, frequency, buckling, dynamic and nonlinear studies. This section will only address standard restraint types, namely Fixed Geometry and Immovable (No translation).

The Immovable (No translation) option is displayed for Sheet Metal parts.

Fixed: For solids, this restraint type sets all translational degrees of freedom to zero. For shells and beams, it sets the translational and the rotational degrees of freedom to zero. For truss joints, it sets the translational degrees of freedom to zero. When using this restraint type, no reference geometry is needed.

View the illustrated table for the attributes and input needed for this restraint.

Attribute	Value
DOFs restrained for solid meshes	3 translations
DOFs restrained for shells and beams	3 translations and 3 rotations
DOFs restrained for truss joints	3 translations
3D symbol (the arrows are for translations and the discs are for rotations)	
Selectable entities	Vertices, edges, faces and beam joints
Selectable reference entity	N/A
Translations	N/A
Rotations	N/A

Immovable (No translation): This restraint type sets all translational degrees of freedom to zero. It is the same for shells, beams and trusses. No reference geometry is used.

To access the immovable restraint, right-click on Fixtures in the Simulation study tree and select Fixed Geometry. Under Standard, select Immovable (No translation). View the illustrated table for the attributes and input needed for this restraint.

Attribute	Value
DOFs restrained for shell meshes	3 translations
DOFs restrained for beam and truss meshes	3 translations
3D symbol	
Selectable entities	Vertices, edges, faces and beam joints
Selectable reference entity	N/A
Translations	N/A
Rotations	N/A

There are differences for Shells and Beams between Immovable (No translation) and Fixed restraint types.

The Immovable option is not available for Solids.

The Fixture PropertyManager allows you to prescribe zero or non-zero displacements on vertices, edges or faces for use with static, frequency, buckling, dynamic and nonlinear studies.

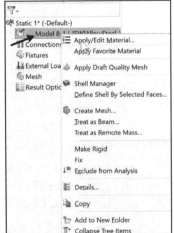

The Fixed Geometry Fixture allows for additional Advanced options: Symmetry, Circular Symmetry, User Reference Geometry, On Flat Faces, On Cylindrical Faces, On Spherical Faces.

Attributes of each option are available in SOLIDWORKS Simulation Help.

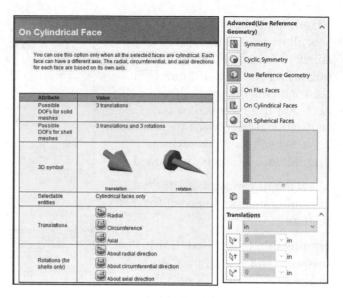

Loads and Restraints in General

Loads and restraints are necessary to define the service environment of the model. The results of analysis directly depend on the specified loads and restraints.

Loads and restraints are applied to geometric entities as features that are fully associative to geometry and automatically adjust to geometric changes.

For example, if you apply a pressure **P** to a face of area **A_1**, the equivalent force applied to the face is **PA_1**.

If you modify the geometry such that the area of the face changes to **A_2**, then the equivalent force automatically changes to **PA_2**. Re-meshing the model is required after any change in geometry to update loads and restraints.

The types of loads and restraints available depend on the type of the study. A load or restraint is applied by the corresponding Property

Manager accessible by right-clicking the Fixtures or External Loads folder in the Simulation study tree, or by clicking Simulation, Loads/Fixture.

Loads: At least one of the following types of loading is required:

- Concentrated force.

- Pressure.

- Prescribed nonzero displacements.

- Body forces (gravitational and/or centrifugal).

- Thermal (define temperatures or get the temperature profile from thermal analysis).

- Imported loads from SOLIDWORKS Motion.

- Imported temperature and pressure from Flow Simulation.

In a linear static thermal stress analysis for an assembly, it is possible to input different temperature boundary conditions for different parts.

Under the External Loads folder you can define Remote Load/Mass and Distributed Mass. In the Remote Load/Mass you define Load, Load/Mass or Displacement. Input values are required for the Remote Location and the Force.

By default, the Location is set to x=0, y=0, z=0. The Force is set to $F_x=0$, $F_y=0$, $F_z=0$. The Force requires you to first select the direction and then enter the value.

Meshing in General

Meshing splits continuous mathematical models into finite elements. The types of elements created by this process **depend on the type of geometry meshed**. SOLIDWORKS Simulation offers three types of elements:

- **Solid elements** - solid geometry.

- **Shell elements** - surface geometry.

- **Beam elements** - wire frame geometry.

In CAD terminology, "Solid" denotes the type of geometry. In FEA terminology, "Solid" denotes the type of element used to mesh the solid CAD Geometry.

Meshing Types

Meshing splits continuous mathematical models into finite elements. Finite element analysis looks at the model as a network of interconnected elements.

Meshing is a crucial step in design analysis. SOLIDWORKS Simulation automatically creates a mixed mesh of:

- **Solid**: The Solid mesh is appropriate for bulky or complex 3D models. In meshing a part or an assembly with solid elements, Simulation generates one of the following types of elements based on the active mesh options for the study:

  - Draft quality mesh. The automatic mesher generates linear tetrahedral solid elements **(4 nodes)**.

  - High quality mesh. The automatic mesher generates parabolic tetrahedral solid elements **(10 nodes)**.

Linear elements are also called first-order, or lower-order elements. Parabolic elements are also called second-order, or higher-order elements.

A linear tetrahedral element is defined by four corner nodes connected by six straight edges.

Linear solid element

A parabolic tetrahedral element assigns 10 nodes to each solid element: four corner nodes and one node at the middle of each edge (a total of six mid-side nodes).

In general, for the same mesh density (number of elements), parabolic elements yield better results than linear elements because:

- They represent curved boundaries more accurately.

- They produce better mathematical approximations. However, parabolic elements require greater computational resources than linear elements.

Parabolic solid element

For structural problems, each node in a solid element has three degrees of freedom that represent the translations in three orthogonal directions.

SOLIDWORKS Simulation uses the X, Y, and Z directions of the global Cartesian coordinate system in formulating the problem.

- **Shell**: Shell elements are suitable for thin parts (sheet metal models). Shell elements are 2D elements capable of resisting membrane and bending loads. When using shell elements, Simulation generates one of the following types of elements depending on the active meshing options for the study:

Linear triangular element

  - **Draft quality mesh**. The Draft quality mesher generates linear triangular shell elements Specifies 4 corner nodes for each solid element and 3 corner nodes for each shell element.

  - **High quality mesh**. The High quality mesher generates parabolic triangular shell elements. Sets the number of integration points for the Jacobian ratio calculations. The Jacobian ratio calculations consider the Gaussian points located within each element. Select 4, 16, & 29 Jacobian points, or At Nodes. Default Jacobian points: 16.

Parabolic triangular element

A linear triangular shell element is defined by three corner nodes connected by three straight edges.

A parabolic triangular element is defined by three corner nodes, three mid-side nodes, and three parabolic edges.

The Shell Definition PropertyManager is used to define the thickness of thin and thick shell elements. The program automatically extracts and assigns the thickness of the sheet metal to the shell. You cannot modify the thickness. You can select between the thin shell and thick shell formulations. You can also define a shell as a composite for static, frequency, and buckling studies. In general use thin shells when the thickness-to-span ratio is less than 0.05.

SW 2020 Version   SW 2021, 2022, 2024 version

 Surface models can only be meshed with shell elements.

- **Beam or Truss**: Beam or Truss elements are suitable for extruded or revolved objects and structural members with constant cross-sections. Beam elements can resist bending, shear, and torsional loads. The typical frame shown is modeled with beams elements to transfer the load to the supports. Modeling such frames with truss elements fails since there is no mechanism to transfer the applied horizontal load to the supports.

💡 A truss is a special beam element that can resist axial deformation only.

Beam elements require defining the exact cross section so that the program can calculate the moments of inertia, neutral axes and the distances from the extreme fibers to the neutral axes. The stresses vary within the plane of the cross-section and along the beam. A beam element is a line element defined by two end points and a cross-section.

Consider a 3D beam with cross-sectional area (A) and the associated mesh. Beam elements can be displayed on actual beam geometry or as hollow cylinders regardless of their actual cross-section.

Beam elements are capable of resisting axial, bending, shear, and torsional loads. Trusses resist axial loads only. When used with weldments, the software defines cross-sectional properties and detects joints.

Mesh on cyclinders and beam geometry

Beam and truss members can be displayed on actual beam geometry or as hollow cylinders regardless of their actual cross-sectional shape.

A Beam element has 3 nodes (one at each end) with 6 degrees of freedom (3 translational and 3 rotational) per node plus one node to define the orientation of the beam cross section.

A Truss element has 2 nodes with 3 translational degrees of freedom per node.

The accuracy of the solution depends on the quality of the mesh. In general, the finer the mesh the better the accuracy.

💡 A compatible mesh is a mesh where elements on touching bodies have overlaying nodes.

The curvature-based mesher supports multi-threaded surface and volume meshing for assembly and multi-body part documents. The standard mesher supports only multi-threaded volume meshing.

It is possible to mesh a part or assembly with a combination of solids, shells and beam elements (mixed mesh) in SOLIDWORKS Simulation.

SOLIDWORKS Simulation Meshing Tips

SOLIDWORKS Simulation Help lists the following Meshing tips that you should know for the CSWA-S exam.

- When you mesh a study, the SOLIDWORKS Simulation meshes all unsuppressed solids, shells and beams:

- Use Solid mesh for bulky objects.

- Use Shell elements for thin objects like sheet metals.

- Use Beam or Truss elements for extruded or revolved objects with constant cross-sections.

- Simplify structural beams to optimize performance in Simulation to be modeled with beam elements. The size of the problem and the resources required are dramatically reduced in this case. For the beam formulation to produce acceptable results, the length of the beam should be 10 times larger than the largest dimension of its cross section.

- Compatible meshing (a mesh where elements on touching bodies have overlaying nodes) is more accurate than incompatible meshing in the interface region. Requesting compatible meshing can cause mesh failure in some cases. Requesting incompatible meshing can result in successful results. You can request compatible meshing and select Re-mesh failed parts with incompatible mesh so that the software uses incompatible meshing only for bodies that fail to mesh.

- Check for interferences between bodies when using a compatible mesh with the curvature-based mesher. If you specify a bonded contact condition between bodies, they should be touching. If interferences are detected, meshing stops, and you can access the Interference Detection PropertyManager to view the interfering parts. Make sure to resolve all interferences before you mesh again.

- If meshing fails, use the Failure Diagnostics tool to locate the cause of mesh failure. Try the proposed options to solve the problem. You can also try different element size, define mesh control, or activate Enable automatic looping for solids.

- The SOLIDWORKS Simplify utility lets you suppress features that meet a specified simplification factor. In the Simulation study tree, right-click Mesh and select Simplify Model for Meshing. This displays the Simplify utility.

- It is good practice to check mesh options before meshing. For example, the Automatic transition can result in generating an unnecessarily large number of elements for models with many small features. The high quality mesh is recommended for most cases. The Automatic looping can help solve meshing problems automatically, but you can adjust its settings for a particular model. The Curvature-based mesher automatically uses smaller element sizes in regions with high curvature.

- To improve results in important areas, use mesh control to set a smaller element size. When meshing an assembly with a wide range of component sizes, default meshing results in a relatively coarse mesh for small components. Component mesh control offers an easy way to give more importance to the selected small components. Use this option to identify important small components.

- For assemblies, check component interference. To detect interference in an assembly, click Tools, Interference Detection. Interference is allowed only when using **shrink fit**. The Treat coincidence as interference and Include multi-body part interferences options allow you to detect touching areas. These are the only areas affected by the global and component contact settings.

Use the mesh and displacement plots to calculate the distance between two nodes using SOLIDWORKS Simulation.

The Global element size parameter provides the ability to **set the global average element size**. SOLIDWORKS suggests a default value based on the model volume and surface area. This option is only available for a standard mesh.

The Ratio value in Mesh Control provides the geometric growth ratio from one layer of elements to the next. To access Mesh Control, right-click the Mesh folder in the Simulation study tree and click Apply Mesh Control.

Running the Study

When you run a study, Simulation calculates the results based on the specified input for materials, restraints, loads and mesh.

Set the default plots that you want to see in your Simulation Study tree under **Simulation**, **Options** from the Main menu.

When you run one or multiple studies, they run as background processes.

In viewing the results after running a study, you can generate plots, lists, graphs, and reports depending on the study and result types.

🔅 If you modify the study (force, material, etc.) you only need to **re-run the study** to update the results. You do not need to re-mesh unless you modified contact conditions.

Displacement Plot - Output of Linear Static Analysis

The Displacement Plot PropertyManager allows you to plot displacement and reaction force results for static, nonlinear, dynamic, drop test studies, or mode shapes for bucking and frequency studies. By default, directions X, Y and Z refer to the **global coordinate system**.

If you choose a reference geometry, these directions refer to the selected reference entity. Displacement components are:

UX = Displacement in the X-direction.

UY = Displacement in the Y-direction.

UZ = Displacement in the Z-direction.

URES = Resultant displacement.

RFX = Reaction force in the X-direction.

RFY = Reaction force in the Y-direction.

RFZ = Reaction force in the Z-axis.

RFRES = Resultant reaction force.

The **Probe** function allows you to query a plot and view the values of plotted quantities at **defined nodes or centers of elements**. When you probe a mesh plot, Simulation displays the node or element number and the global coordinates of the node. When you probe a result plot, SOLIDWORKS Simulation displays the node or element number, the value of the plotted result, and the global coordinates of the node or center of the element. For example, in a nodal stress plot, the node number, the stress value, and the global x, y, and z coordinates appear.

Adaptive Methods for Static Studies

Adaptive methods help you obtain an accurate solution for static studies. There are two types of adaptive methods: h-adaptive and p-adaptive method.

The concept of the h-method (available for solid part and assembly documents) is to use smaller elements (increase the number of elements) in regions with high relative errors to improve accuracy of the results. After running the study and estimating errors, the software automatically refines the mesh where needed to improve results.

The p-adaptive method (available for solid part and assembly documents) increases the polynomial order of elements with high relative errors. The p-method does not change the mesh. It changes the order of the polynomials used to approximate the displacement field using a unified polynomial order for all elements. See SOLIDWORKS Help for additional information.

Understand how to select either p-adaptive or h-adaptive methods for the CSWA- S exam.

Table 1 displays the User Interface updates for Contact terms between SOLIDWORKS 2020 and newer versions. The CSWA-S exam uses the older contact terms. The older terms are used in this project.

Table 1: User Interface Updates for Contact Terms

2020	2021
Contacts	Interactions
Contact Visualization Plot	Interaction Viewer
Contact set	Local interaction
Component contact	Component interaction
Global contact	Global interaction
Allow penetration	Free
No penetration	Contact
Incompatible mesh	Independent mesh
Compatible mesh	Continuous mesh
Compatible bonding	Enforce continuous mesh between touching faces

Sample Exam Questions

The questions below are examples of what to expect on the certification exam. The multiple-choice questions should serve as a check for basic knowledge of the exam materials. **Note**: The Official Guide to SOLIDWORKS Associate Exams provides over 90 questions on the CSWA-S with additional models, exercises and tips.

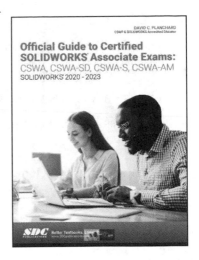

1. What is the Modulus of Elasticity?

- The slope of the Deflection-Stress curve.

- The slope of the Stress-Strain curve in its linear section.

- The slope of the Force-Deflection curve in its linear section.

- The first inflection point of a Strain curve.

2. What is Stress?

- A measure of power.

- A measure of strain.

- A measure of material strength.

- A measure of the average amount of force exerted per unit area.

3. Which of the following assumptions are true for a static analysis in SOLIDWORKS Simulation with small displacements?

- Inertia effects are negligible and loads are applied slowly.

- The model is not fully elastic. If loads are removed, the model will not return to its original position.

- Results are proportional to loads.

- All the displacements are small relative to the model geometry.

4. What is Yield Stress?

- The stress level beyond which the material becomes plastic.

- The stress level beyond which the material breaks.

- The strain level above the stress level which the material breaks.

- The stress level beyond the melting point of the material.

5. A high quality Shell element has _____ nodes.

- 4

- 5

- 6

- 8

6. Stress σ is proportional to _____ in a Linear Elastic Material.

- Strain.

- Stress.

- Force.

- Pressure.

7. The Elastic Modulus (Young's Modulus) is the slope defined as _____ divided by
_____.

- Strain, Stress.

- Stress, Strain.

- Stress, Force.

- Force, Area.

8. Linear static analysis assumes that the relationship between loads and the induced
response is _____.

- Flat.

- Linear.

- Doubles per area.

- Translational.

9. In SOLIDWORKS Simulation, the Factor of Safety (FOS) calculations are based on one of the following failure criteria.

- Maximum von Mises Stress.

- Maximum shear stress (Tresca).

- Mohr-Coulomb stress.

- Maximum Normal stress.

10. The Yield point is the point where the material begins to deform at a faster rate than at the elastic limit. The material behaves _____ in the Plastic Range.

- Flatly.

- Linearly.

- Non-Linearly.

- Like a liquid.

11. What are the Degrees of Freedom (DOFs) restrained for a Solid?

- None.

- 3 Translations.

- 3 Translations and 3 Rotations.

- 3 Rotations.

12. What are the Degrees of Freedom (DOFs) restrained for Truss joints?

- None.

- 3 Translations.

- 3 Translations and 3 Rotations.

- 3 Rotations.

13. What are the Degrees of Freedom (DOFs) restrained for Shells and Beams?

- None.

- 3 Translations.

- 3 Translations and 3 Rotations.

- 3 Rotations.

14. Which statements are true for Material Properties using SOLIDWORKS Simulation?

- For solid assemblies, each component can have a different material.

- For shell models, each shell cannot have a different material and thickness.

- For shell models, the material of the part is used for all shells.

- For beam models, each beam cannot have a different material.

15. A Beam element has _____nodes (one at each end) with _____degrees of freedom per node plus_____ node to define the orientation of the beam cross section.

- 6, 3, 1

- 3, 3, 1

- 3, 6, 1

- None of the above.

16. A Truss element has _____ nodes with _____ translational degrees of freedom per node.

- 2, 3

- 3, 3

- 6, 6

- 2, 2

17. In general, the finer the mesh the better the accuracy of the results.

- True.

- False.

18. How does SOLIDWORKS Simulation automatically treat a Sheet metal part with uniform thickness?

- Shell.

- Solid.

- Beam.

- Mixed Mesh.

19. Use the mesh and displacement plots to calculate the distance between two _____ using SOLIDWORKS Simulation.

- Nodes.
- Elements.
- Bodies.
- Surfaces.

20. Surface models can only be meshed with _____ elements.

- Shell.
- Beam.
- Mixed Mesh.
- Solid.

21. The shell mesh is generated on the surface (located at the mid-surface of the shell).

- True.
- False.

22. In general, use Thin shells when the thickness-to-span ratio is less than _____.

- 0.05
- .5
- 1
- 2

23. The model (a rectangular plate) has a length to thickness ratio of less than 5. You extracted its mid-surface to use it in SOLIDWORKS Simulation. You should use a _____.

- Thin Shell element formulation.
- Thick Shell element formulation.
- Thick or Thin Shell element formulation, it does not matter.
- Beam Shell element formulation.

24. The model, a rectangular sheet metal part, uses SOLIDWORKS Simulation. You should use a:

- Thin Shell element formulation.

- Thick Shell element formulation.

- Thick or Thin Shell element formulation, it does not matter.

- Beam Shell element formulation.

25. The Global element size parameter provides the ability to set the global average element size. SOLIDWORKS Simulation suggests a default value based on the model volume and _____ area. This option is only available for a standard mesh.

- Force.

- Pressure.

- Surface.

- None of the above.

26. A remote load applied on a face with a Force component and no Moment can result in: Note: Remember (DOFs restrain).

- A Force and Moment of the face.

- A Force on the face only.

- A Moment on the face only.

- A Pressure and Force on the face.

27. There are _____ DOFs restrained for a Solid element.

- 3

- 1

- 6

- None

28. There are _____ DOFs restrained for a Beam element.

- 3

- 1

- 6

- None

29. What best describes the difference(s) between a Fixed and Immovable (No translation) boundary condition in SOLIDWORKS Simulation?

- There are no differences.

- There are no difference(s) for Shells but it is different for Solids.

- There is no difference(s) for Solids but it is different for Shells and Beams.

- There are only differences(s) for a Static Study.

30. Can a non-uniform pressure or force be applied on a face using SOLIDWORKS Simulation?

- No.

- Yes, but the variation must be along a single direction only.

- Yes. The variation can be in two directions and is described by a binomial equation.

- Yes, but the variation must be linear.

31. You are performing an analysis on your model. You select five faces, 3 edges and 2 vertices and apply a force of 20lbf. What is the total force applied to the model using SOLIDWORKS Simulation?

- 100lbf

- 1600lbf

- 180lbf

- 200lbf

32. Yield strength is typically determined at _____ strain.
- 0.1%

- 0.2%

- 0.02%

- 0.002%

33. There are four key assumptions made in Linear Static Analysis: 1. Effects of inertia and damping are neglected, 2. The response of the system is directly proportional to the applied loads, 3. Loads are applied slowly and gradually, and_____ .

- Displacements are very small. The highest stress is in the linear range of the stress-strain curve.

- There are no loads.

- Material is not elastic.
- Loads are applied quickly.

34. How many degrees of freedom does a physical structure have?

- Zero.
- Three - Rotations only.
- Three - Translations only.
- Six - Three translations and three rotational.

35. What criteria are best suited to check the failure of brittle material in SOLIDWORKS Simulation?

- Maximum von Mises Stress and Maximum Shear Stress criteria.
- Maximum von Misses Stress and Maximum Shear Strain criteria.
- Maximum von Misses Stress and Maximum Normal Stress criteria.
- Mohr-Coulomb Stress and Maximum Normal Stress criteria.

36. You are performing an analysis on your model. You select three faces and apply a force of 40lb. What is the total force applied to the model using SOLIDWORKS Simulation?

- 40lb
- 20lb
- 120lb
- Additional information is required.

37. A material is orthotropic if its mechanical or thermal properties are not unique and independent in three mutually perpendicular directions.

- True
- False

38. An increase in the number of elements in a mesh for a part will:

- Decrease calculation accuracy and time.

- Increase calculation accuracy and time.

- Have no effect on the calculation.

- Change the FOS below 1.

39. SOLIDWORKS Simulation uses the von Mises Yield criteria to calculate the Factor of Safety of many ductile materials. According to the criteria:

- Material yields when the von Mises stress in the model equals the yield strength of the material.

- Material yields when the von Mises stress in the model is 5 times greater than the minimum tensile strength of the material.

- Material yields when the von Mises stress in the model is 3 times greater than the FOS of the material.

- None of the above.

40. SOLIDWORKS Simulation calculates structural failure on:

- Buckling.

- Fatigue.

- Creep.

- Material yield.

41. Apply a uniform total force of 200lb on two faces of a model. The two faces have different areas. How do you apply the load using SOLIDWORKS Simulation for a Linear Static Study?

- Select the two faces and input a normal to direction force of 200lb on each face .

- Select the two faces and a reference plane. Apply 100lb on each face.

- Apply equal force to the two faces. The force on each face is the total force divided by the total area of the two faces.

- None of the above.

42. Maximum and Minimum value indicators are displayed on Stress and Displacement plots in SOLIDWORKS Simulation for a Linear Static Study.

- True.
- False.

43. What SOLIDWORKS Simulation tool should you use to determine the result values at specific locations (nodes) in a model using SOLIDWORKS Simulation?

- Section tool.
- Probe tool.
- Clipping tool.
- Surface tool.

44. What criteria are best suited to check the failure of ductile materials in SOLIDWORKS Simulation?

- Maximum von Mises Strain and Maximum Shear Strain criteria.
- Maximum von Misses Stress and Maximum Shear Stress criteria.
- Maximum Mohr-Coulomb Stress and Maximum Mohr-Coulomb Shear Strain criteria.
- Mohr-Coulomb Stress and Maximum Normal Stress criteria.

45. Set the scale factor for plots_____ to avoid any misinterpretation of the results, after performing a Static analysis with gap/contact elements.

- Equal to 0.
- Equal to 1.
- Less than 1.
- To the Maximum displacement value for the model.

FEA Modeling Section

Tutorial FEA Model 8-1

An exam question in this category could read:

In the figure displayed, what is the vertical displacement in the Global Y direction in (inches) at the location of the dot? Calculate your answer. Decimal place: 3.

Given Information:

Material: Alloy Steel

Elastic modulus = $2.1e11$ N/m^2

Poisson's ratio = 0.28

$F_1 = 200$lbf

Use the default high quality element size to mesh.

Use the models from the CSWA-S model folder for this section.

Let's start.

1. **Open** Model 8-1 from the Chapter 8 folder.

The bar that you opened was created on the Front Plane.

The upper left corner of the rectangle is located at the origin. This simplifies the actual deformation of the part. The height dimension references zero in the Global Y direction.

Split Lines were created to provide the ability to locate the needed Joints in this problem.

To add the force at the center of the beam, a split line in the shape of a small circle on the top face of the right end of the beam is used.

Proper model set up is very important to obtain the correct mesh and final results.

Create a Static SOLIDWORKS Simulation Study. Set Simulation Options.

2. **Add-In** SOLIDWORKS Simulation.

3. Click the **Simulation** tab in the CommandManager.

4. Click **New Study** from the drop-down menu.

5. Accept the default name (Static #).

   Click **OK** ✓ from the Study PropertyManager.

Set unit system, units, and mesh type.

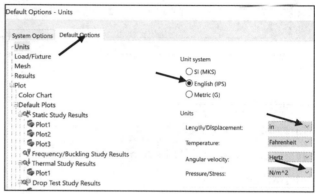

6. Click **Simulation**, **Options** from the Main menu. The System Options - General dialog box is displayed.

7. Click the **Default Options** tab.

Set unit system.

8. Select **English (IPS)**.

Set units.

9. Select **in** for Length/Displacement.

10. Select **(N/m²)** for Pressure/Stress.

Select mesh quality.

11. Click the **Mesh folder**.

12. Select **High** for Mesh quality. View your options. Default Gaussian points: 16.

13. Click **OK** from the Default Options - Mesh dialog box.

Treat the model as a Beam.

14. Right-click **Model 8-1** in the Study.

15. Click **Treat as Beam**. A Joint group folder is created.

16. Right-click the **Joint group** folder. A joint is identified at the free end of a structural member or at the intersection of two or more structural members.

17. Click **Edit**. The Edit Joints PropertyManager is displayed. All is selected by default. The software considers all structural members when calculating joints. Joint 1 and Joint 2 are displayed in the Results box.

18. Click **OK** ✓ from the Edit Joints PropertyManager.

The Results box displays the joints that the software finds in the model. The software highlights using pink color, the joints where two or more structural members intersect.

Set Fixed Geometry as Fixture type.

19. Right-click the **Fixtures** folder.

20. Click **Fixed Geometry**. The Fixture PropertyManager is displayed. View your options. Fixed Geometry is selected by default.

21. Click the **joint on the left side** of the beam. Joint<1,1> is displayed in the Joints box.

22. Click **OK** ✓ from the Fixture PropertyManager. Fixed-1 is created in the Static Study.

Set the Force at the end of the beam.

23. Right-click the **External Loads** folder.

24. Click **Force**. The Force/Torque PropertyManager is displayed.

25. Click the **Joints** option.

26. Click the **joint on the right side** of the beam. Joint<2,1> is displayed in the Joints box.

27. Click **inside** the Face, Edge, Plane for Direction 1 box.

28. Click the **end face** of the beam as the plane for direction as illustrated.

29. Select **English (IPS)** for units.

30. Click the **Along Plane Dir 1 Force** box.

31. Enter **200**lbf Along Plane Direction1. The arrow is displayed downwards; reverse direction if needed.

32. Click **OK** ✓ from the Force/Torque PropertyManager. Force-1 is created in the Static Study.

Meshing generates 3D tetrahedral solid elements, 2D triangular shell elements, and 1D beam elements. A mesh consists of one type of elements unless the mixed mesh type is specified. Solid elements are naturally suitable for bulky models. Shell elements are naturally suitable for modeling thin parts (sheet metals), and beams and trusses are suitable for modeling structural members.

Apply the standard default setting for the mesh. Mesh and run the study.

33. Right-click the **Mesh** folder.

34. Click **Mesh and Run**. View the results.

35. Double-click the
**Displacement1 (-Res
disp-)** folder. View the
results.

Set plot results. Set number
format to floating. Set
number of decimals to 3.
Display URES: Resultant
Displacement.

36. Right-click the
**Displacement1 (-Res
disp-)** folder.

37. Click **Edit Definition**. The
Displacement plot FeatureManager
is displayed.

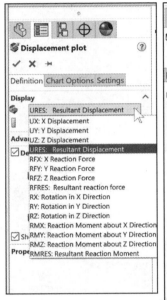

38. Click the **Definition tab**.

39. Click the **Display drop-down**
arrow.

40. Select **URES: Resultant
Displacement**.

41. Click the **Chart Options** tab.

42. Click the **show max annotation**
box.

43. Select **Floating** from the number
format box.

44. Select **3** for number of decimal
places.

45. Click **OK** from the Displacement
plot FeatureManager. View the results in the graphic
window.

Calculate the displacement at 7cm.

46. Click **List Stress, Displacement and Strain** from the Results Advisor drop-down menu. The List Results PropertyManager is displayed.

47. Select **Displacement** for Quantity.

48. Select **UY: Y Displacement** for Component. Select **inches** for units.

49. Click **OK** ✓ from the List Results PropertyManager. The List Results dialog box is displayed.

50. Select **inches** for units.

51. **View** the results.

52. **Scroll down** until you see values around 70mm for the distance along the X direction. See the value of displacement (UY) (in). Click **Close**.

To find the exact value at 70mm, use linear interpolation.

This method is shown below and uses the values greater and less than the optimal one to find the actual displacement.

$$\frac{X_U - X_L}{UY_U - UY_L} = \frac{X_U - X_O}{UY_U - UY_O}$$

$$\frac{71.875 - 68.75}{-0.0272206 - -0.0250617} = \frac{71.875 - 70}{-0.02718 - UY_O}$$

$$UY_O = 0.0258$$

At the distance of 7cm (70mm) the displacement is found to be approximately 0.026in.

The correct answer is **B**.

A = 0.034in

B = **0.026**in

C = 0.043in

D = 0.051in

Note: In SOLIDWORKS 2024, the Mesh Quality Plot Folder is displayed in the Study tree. Default Mesh quality - **High, Jacobian points: 16**.

The purpose of this chapter is not to educate a new or intermediate user on SOLIDWORKS Simulation, but to cover and to inform you on the types of questions, layout and what to expect when taking the CSWA-S exam.

Tutorial FEA Model 8-2

Below is a second way to address the first problem (Tutorial FEA Model 8-1) with a different model, using the **Study Advisor** and the **Probe** tool.

In the figure displayed, what is the vertical displacement in the Global Y direction in (inches) at the location of the red dot? Decimal place: 3.

Given Information:

Material: Alloy Steel

Elastic modulus = 2.1e11 N/m^2

Poisson's ratio = 0.28

F$_1$ = 200lbf

Use the default high quality element size to mesh.

Use the models from the CSWA-S model folder for this section.

Let's start.

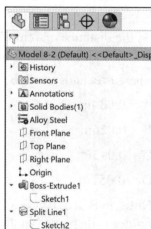

1. **Open** Model 8-2 from the Chapter 8 folder.

The bar that you opened was created on the Front Plane. The bar was created so that the origin is the point at which the force is applied. A Spline line is created across the part at a distance of 70mm (7cm) from the end that is to be fixed.

Create a Static SOLIDWORKS Simulation Study. Set Simulation Options.

2. **Add-In** SOLIDWORKS Simulation.

3. Click the **Simulation** tab in the CommandManager.

4. Click **New Study** from the drop-down menu.

5. Accept the default name (Static #). Click **OK** ✓ from the Study PropertyManager.

6. Click **Simulation**, **Options** from the Main menu. The System Options - General dialog box is displayed.

7. Click the **Default Options** tab.

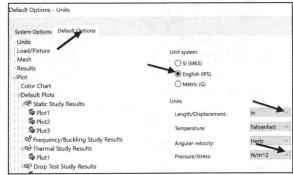

Set unit system, units and mesh quality.

8. Select **English (IPS)** for unit system.

9. Select **in** for Length/Displacement for units.

10. Select **N/m²** for units.

Set mesh quality.

11. Click the **Mesh folder**.

12. Select **High** for Mesh quality.

13. Click **OK** from the Default Options - Mesh dialog box.

Set Fixture type. Use the Study Advisor. Set as Fixed Geometry.

14. Right-click the **Fixtures** folder from the Simulation study tree.

15. Click **Fixtures Advisor**. The Simulation Advisor is displayed to the right of the Graphics area.

16. Click **Add a fixture**. The Fixture PropertyManager is displayed.

17. Select the **left face** as illustrated.

18. Set as **Fixed Geometry**.

19. Click **OK** ✓ from the Fixture PropertyManager.

Set the Force.

20. Right-click the **External Loads** folder from the Simulation study tree.

21. Select **Remote Load/Mass**. The Remote Loads/Mass PropertyManager is displayed. The Remote Loads/Mass PropertyManager allows you to apply remote loads, remote masses, and remote displacements for static, nonlinear static, and topology studies.

22. Select the **top face of the part on each side of the Split lin**e - for Select faces.

23. Set the Reference Coordinate system to **Global**.

24. Leave all of the location boxes at **zero** value.

25. Select the **Translational Components** box.

26. Select **lbf** for Force unit.

27. Select **in** for Translational unit.

28. Click the **Y-Direction** button.

29. Enter **200 in Y-Direction**. The force is pointing downward. Reverse direction if necessary.

30. Click **OK** ✓ from the Remote Loads/Mass PropertyManager. -200 lbf is displayed in the Remote load folder.

Meshing generates 3D tetrahedral solid elements, 2D triangular shell elements, and 1D beam elements. A mesh consists of one type of elements unless the mixed mesh type is specified. Solid elements are naturally suitable for bulky models. Shell elements are naturally suitable for modeling thin parts (sheet metals), and beams and trusses are suitable for modeling structural members.

Use the standard default setting for the mesh. Mesh and run the study.

31. Right-click the **Mesh** folder.

32. Click **Mesh and Run**. View the results.

33. Click the **Play Animation** button. View the model. Explore the below options in the Simulation Advisor dialog box.

34. Click **Stop**.

35. Double-click the **Displacement1 (-Res-disp1)** folder.

Set plot results. Set number format to floating. Set number of decimals to 3. Display URES: Resultant Displacement.

36. Right-click the **Displacement1 (-Res-disp1)** folder.

37. Click **Edit Definition**. The Displacement plot FeatureManager is displayed.

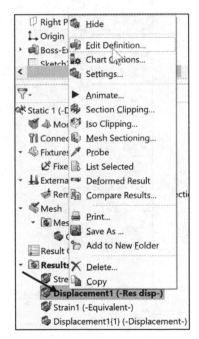

38. Click the **Definition tab**. View your options.

39. Click the **Display drop-down** arrow.

40. Select **URES: Resultant Displacement**.

41. Click the **Chart Options** tab. View your options.

42. Click the **show max annotation** box.

43. Select **Floating** from the Number format box.

44. Select **3** for number of decimal places.

45. Click **OK** from the Displacement plot FeaturcManager. View the results in the graphic window.

Locate the displacement at 7cm using the Probe tool.

46. Click the **Displacement1 (-Res disp-)** Results folder.

47. Click **Probe** from the Plot Tools drop-down menu.

48. Click the **Split line** at 7cm on the model as illustrated. View the results in the graphics window.

49. Click **OK** from the Probe Results PropertyManager.

At the distance of 7cm (70mm) the displacement is found to be approximately 0.026in.

The correct answer is **B**.

A = 0.034in

B = 0.026in

C = 0.043in

D = 0.051in

Sample CSWA-S part

In the CSWA-S exam:

- Comprehend Remote loads/mass to apply an external load.

- Knowledge in defining coordinates of the reference, node location, remote forces, and remote translations.

- Awareness of how to apply Fixed Geometry, Fixed Hinge, and Advanced Fixtures.

- Familiarity with how to apply h-adaptive or p-adaptive method during a mesh.

- Understanding to apply the German material standard (DIN).

- Ability to apply different materials to various components in an assembly.

Tutorial FEA Model 8-3

An exam question in the Solid category could read:

In the figure displayed, what is the maximum URES: Resultant Displacement in mm on the annular face of the model? The three holes are fixed.

Decimal place: 3.

<u>**Given Information:**</u>

Material: Alloy Steel

A normal force **F₁** is applied to the annular face. $F_1 = 3000lbf$.

The three holes are **fixed**.

Use the default high quality element size to mesh.

Use the models from the Chapter 8 folder for this section.

Let's start.

1. **Open** Model 8-3 from the Chapter 8 folder.

Create a Static SOLIDWORKS Simulation Study. Set Simulation Options.

2. **Add-In** SOLIDWORKS Simulation.

3. **Create** a new Static Study. Accept the default name (Static #).

4. Click **Simulation**, **Options** from the Main menu.

5. Click the **Default Options** tab.

Set unit system, units and mesh quality.

6. Select **English (IPS)** for unit system. Select **in** for units.

7. Select **psi** for pressure/stress units.

8. Click the **Mesh folder**.

9. Select **High** for mesh quality.

10. Click **OK** from the Default Options - Mesh dialog box.

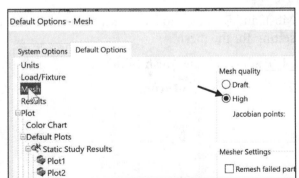

Apply Fixed Geometry to three cylindrical faces.

11. Right-click the **Fixtures** folder.

12. Click **Fixed Geometry**. The Fixture PropertyManager is displayed.

13. Select the **three cylindrical faces** of the hole pattern as illustrated.

14. Click **OK** ✓ from the Fixtures PropertyManager. Fixed-1 is created.

Apply an External Load.

15. Right-click the **External Loads** folder.

16. Click **Force**. The Force/Torque PropertyManager is displayed.

17. Click the **face** of the model as illustrated. Face<1> is displayed in the Faces and Shell edges for Normal Force box.

18. Select **English (IPS)** for units.

19. Enter **3000**lbf normal to the annular face.

20. Click **OK** ✓ from the Force/Torque PropertyManager. Force-1 is created in the Study tree.

Mesh and Run the model. Use the standard default setting for the mesh.

21. Right-click the **Mesh** folder.

22. Click **Mesh and Run**.

23. **View** the results.

Set plot results. Set number format to floating. Set number of decimals to 3. Display URES: Resultant Displacement in mm.

24. Right-click the **Displacement1 (-Res disp-)** folder.

25. Click **Edit Definition**. The Displacement plot FeatureManager is displayed.

26. Click the **Definition tab**.

27. Select **mm** for units.

28. Click the **Chart Options** tab.

29. Click the **show max annotation** box.

30. Select **Floating** from the Number format box.

31. Select **3** for number of decimal places.

32. Click **OK** from the Displacement plot FeatureManager. View the results in the graphic window.

Note: Using SW 2024, the answer is 1.661. Using older versions, the answer will vary but be within 1% for the exam number.

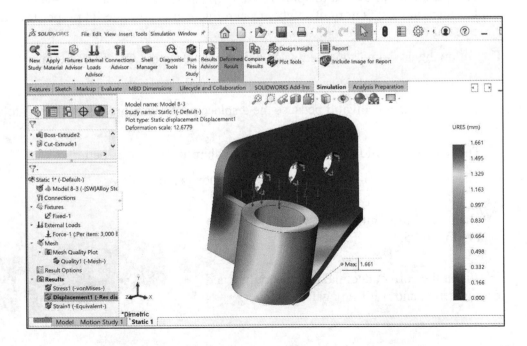

The correct answer is **±1%** of this value.

The correct answer is **C**.

A = 1.112mm

B = 1.014mm

C = 1.661mm

D = 1.734mm

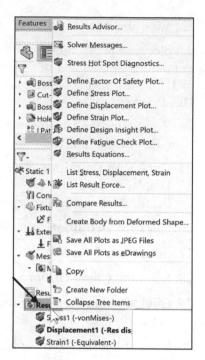

Displacement components are: UX = Displacement in the X-direction, UY = Displacement in the Y-direction, UZ = Displacement in the Z-direction, URES = Resultant displacement.

After you calculate displacement or other parameters in a Simulation Study, the CSWA-S exam will also deliver a series of successive questions to test the understanding of the results.

In the first question you calculate Displacement; in the second question you calculate Resultant Force. Do not re-mesh and re-run. Create the required parameter in the Results folder.

In the third question, you are asked to determine if the results are valid or invalid. If the materials yield strength was passed, then the results are invalid.

Use the Define Factor of Safety Plot to determine if your results are valid or invalid. The CSWA-S exam requires you to apply Finite Element Method theory and review displacement values, factory of safety, mesh refinement, and material properties such as yield strength.

The following are some statements you will encounter in the CSWA-S exam:

- The results are invalid because the material's yield strength was passed.

- The results area is invalid because the displacement was more than ½ the plate's thickness.

- The results are valid as they are, even if mesh refinement was better.

- The results are invalid because a dynamic study is required.

In general use Thin Shells when thickness to span ratio < 0.05.

Input the correct decimal places (trailing and leading zeros) and if it's a negative number.

Tutorial FEA Model 8-4

An exam question in the Sheet Metal category could read:

In the figure displayed, what is the **maximum UX: X displacement** in millimeters?

Decimal place: 4.

<u>**Given Information:**</u>

Material: Alloy Steel.

A normal force of **400N** is applied to the **inside** face as illustrated.

The thickness of the material is **0.15in**.

The right edge as illustrated and two circular edges are **immovable**.

Use the default high quality element size to mesh.

Let's start.

You need to define the thickness. The Fixture option Immovable is added for shell models. Thin models created with no Sheet Metal feature require you to define the use of **Shell elements** in SOLIDWORKS Simulation.

Models created with the Sheet Metal feature automatically create Shell elements in SOLIDWORKS Simulation.

1. **Open** Model 8-4 from the Chapter 8 folder.

Create a Static SOLIDWORKS Simulation Study. Set Simulation Options.

2. **Add-In** SOLIDWORKS Simulation.

3. Click the **Simulation** tab in the CommandManager.

4. Click **New Study** from the drop-down menu.

5. Accept the default name (Static #).
   Click **OK** ✓ from the Study PropertyManager.

6. Click **Simulation**, **Options** from the Main menu. The System Options - General dialog box is displayed.

7. Click the **Default Options** tab.

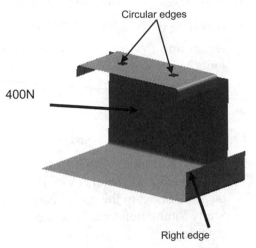

Circular edges

400N

Right edge

Set unit system, units and mesh quality.

8. Select **SI (MKS)** for unit system.

9. Select **mm** for length/Displacement units.

10. Select **N/m²** for Pressure/Stress units.

11. Click the **Mesh** folder.

12. Click **High** for Mesh quality.

13. Click **OK** from the Default Options – Mesh dialog box. A check mark is displayed next to the model name in the Study Simulation tree. Note the comment.

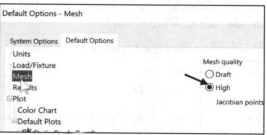

The Shell Definition PropertyManager is used to define the thickness of thin and thick shell elements. You can also define a shell as a composite for static, frequency, and buckling studies.

There are three shell types in a linear static study:

- Thin: In general use thin shells when the thickness-to-span ratio is less than 0.05.

- Thick: Use the thick shells when the thickness-to-span ratio is greater than 0.05.

- Composite: Available only for surface geometries in static, frequency and buckling studies. Defines the shell as a composite laminate.

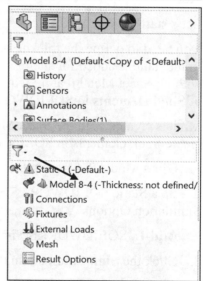

Define Shell thickness.

14. Right-click **Model 8-4** in the Study tree.

15. Click **Edit Definition** from the drop-down menu. The Shell Definition PropertyManager is used to define the thickness of thin and thick shell elements. You can also define a shell as a composite for static, frequency, and buckling studies.

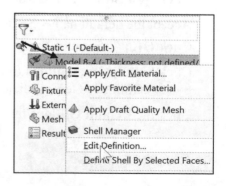

16. Click **Thin** for Type.

17. Enter **0.15in** for shell thickness.

18. Click **OK** ✓ from the Shell Definition plot PropertyManager.

Apply Fix Geometry Fixtures. The right edge as illustrated and two circular edges are immovable. This restraint type sets all translational degrees of freedom to zero. It is the same for shells, beams and trusses. No reference geometry is used.

19. Right-click **Fixtures** folder.

20. Click **Fixed Geometry**. The Fixture PropertyManager is displayed.

21. Click the **Immovable (No translation)** option.

22. Select the **edge** and the **two circular edges** of the model as illustrated.

23. Click **OK** ✓ from the Fixture PropertyManager. Immovable-1 is created.

Immovable (No translation):

Apply an External Normal Load (Force) of 400N to the inside face of the model.

24. Right-click the **External Loads** folder from the study.

25. Click **Force**. The Force/Torque PropertyManager is displayed.

26. Click the **inside face** of the model as illustrated.

27. Select **Normal** for Force component.

28. Select **SI** for units.

29. Enter **400N** for force value.

30. Click **OK** ✓ from the Force/Torque PropertyManager.

Mesh and Run the model. Use the standard default setting for the mesh.

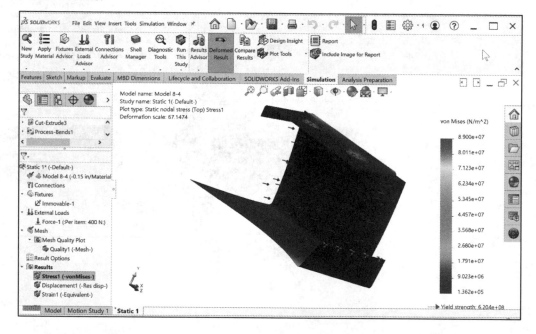

31. Right-click the **Mesh** folder.

32. Click **Mesh and Run**. **Note**: Large displacement is used. View the results.

Set plot results. Set number format to floating. Set number of decimals to 4. Calculate maximum global displacement in X. Change units to mm.

33. Double-click the **Displacement1 (-Res disp-)** folder.

34. Right-click the **Displacement1 (-Res disp-)** folder.

35. Click **Edit Definition**. The Displacement plot FeatureManager is displayed.

36. Click the **Definition tab**.

37. Select **UX: X Displacement** for display.

38. Select **mm** for units.

39. Click the **Chart Options** tab.

40. Click the **show max annotation** box.

41. Select **Floating** from the Number format box.

42. Select **4** for number of decimal places.

43. Click **OK** from the Displacement plot FeatureManager. View the results in the graphic window.

Note: Using SW 2024, the answer is 0.4382. Using older versions, the answer will vary but be within 1% for the exam number.

The maximum displacement in X, UX = 0.4382mm.

The correct answer is **±1%** of this value.

The correct answer is **D**.

A = 0.4000mm

B = 0.4120mm

C = 1.6559mm

D = 0.4382mm

When you enter a value in the CSWA-S exam, include the required number of decimal places and leading and trailing zeroes.

Tutorial FEA Model 8-4 Part 2

An exam question in the Understanding your Knowledge category could read:

Given Information:

From the study you just completed, can you modify the applied force from **400N** to **4000N** without creating a new mesh and still calculate the correct maximum von Mises stress?

Think about the problem. Think about the model.

Answer: In SOLIDWORKS Simulation you need to only select **Run**.

In a Yes/No question, the answer would be: **Yes**.

Screen shot from the exam

Tutorial FEA Model 8-4 Part 3

An exam question in the understanding your knowledge category could read:

Given Information:

From the study you just completed with a force of 4000N applied, are the results for the maximum von Mises stress valid?

Answer: The results are not valid because the maximum von Mises stress (900 MPa) is greater than the Yield strength (620 MPa) of the material. In a Yes/No question, the answer would be: **No**.

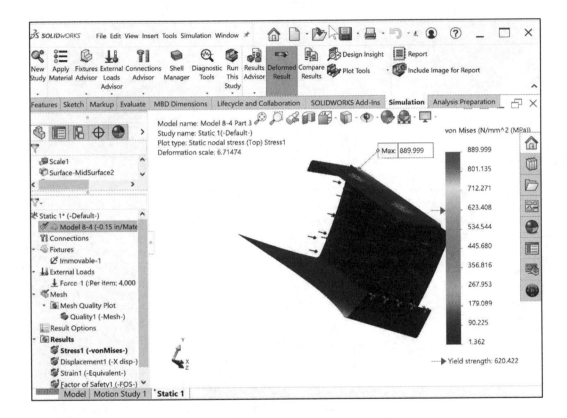

Notes:

Project 9

SOLIDWORKS and the 3DEXPERIENCE Platform

Below are the desired outcomes and usage competencies based on the completion of Project 9.

Project Desired Outcomes:	Usage Competencies:
• Utilize and understand SOLIDWORKS and the **3D**EXPERIENCE platform.	• Obtain your **3D**EXPERIENCE login credentials. • Download the needed Apps from the **3D**EXPERIENCE platform.
• Knowledge of Apps and tools on the platform using a single modeling environment inside of SOLIDWORKS.	• Create a Bookmark and save to the platform. • View and use the auto generated embedded PLM model properties from the platform. • Use Bookmarks (links) to delete entire groups of data.
• Understand new Search tools.	• Create and apply metadata and 6W Tags.

Notes:

Project 9 – SOLIDWORKS and the 3DEXPERIENCE Platform

Project Objective

Project 9 offers a look at SOLIDWORKS and the **3D**EXPERIENCE platform. The project is written to answer a few keys questions for the new user. What is the **3D**EXPERIENCE platform? How does it work within my SOLIDWORKS session?

The platform enables you to collaborate with students and teams on and off campus through a browser in a secure cloud environment. Think of the **3D**EXPERIENCE platform simply as a SOLIDWORKS Add-In on your desktop. Through the Task Pane, students can manage, revise, and share information. Experience a single modeling environment. Access the platform within the SOLIDWORKS Task Pane.

Detailed steps are provided to obtain a **3D**EXPERIENCE ID (login credentials) to access the platform.

Nine step-by-step PDF tutorials with models are provided to practice and understand the tools and advantages of SOLIDWORKS and the platform.

On the completion of the project, you will be able to:

- Obtain a **3D**EXPERIENCE ID to access the platform.

- Grasp **3D**EXPERIENCE platform terminology.

- Launch the SOLIDWORKS App from the **3D**EXPERIENCE platform.

- Use the new Save and Open tools.

- Identify and apply the Overwrite Protection tools.

- Create Bookmarks and a Collaborative space.

- Identify the auto generated embedded Product Lifecycle Management (PLM) model properties.

- Handle complex model relations in a Graph or Datagrid view.

- Apply the Revision and Lifecycle Maturity tools.

- Discuss and create a 3DDrive.

- Apply improved search tools (Searchable metadata and 6W Tags).

- Examine and utilize 3DPlay.

- Investigate and explore 3DSwym.

3DEXPERIENCE Platform

What is the **3D**EXPERIENCE platform? How does it work within my SOLIDWORKS session?

The **3D**EXPERIENCE platform provides downloadable content (DLC) from a secure cloud location within your SOLIDWORKS session. The platform enables you to collaborate with students and teams on and off campus through a browser.

Think of the **3D**EXPERIENCE platform simply as a SOLIDWORKS Add-In on your desktop. Through the Task Pane, students can manage, revise, and share information. Experience a single modeling environment. Access the platform within the SOLIDWORKS Task Pane.

New Save Tools

Students need to better understand how to save files, where to save files, and create revisions, especially when using assemblies with common reference component names.

The platform provides additional tools and information inside your SOLIDWORKS session about the state of a part, component, assembly or drawing. The traditional save commands are still available.

When connected to the platform, there are many new save options available to the user. These options ensure that the latest revision and updates are being saved to the correct Bookmarks (folders) in the cloud.

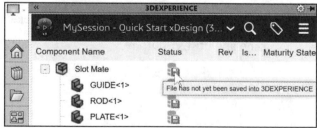

The Status is displayed in the SOLIDWORKS Task Pane (MySession Panel) and in the FeatureManager.

Secure links, not files, can be created to share information inside/outside of your team.

Use the Share with users tab to share files with members in your Collaborative space.

Use Share by link to share a link (from the platform) to anyone outside of your Collaborative space. **Note**: 3DDrive auto-synchs to your local hard drive.

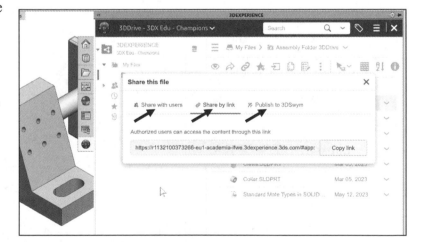

In this project, review and address numerous Save options: Save to this PC, Save to 3DEXPERIENCE, Save Active Window, Save with Options, etc. View several Product Lifecycle Management (PLM) attributes of a model.

In a short step-by-step PDF tutorial, open a SOLIDWORKS part that has not been previously uploaded or saved to the **3D**EXPERIENCE platform. Lock the part. Modify the part. Save the modified SOLIDWORKS part to the **3D**EXPERIENCE platform. Create a New Revision, Unlock the part and view the maturity state of the part. Save the part to a Collaborative space in the MySession panel. This will all be done all within a SOLIDWORKS session. Experience a single modeling environment.

New Open Tools

Most people use the traditional Windows folder structure to open their SOLIDWORKS models.

You can still do this. A new option is to open any model directly from the platform in your cloud storage space, within a SOLIDWORKS session.

Use the Open from 3DEXPERIENCE command as illustrated in the Open dialog box.

You can also use the Open command directly in the SOLIDWORKS Task Pane in the MySession Panel.

These and more open options provide key SOLIDWORKS properties for all models in your space: **Revision**, **Modification date**, **Creation date**, **Owner name**, **Maturity state**, **Lock**, **UnLock**, etc.

Auto generated embedded model properties ensure academic integrity, contact identification, and improved search capabilities.

You still can utilize the standard Quick Filters: **Filter Assemblies (*.asm)**, **Filter Parts (*.prt)**, **Filter drawings (*.drw)** and **Filter Top-Level Assemblies**.

There are new search filters to help in the open command. You can now search by name, part number, Bookmark and My Content.

Note: Think of a Bookmark Workspace as a directory which can contain Bookmark folders. Bookmark folders are used to further categorize bookmarks like a folder tree structure (Windows Explorer). One of the best, simple practices for SOLIDWORKS data is to create bookmarks upon the initial save to the **3D**EXPERIENCE platform. Use bookmarks to delete entire groups of data. For assemblies within the Bookmark folder, there is an option to delete the entire structure of the assembly and all reference components.

Double-click any model icon in the Open dialog box, (I selected the ROBOT icon in this example). The model opens directly in your SOLIDWORKS session. View the provided model information in the MySession panel.

In a short step-by-step PDF tutorial, open a SOLIDWORKS assembly that has not been previously uploaded or saved to the **3D**EXPERIENCE platform. Create Bookmarks (links) to delete entire groups of data. View the structure option to delete the assembly document and all reference components. Apply the Collaborative Lifecycle App.

Overwrite Protection

The platform provides visual properties in the SOLIDWORKS Task Pane Lock/UnLock, Owners Name, Maturity state (Lifecycle state of the family, for example: Private, In Work, Frozen, Released, Obsolete), Description, Revision, etc. The model properties are key to prevent other members from accessing the model when someone else on the team is working on the document.

In my experience, Junior and Senior Design classes have a very hard time using any type of Product data management (PDM) software. PDM is not normally taught in schools, and most students do not realize how important it is. By the end of the semester, they start realizing the power of PDM in a design organization.

If a class or team does use a PDM software package, it changes semester to semester depending on volunteers, and what they have been exposed to in the past. Simply, no student wants to do PDM.

Product Lifecycle Management (PLM) cloud Apps connect directly to SOLIDWORKS through the Task Pane. The auto generated embedded model properties (Component Name, Status (Displays an icon that represents the status of a given revision, saved or not saved to the platform), Locked status (Locked or Unlocked), Rev (Revision #), Is Last Revision (Yes/No), Maturity State (Lifecycle state of the family, for example: Private, In Work, Frozen, Released, Obsolete), Description, File Name (Displays the PLM external ID of the object for reference objects and the PLM external ID of the connection for instances) and Type (Physical Product or Drawing).

In a short step-by-step PDF tutorial, open a SOLIDWORKS assembly that has not been previously uploaded or saved to the **3D**EXPERIENCE platform.

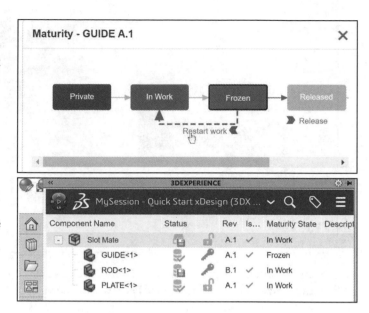

The following options in the SOLIDWORKS MySession, Lifecycle tab are reviewed: **Save Active Window**, **Save Active Window As New**, **Reload from Server**, **Replace By Revision** and **Maturity**.

Change Maturity state of a component, assembly and several components with different maturity status.

Complex Model Relations

Saving large assemblies or drawings can be a difficult task. An understanding of file management is required for the assembly or drawing document with their reference components.

Did you ever experience a broken reference in an assembly or drawing?

The SOLIDWORKS **3D**EXPERIENCE platform provide tools to ensure correct references are opened and saved in an assembly or drawing.

In a short step-by-step PDF tutorial, open a SOLIDWORKS assembly that has not been previously uploaded or saved to the **3D**EXPERIENCE platform.

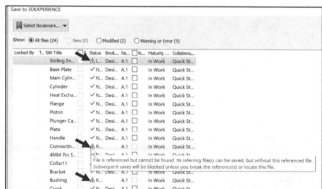

View all references in an assembly in a Graph and Datagrid view.

Open a component and view the auto generated embedded model properties. View the Active Member, Parents (Sub-assemblies), children, Bookmark locations and more, all inside of your SOLIDWORKS desktop session.

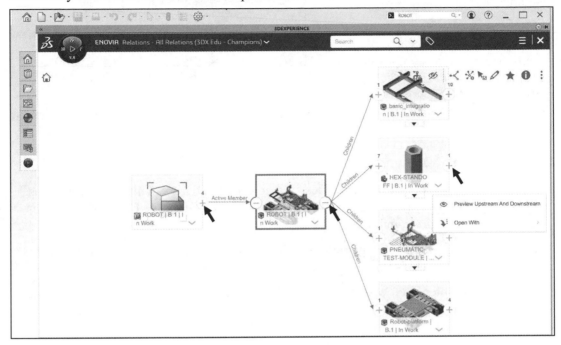

Handling Revisions

Revisions now have their own space in the MySession panel in the **3D**EXPERIENCE platform.

Students no longer need to use the Save As command to create a new revision.

The platform addresses revisions (when needed) and comments that are embedded in the document history. Key document attributes are always available.

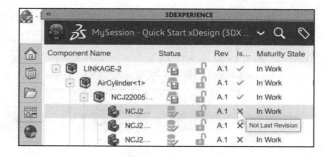

In a short step-by-step PDF tutorial open a SOLIDWORKS assembly that has not been previously uploaded or saved to the **3D**EXPERIENCE platform. Use the New Revision tool in the Action bar. Change the revision status of a component in the assembly after a team design review.

Storage Space - 3DDrive

The **3D**EXPERIENCE platform provides an App called 3DDrive. Think of 3DDrive as another online storage space such as OneDrive, Dropbox, or Google Drive to keep files organized and shared through folders.

Are there other free cloud storage options? Yes, but having data stored with Dassault Systèmes that understands your content is the differentiator.

3DDrive is fully integrated within your SOLIDWORKS desktop and is auto-synched to Windows Explorer. You can access 3DDrive three different ways: through Windows Explorer, directly in SOLIDWORKS or from the **3D**EXPERIENCE platform.

3DDrive provides the ability to store, share and manage files inside and outside of your class or team through a link using any device (phone, tablet, PC, etc.). View embedded design comments during the lifecycle process of the document. When the sharing of the link is finished, you can remove the outside person. No file is provided. This ensures security on the platform of your design.

In a short step-by-step PDF tutorial, install **3D**EXPERIENCE Drive for Windows directly from your SOLIDWORKS desktop. The following items are covered in 3DDrive: **upload**, **download**, **view properties**, **share, roles**, **embed comments**, **links**, **3DSwym**, **members**, **revoke access**, **delete files** and **restore files**. **Note**: Anything you discard in 3DDrive will end up in the trash bin awaiting a final deletion. This gives you a second chance to consider whether you really want to delete a file.

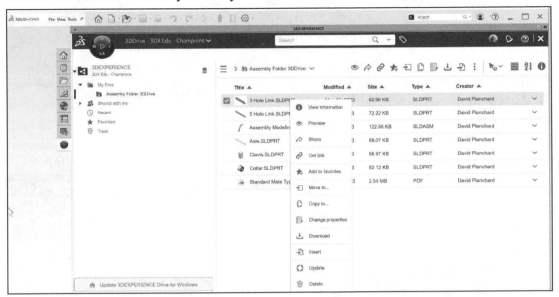

3DEXPERIENCE Platform Search Tools

Using the traditional Windows Explorer search tools are limited in a CAD environment. Conventional Save tools in SOLIDWORKS do not capture many key PLM attributes.

Open a file (using the Browser) within SOLIDWORKS. To open a file in the Browser requires searching for the file. No Apps are needed. It should remind you of using Google or any other search engine.

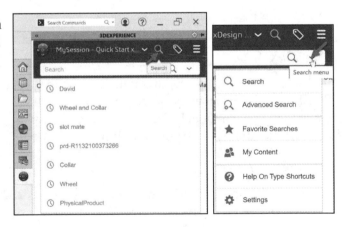

View the searchable metadata that is automatically generated and embedded in each model.

Search by Description, Name, Owner, Creation date, Modification date, type, etc.

Create and use 6W Tags (What, Who, When, Where, How and Why).

Perform source searches using Collaborative spaces, Communities, 3DDrive and more. Search by Type: Physical Product, 3D Shapes, Documents, 2D Component Templates and 3D Annotations.

In a short step-by-step PDF tutorial, open a file (using the Browser) within SOLIDWORKS. View the searchable metadata that is automatically generated and embedded in each model. Create a 6W tag. Apply the advanced search features to demonstrate the power of the platform.

3DPlay

3DPlay is the universal player and viewer on the **3D**EXPERIENCE platform. Use 3DPlay to view 3D models, image files, PDFs, 2D drawings, PowerPoint presentations, and assembly instruction files.

For design reviews, 3DPlay provides the ability to explode and collapse components in the assembly while making hand or typed markups and measurements. There are various Section tools to view the internals of a complex assembly.

When the 3DPlay link is shared with people outside of your team/platform via another App called 3DDrive, 3DPlay is INCLUDED. This provides someone who has no roles and isn't even associated with your platform the ability to view the link and even mark it up during the review or quote process.

When the sharing of the link is finished, you can remove the outside person. No file is provided. This ensures security on the platform of your design.

In a short step-by-step PDF tutorial, open a SOLIDWORKS assembly that has not been previously uploaded or saved to the platform. View the 3DEXPERIENCE Add-Ins on your SOLIDWORKS desktop.

Access the 3DPlay Widget. View your default or created Dashboard on the platform. View and explore various tools in the Action bar.

3DSwym

The 3DSwym App is the Social and Collaborative App on the platform for students, engineers, scientists and other professionals to view, participate, and create in Communities of common interests.

The 3DSWymer Role provides the ability to connect and collaborate in a cloud environment on the **3D**EXPERIENCE Platform. A Role is a collection of Apps. The 3DSwym App is contained within the **3DSwymer Role**.

Communities can engage and establish new relationships through likes, comments, file exchanges, shares, Wikis, Wedos, and more.

Communities can be private, public or secret.

Access 3DSwym communities from the SOLIDWORKS desktop or directly from your browser.

In a short step-by-step PDF tutorial, access 3DSwym communities from your SOLIDWORKS desktop. Explore four communities: Open COVID19 Community, SOLIDWORKS Forum Community, 3DEXPERIENCE EDU Student Community, 3DEXPERIENCE EDU Academics Community.

In the next section, three tutorials are provided: **3DEXPERIENCE Lesson 1: Getting Started with SOLIDWORKS and the Platform, 3DEXPERIENCE Lesson 2: SOLIDWORKS Save and Revision**, and **3DEXPERIENCE Lesson 3: SOLIDWORKS and Lifecycle Maturity States**. PDFs of these tutorials and six additional tutorials and models are provided in the project folder.

3DEXPERIENCE Lesson 1: Getting Started with SOLIDWORKS and the Platform

In this lesson, close SOLIDWORKS on your desktop before you begin.

Work through the steps provided to obtain your 3DEXPERIENCE ID (login credentials) to access the platform.

Install the Design with SOLIDWORKS App. This provides the ability to upload, store, manage, and share files.

Open a SOLIDWORKS assembly and view the 3DEXPERIENCE Add-In on your SOLIDWORKS desktop.

The MySession panel displays the two-way communication between SOLIDWORKS running on your computer and the **3D**EXPERIENCE platform running in the cloud in a single modeling environment.

There are three different user types when using the **3D**EXPERIENCE platform: Administrator, Member (student), and Administrator/Member (educator). This lesson is written for an individual Member, who has permission, with their own computer. For computer lab installation, contact your Value Added Reseller (Partner).

Note: This lesson is not for SOLIDWORKS Connected "3DEXPERIENCE SOLIDWORKS".

Before you begin: You need to purchase a 3DEXPERIENCE education offer from your VAR.

Create and register your 3DEXPERIENCE ID (credentials) to the platform. An internet connection is required.

Enter www.solidworks.com/academic-cloud to create and register a new ID on an academic tenant. In an academic environment, it is strongly recommended to use your school's assigned email address.

Input the requested information.

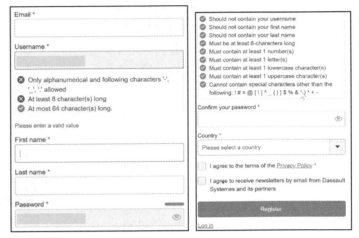

Click **Register**.

You will receive the 3DEXPERIENCE platform Invitation link via email. The email you use must be the same as your 3DEXPERIENCE ID.

Click **Launch your 3DEXPERIENCE platform** from the Invitation email as illustrated.

Bookmark the address. This is your unique link to the **3D**EXPERIENCE platform.

Store the Invitation email in a safe location.

Log in with your 3DEXPERIENCE ID.

Review and **close** the Welcome box.

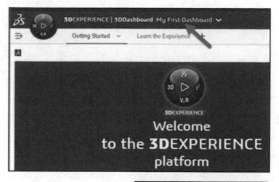

At initial login, you are provided with a general 3DEXPERIENCE Dashboard. The name of the Dashboard is My First Dashboard. Dashboards organize Apps and Widgets. A Widget is an extension of the App that allows you to use it in the Dashboard.

Depending on your license, (academic, student, maker, or commercial) you will have a default Dashboard. For academics, your IT administrator or instructor will create a default Dashboard for a class.

Click the **Menu button** as illustrated to view the Dashboards and cockpits list. This displays a list of the Dashboards that you have access to.

In the example, My First Dashboard is displayed. I created other Dashboards. You may see additional Dashboards.

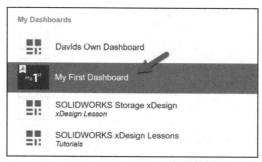

The Compass is a means of launching your Apps in the **3DEXPERIENCE** platform and inside of a SOLIDWORKS session.

Click the **center** of the Compass as illustrated.

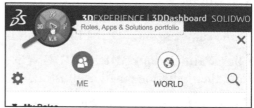

A 3DDashboard is displayed. The panel displays the Roles, Cockpits, My Favorite Apps and Apps that you have been assigned or purchased.

Install the Design with SOLIDWORKS App. The Design with SOLIDWORKS App provides access to use the **3DEXPERIENCE** platform from within SOLIDWORKS. This provides the ability to upload, store, manage, and share files.

Note: This lesson is not for SOLIDWORKS Connected "3DEXPERIENCE SOLIDWORKS".

Drag the slider bar downward until you see the Design with SOLIDWORKS App.

Click **Design with SOLIDWORKS**.

If this is the first time you installed any applications from the platform, install the 3DEXPERIENCE Launcher.

Click **Download 3DEXPERIENCE Launcher** from the dialog box. The 3DEXPERIENCE Launcher will be downloaded to your browser setting location.

Run the downloaded application.

The Welcome to the 3DEXPERIENCE Launcher Setup Wizard is displayed. Install the Launcher.

Click **Next**.

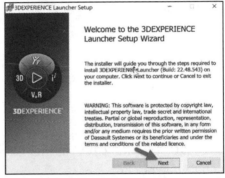

Select the location of the downloaded files.

Click **Install**.

When completed, click **Finish**.

When finished, the Welcome to the 3DEXPERIENCE launcher dialog box is displayed.

Click **Continue**.

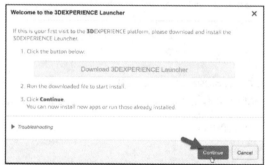

Click **Install All Roles Containing Design with SOLIDWORKS**.

The 3DEXPERIENCE Launcher is displayed in the Graphics area.

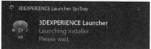

This starts the process of the download and installation of the SOLIDWORKS connector for the platform.

Click **Next>**.

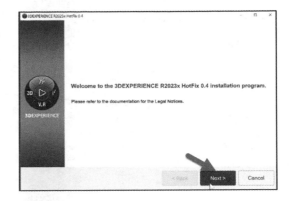

Select the location of the downloaded files.

Click **Next>**.

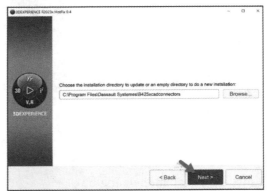

Select the location for your vault (PDM) view.

This needs to be a local disk drive on your computer. The default is C:\3DEXPERIENCE.

Click **Next>**.

Select language.

Click **Next>**.

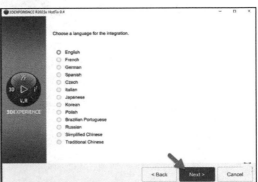

Install the files on your hard drive.

Click **Install>**.

Once completed, Click **OK**.

Close the Installation dialog box.

Click **Close**.

Close all running programs.

Restart your computer.

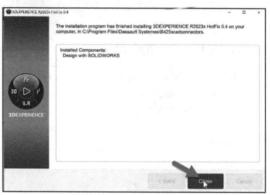

Log in to the **3D**EXPERIENCE platform. Use your unique link from your Invitation email to the **3D**EXPERIENCE platform.

Enter your **3D**EXPERIENCE ID.

Launch the Design with SOLIDWORKS App.

Click the **center** of the Compass.

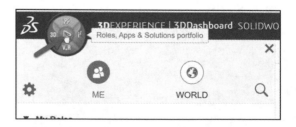

Drag the slider bar downward under My Apps until you see the Design with SOLIDWORKS App.

Click the **Design with SOLIDWORKS** App.

SOLIDWORKS is opened and you are connected to the **3D**EXPERIENCE platform.

Close the Welcome - SOLIDWORKS dialog box.

Click **Close**.

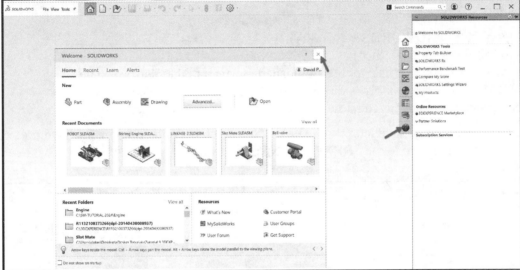

Click the **3DEXPERIENCE** icon in the Task Pane. If you do not see the 3DEXPERIENCE icon, select 3DEXPERIENCE from the SOLIDWORKS Add-Ins.

Read the message.

Click **Accept All**.

The MySession panel displays the two-way communication between SOLIDWORKS running on your computer and the **3D**EXPERIENCE platform running in the cloud.

Note: The default storage space, "Collaborative space" is called Common Space.

The illustration displays my Collaborative space called: Quick Start xDesign.

Open the SOLIDWORKS assembly (provided in the Project 9 folder) that has not been uploaded or saved to the platform. The SOLIDWORKS assembly is displayed in the Graphics area. The assembly and components are displayed in the FeatureManager and Task Pane.

The Status, orange save icons, indicate that the assembly and components have not been uploaded and saved to the **3D**EXPERIENCE platform.

Below is an illustration of the assembly after it is saved to the **3D**EXPERIENCE platform.

Note: Once the 3DEXPERIENCE Launcher has been installed, your computer may display the 3DPassport dialog box in the desktop. This is an easy way to log in to the 3DEXPERIENCE platform before starting a SOLIDWORKS session.

Close all models.

Exit SOLIDWORKS.

We are finished with this lesson. In the next lesson explore saving and revision of a SOLIDWORKS document to the **3D**EXPERIENCE platform.

3DEXPERIENCE Lesson 2: SOLIDWORKS Save and Revision

The entire lesson is within the SOLIDWORKS environment connected to the **3D**EXPERIENCE platform. Start a SOLIDWORKS session.

Open a SOLIDWORKS part that has not been previously uploaded or saved to the **3D**EXPERIENCE platform. View the 3DEXPERIENCE Add-In on your SOLIDWORKS desktop.

Review and address numerous Save options: **Save to this PC**, **Save to 3DEXPERIENCE**, **Save Active Window**, **Save with Options**, etc.

View several Product Lifecycle Management (PLM) attributes (Bookmark, Revision, Maturity State, Collaborative Space, and Reservation Owner) of the part.

Lock the part. Modify the SOLIDWORKS part. Save the modified SOLIDWORKS part to the **3D**EXPERIENCE platform. Create a New Revision, Unlock the part and view the maturity state of the part. Save the model to your Collaborative space in the MySession panel.

Collaborative space is defined by an administrator (instructor or IT personnel) or the user, if the user has access rights. The default Collaborative space for a student is named Common Space.

Create and use Bookmarks, not new Collaborative spaces, to save different assemblies with common referenced components for your classes, teams and organizations. Bookmarks provide a file/folder structure similar to Windows Explorer.

A part is used in this lesson to simply demonstrate the power of Collaborative space, Bookmarks and revisions with product lifecycle management attributes. In the next lesson, you will explore an assembly.

Before we start, there are a few items that you need to know.

In this lesson, use your assigned Collaborative space on the **3D**EXPERIENCE platform. An internet connection is required.

3DEXPERIENCE Launcher needs to be installed.

This lesson is not for SOLIDWORKS Connected "3DEXPERIENCE SOLIDWORKS".

The **3D**EXPERIENCE platform is browser driven. Your existing cookies and cache determine what you will see on your computer desktop or during a SOLIDWORKS login.

A full installation of SOLIDWORKS 2019 SP0 or later is required.

Start a SOLIDWORKS session from your desktop.

Double-click the **SOLIDWORKS** icon.

View the illustration below. Depending on your system setup, cookies, and cache, it will be different.

Input the requested information.

Click **Accept All**.

The Welcome - SOLIDWORKS dialog box is displayed.

You are logged into the **3D**EXPERIENCE platform.

Close the Welcome dialog box.

Click the **3DEXPERIENCE** icon in the Task Pane. The MySession panel is displayed. This displays the two-way communication between SOLIDWORKS running on your machine and the **3D**EXPERIENCE platform running in the cloud. Experience a single modeling environment.

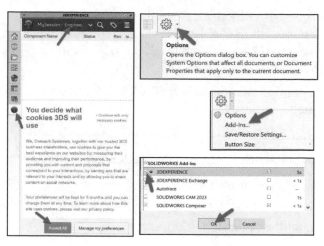

The default Collaborative space for a student is named Common Space. The illustrated Collaborative space is named Engineering Design.

Note: If you do not see the 3DEXPERIENCE icon, click the Options drop-down arrow, click Add-Ins, check the 3DEXPERIENCE box, click OK, from the SOLIDWORKS Main menu.

Click **Accept All**.

View your available Roles, Cockpits, My Favorite Apps, and Apps.

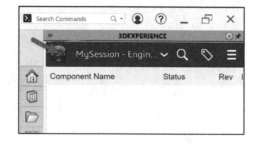

Click the **center of the Compass** in the MySession panel as illustrated.

Drag the slider bar downward to view your available Roles, Cockpits, My Favorite Apps, and Apps. The display will vary depending on your license and Role.

Return to the Compass in the MySession panel.

Click **Close**.

Open the SOLIDWORKS part Cosmetic Thread 7-6 from the Project 9 folder that has not been previously uploaded or saved to the **3D**EXPERIENCE platform.

Expand the Task Pane. The MySession panel displays a tree view of the active file and a list of commands that can be accessed through the Action bar and context menu.

The MySession panel displays the **Component Name**, **Status** (Displays an icon that represents the status of a given revision, saved or not saved to the platform), **Locked status** (Locked or Unlocked), **Rev** (Revision #), **Is Last Revision** (Yes/No), **Maturity State** (Lifecycle state of the family, for example: Private, In Work, Frozen, Released, Obsolete), **Description**, **File Name** (Displays the PLM external ID of the object for reference objects and the PLM external ID of the connection for instances) and **Type** (Physical Product or Drawing).

At this time, they are blank. You opened a SOLIDWORKS part that has not been previously uploaded or saved to the platform.

The Action bar provides the ability to upload files to the **3D**EXPERIENCE platform. There are other tools that can be accessed. The Lifecycle tab is selected by default.

There are many new save options in SOLIDWORKS when using the 3DEXPERIENCE platform. Let's explore a few.

Right-click the **orange Save** icon in the MySession panel. There are two save options to the **3D**EXPERIENCE platform.

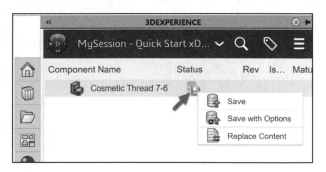

Save: Opens the Save dialog box to save a file to a Collaborative space (Common space for a student) on the **3D**EXPERIENCE platform. This command does not provide the ability to select a Bookmark location. It is highly recommended to save files in a Bookmark. Think of a Bookmark Workspace as a directory which can contain Bookmark folders. Bookmark folders are used to further categorize bookmarks like a folder tree structure (Windows Explorer).

Use Bookmarks to delete entire groups of data or just a single file. For assemblies within the bookmark folder, there is an option to delete the entire structure of the assembly and all reference components. Another option to delete files (Physical Products) from the platform is to use the Collaborative Lifecycle App.

Save with Options: One of the best simple practices for SOLIDWORKS data is to create bookmarks upon the initial save to the **3D**EXPERIENCE platform. Choosing "Save With Options" provides you the choice of existing Bookmark folders or to create a new Bookmark to contain the file references.

Expand Lifecycle (PLM operations to files using the **3D**EXPERIENCE platform in the Action bar) as illustrated.

There are two Save options displaycd to the 3DEXPERIENCE platform.

Save Active Window: Uploads the files to the **3D**EXPERIENCE platform. One of the best simple practices for SOLIDWORKS data is to create bookmarks upon the initial save to the **3D**EXPERIENCE platform. Choosing the "Save Active Window" provides the user the choice of existing bookmark folders or to create a new bookmark to contain the file references.

Save Active Window as New: Saves a copy of existing files with new names to the **3D**EXPERIENCE platform. Not available for the first time.

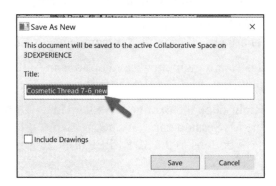

Click **File** from the SOLIDWORKS Main menu. View the save options.

Save to This PC: Saves the document to your local hard drive in the folder which was last opened from.

Save to 3DEXPERIENCE: Opens the Save as New dialog box to save the file to a Collaborative space on the **3D**EXPERIENCE platform. This command does not provide the ability to select a Bookmark location.

Save As: Saves the document to a new file name that becomes the active document without saving the original document.

Save All: Saves all the open files that have been modified since they were last saved.

Save As Template: **3D**EXPERIENCE users can create a template for the **3D**EXPERIENCE platform and save it directly from SOLIDWORKS.

View the 3DEXPERIENCE files that were saved on your computer.

Click the **3DEXPERIENCE files on This PC** icon in the Task Pane. View the saved files in the **3D**EXPERIENCE platform and their Collaborative space locations. An example is displayed.

Save the model, Cosmetic Thread 7-6 to your local hard drive.

Click **Save to This PC** from the SOLIDWORKS Menu bar.

Save the model to the **3D**EXPERIENCE platform. Use the Save Active Window command in the Action bar.

Click **Save Active Window** from the Lifecycle tab in the Action bar.

The file is temporarily being saved to a local cache area. The platform is checking the model. **Note**: for an assembly, the platform is checking for out of date components, modified components from the last save to the platform, different revisions, missing components, etc.

The Save to **3D**EXPERIENCE dialog box and PLM attributes are displayed. The PLM attributes include Bookmark, Revision, Maturity State, Collaborative Space, and Reservation Owner, etc.

Save the model. Use your existing Collaborative space. Create a new Bookmark.

One of the best simple practices for SOLIDWORKS data is to create Bookmarks upon the initial save to the **3D**EXPERIENCE platform.

Click the **Select Bookmark** down arrow.

Click **Select Bookmark** (Workspace). The Select a Bookmark dialog box is displayed.

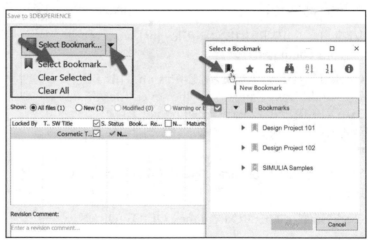

Select the **Bookmarks** box as illustrated.

Click the **New Bookmark** icon. The below illustration displays my Bookmarks.

Enter **Design Project 103** for New Bookmark Title.

Select your Collaborative space. The below illustration displays a few of my created Collaborative spaces.

Click **OK**.

Bookmark Design Project 103 is created.

Select Design Project 103.

Click **Apply**.

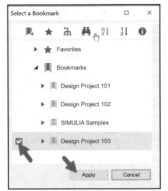

Hover over the Bookmarks column. Design Project 103 is displayed.

Enter **Revision Comment**.
Note: The Revision Comment box was removed in the latest **3D**EXPERIENCE hot fix.

Enter **Initial Save**.

Click **Save**. The model is saved to your Collaborative space and Bookmarked on the **3D**EXPERIENCE platform.

You are returned back to your SOLIDWORKS session.

Expand the Task Pane area. View the auto generated embedded model properties. Use these properties for enhanced search options.

View the columns in the MySession panel: **Component Name** (Cosmetic Thread 7-6), **Status** (Saved to the platform), **Locked Status** (Unlocked), **Rev** (A.1), **Is Last Revision** (Yes), **Maturity State** (In Work), **File Name** (Cosmetic Thread 7-6), **Collaborative space** (Engineering Design), **Name** (prd-R1132100373266), **Type** (Physical Product), **Convert status** (Yes). **Note**: A Physical Product is a configuration of a part or assembly that can be manufactured. When you save a .sldprt or .sldasm file to a Collaborative space, each configuration becomes a Physical Product with a name that you can find in searches.

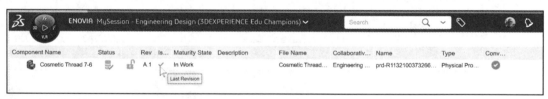

Explore the Status column. View Properties.

Right-click **Status** from the MySession panel.

Click **Properties**. View the auto generated embedded model properties.

Close the dialog box. Return to the MySession panel.

Click **Close**.

Right-click the **Not Locked** icon in the MySession panel. View your options.

Click Lock. The lock process displays a green key icon. You have full read write access. It also prevents any other team member to make changes to the file. This will reduce duplicated work and ensure that you and your team are working on the most updated file/files.

Modify the Cosmetic Thread length of the model.

Right-click Cosmetic Thread from the FeatureManager.

Click **Edit Feature** from the Pop-up menu. The Cosmetic Thread PropertyManager is displayed.

Enter **1.250in** for Depth.

Click **OK**.

The Status icon in the MySession panel informs you that the file was modified. The local copy (on your desktop) is different than the saved copy on the platform.

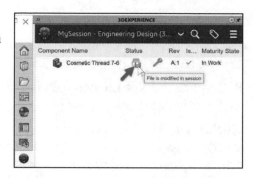

Save and update the model to the **3D**EXPERIENCE platform in your Bookmark.

Create a New Revision.

Click **Save Active Window** from the Lifecycle tab in the Action bar.

The Save to 3DEXPERIENCE dialog box is displayed.

Check the **New Revision** box.
Note: Locked By, displays the first and last name of the person that has locked the corresponding family on the **3D**EXPERIENCE platform.

Enter a Revision comment.
Note: The Revision Comment box was removed in the latest **3D**EXPERIENCE hot fix.

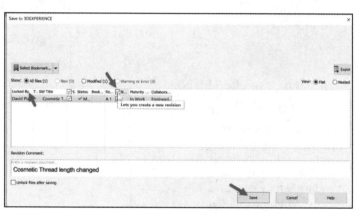

Return to your SOLIDWORKS session.

Click **Save**.

View the updated MySession panel. Status: updated. Revision: B.1.

Unlock the model. This provides model access to people on your team or other groups with the latest changes.

Right-click the green key icon.

Click **Unlock**.

Check the Revision history of the model.

Click **Cosmetic Thread 7-6** in the MySession panel.

Click **Revisions** under the Lifecycle tab in the Action bar as illustrated.

View the Revisions dialog box. The box displays the history revision information on the document. Revision A.1 and B.1 are displayed. Maturity State: In Work. Creation and Modification Dates are displayed. Close the Revisions dialog box.

Click **Close**.

View Lifecycle information directly in the FeatureManager design tree.

Expand the FeatureManager as illustrated. View the provided information.

Close the part.

This entire lesson was within the SOLIDWORKS environment connected to the 3DEXPERIENCE platform.

We are finished with the lesson. In the next lesson, explore **3DEXPERIENCE Lesson 4: SOLIDWORKS and Lifecycle Maturity States**.

3DEXPERIENCE Lesson 4: SOLIDWORKS and Lifecycle Maturity States

Product Lifecycle Management (PLM) is not normally taught in schools, and most students do not realize how important it is.

Product Lifecycle Management (PLM) cloud Apps connect directly to SOLIDWORKS through the Task Pane. The auto generated embedded model properties provide students awareness to implement sound engineering practices.

In this lesson, you will investigate the following options in the SOLIDWORKS MySession, Lifecycle tab: Save Active Window, Save Active Window As New, Reload from Server, Replace By Revision and Maturity.

Use the Collaborative Lifecycle App in the Action bar on a SOLIDWORKS assembly. Save the assembly to the 3DEXPERIENCE platform in a Bookmark. Bookmarks provides a file/folder structure similar to Windows Explorer.

Lock a component. Modify a component. Create a new revision. Save the latest changes to the platform.

Change Maturity state of a component, assembly and several components with different maturity status. Maturity states provide information about the content's completion stage and its validity.

Change the Revision of the assembly. Note: Verify with your administrator/instructor that your credentials match the Collaborative space of the object you want to change the maturity of.

Before we start, there are a few items that you need to know.

In this lesson, create a Collaborative space on the 3DEXPERIENCE platform or use your default Collaborative space named "Common Space".

3DEXPERIENCE Launcher needs to be installed. **Note**: This lesson is not for SOLIDWORKS Connected "3DEXPERIENCE SOLIDWORKS".

The **3D**EXPERIENCE platform is browser driven. Your existing cookies and cache determine what you will see on your computer desktop or during a SOLIDWORKS login. A full installation of SOLIDWORKS 2019 SP0 or later is required.

What is the difference between traditional Product Data Management (PDM) and 3DEXPERIENCE Product Lifecycle Management (PLM)?

Traditional PDM is a system that centralizes product-related data and processes. Engineers use PDM software to track revisions, manage change orders, generate Bills of Materials (BOMs), and more. PDM software helps student teams and companies to bring the designs to final release, reducing redundancy work, increasing product development, and collaboration.

3DEXPERIENCE PLM is a cloud-based platform App in a Collaborative space which enables teams and organizations to control and track large amounts of data from part and assembly files to simulation and CAD/CAM files. Further, PLM addresses the countless files necessary for effective project communications, such as spreadsheets, PDFs, and emails, to name a few; all need to be managed to optimize collaboration and keep product designs on course. PLM is part of the **3D**EXPERIENCE Works portfolio.

Start a SOLIDWORKS session from your desktop.

Double-click the **SOLIDWORKS** icon.

View the illustration below. Depending on your system setup, cookies, and cache, your display will be different.

Log in to the 3DEXPERIENCE platform.

Enter your 3DEXPERIENCE ID.

The SOLIDWORKS - Welcome dialog box is displayed.

You are logged into the 3DEXPERIENCE platform.

Close the Welcome dialog box.

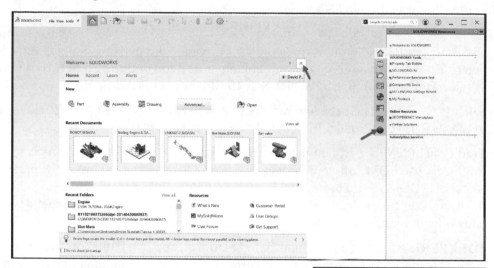

Click the **3DEXPERIENCE** icon in the Task Pane. The MySession panel is displayed. This displays the two-way communication between SOLIDWORKS running on your machine and the **3D**EXPERIENCE platform running in the cloud. If needed, Click Accept All.

Collaborative space is defined by an administrator (instructor or IT personnel) or the user, if the user has access rights. The default Collaborative space for a student is named Common Space. The illustrated Collaborative space is named Engineering Design.

Open the SOLIDWORKS Slot Mate assembly that has not been previously uploaded or saved to the **3D**EXPERIENCE platform.

Expand the Task Pane bar. The MySession panel displays a tree view of the active file and a list of commands that can be accessed through the Action bar and Context menu. The MySession panel displays the two-way communication between SOLIDWORKS running on your machine and the **3D**EXPERIENCE platform running in the cloud.

The MySession panel also displays the **Component Name**, **Status** (Displays an icon that represents the status of a given revision, saved or not saved to the platform), **Locked status** (Locked or Unlocked), **Rev** (Revision #), **Is Last Revision** (Yes/No), **Maturity State** (Lifecycle state of the family, for example: Private, In Work, Frozen, Released, Obsolete), **Description**, **File Name** (Displays the PLM external ID of the object for reference objects and the PLM external ID of the connection for instances) and **Type** (Physical Product or Drawing).

At this time, they are blank. You opened a SOLIDWORKS assembly that has not been previously uploaded or saved to the platform.

The Action bar provides the ability to upload files to the **3D**EXPERIENCE platform. There are other tools that can be accessed. The Lifecycle tab is selected by default.

In this lesson, the Collaborative space on the platform is Engineering Design. Note: Within a Collaborative space, your access determines what you can do depending on whether the Collaborative space is public, protected, or private.

Upload and save the files to the **3D**EXPERIENCE platform. One of the best simple practices for SOLIDWORKS data is to create Bookmarks upon the initial save to the **3D**EXPERIENCE platform. Choosing the "Save Active Window" provides the user the choice of existing Bookmark folders or to create a new Bookmark to contain the file references.

Expand the Lifecycle tab.

Click **Save Active Window**.

The file is temporarily being saved to a local cache area. The platform is checking the model.

The Save to 3DEXPERIENCE dialog box and PLM attributes are displayed. The attributes include Bookmark, Revision, Maturity State, Collaborative Space, and Reservation Owner.

Save the model. Use your existing Collaborative space. Create a new Bookmark. Use Bookmarks, not Collaborative spaces to save different assemblies with common referenced components. Bookmarks provide a folder/file structure similar to Windows.

Click the **Select Bookmark** down arrow.

Click **Select Bookmark**. The Select a Bookmark dialog box is displayed.

Select the **Bookmarks** box as illustrated.

Click the **New Bookmark** icon.
Note: The illustration displays other created Bookmarks.

Enter **Design Project 104** for New Bookmark Title. **Select** your Collaborative space.

Click **OK**. Select **Design Project 104**. Click **Apply to Selected**.

Enter Revision Comment. Save to the 3DEXPERIENCE platform.

Enter **comment** in the Revision Comment box. **Note**: The Revision Comment box was removed in the latest **3D**EXPERIENCE hot fix.

Click **Save**.

The files are directly loaded into your Collaborative space and Bookmarked on the **3D**EXPERIENCE platform.

You are returned to your SOLIDWORKS session.

View the updated MySession panel.

The Maturity State is "In Work". This is the default Lifecycle state after you saved the model to the **3D**EXPERIENCE platform.

Modify the length of the ROD from 100mm to 105mm. Lock the component. This prevents anyone in your Collaborative space from making a modification to the component while you are working on it.

Click **ROD** in the MySession panel.

Hover over the **Lock Status**. At this time, the Lock Status is "Not Locked".

Right-click the **Not Locked** icon.

Click **Lock** from the drop-down menu. View the Lock Status "Locked by me".

Open ROD from the FeatureManager.

Right-click **ROD** from the FeatureManager.

Click **Open Part** in the Context box.

The ROD is displayed in the Graphics window.

Double-click **Base Extrude**.

Click the **100mm** length. The modify dialog box is displayed.

Enter **105mm**.

Return to the assembly.

Save and **Close** the ROD part.

View the results in the MySession panel. The orange save Status icon displayed for the Slot Mate informs you that the local file on your computer is not saved in the platform.

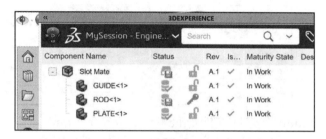

The orange save Status icon displayed for the ROD informs you that the local file is locked and modified, but the modifications are not yet propagated to the platform.

Update the Slot Mate assembly with the ROD change to the platform.

Lock the **Slot Mate** assembly as illustrated.

Right-click the **orange Status** icon of the Slot Mate assembly.

Click **Save with Options**.

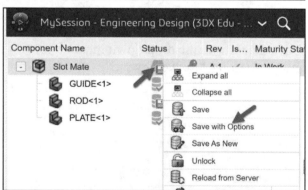

The Save to 3DEXPERIENCE dialog box is displayed. The Slot Mate assembly and ROD are locked by the user.

The Collaborative space in this lesson is Engineering Design.

The Maturity Status is "In Work". The first default Revision is A.1.

Create a new Revision.

Click the **Lets you create a new revision** box.

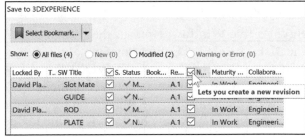

Enter **Revision Comment**. **Note**: The Revision Comment box was removed in the latest **3D**EXPERIENCE hot fix.

Check the **Unlock files** after saving box.

Save the assembly and new revision to the platform.

Click **Save** from the 3DEXPERIENCE dialog box.

View the updated MySession panel. The assembly and reference components are saved to the platform with a new revision. They are all unlocked. Anyone in your Collaborative space has access to the assembly.

Investigate the Lifecycle tab in the Action bar.

Click **Slot Mate** from the MySession panel. The Lifecycle tools are highlighted.

Review the options.

Save Active Window and the **Save Active Window As New**.

Save Active Window: Uploads the files to the **3D**EXPERIENCE platform. One of the best simple practices for SOLIDWORKS data is to create bookmarks upon the initial save to the **3D**EXPERIENCE platform. Choosing the "Save Active Window" provides the user the choice of existing bookmark folders or to create a new bookmark to contain the file references.

Save Active Window as New: Saves a copy of existing files with new names to the **3D**EXPERIENCE platform. Not available for the first time.

Reload from Server. Updates the selected revision in a SOLIDWORKS session with the latest version that exists on the **3D**EXPERIENCE platform.

There are three sub-options: *Reload selected file only (do not reload references), Reload references only (do not reload the selected file), Reload file and references (reload the selected file and all its references).*

Replace By Revision. Opens the Replace by Revision dialog box to replace the revision that is opened in the SOLIDWORKS session. There are two sub-options: *Update Revisions (Opens the Replace by Revision dialog box to replace the revision that is opened in session), Replace by Latest Revision (Updates the revision for the selected object with the latest revision).*

Maturity. Defines the Maturity state of a component as it evolves throughout its lifecycle: Private (documents are only accessible to the owner and are not shared in the Collaborative space), In Work

(documents are initially saved to the In Work state and are accessible in the Collaborative space), Frozen (documents are typically set Frozen while in review to prevent change), Released (documents are typically set to Released when in production. Released documents cannot be modified), Obsolete (documents are typically set to Obsolete when no longer used).

Below are two Maturity graphs. Each box shows a Maturity state. The lines with arrows from one Maturity state to another show which Maturity states the content can move to, from that Maturity state. A right-pointing arrow means the content can be promoted to that Maturity state. A left-pointing arrow means the content can be demoted to that Maturity state. The arrows between Maturity states are called transitions.

Change the Maturity state of the ROD component.

Click **ROD** from the MySession panel. ROD is selected in the Graphics area.

Click **Maturity** from the Lifecycle tab in the Action bar.

The Maturity - ROD B.1 dialog box is displayed. View the Maturity graph.

The highlighted box (In Work) designates the present Maturity state.

Hover over Freeze as illustrated. View the arrow movement from "In Work" to "Freeze".

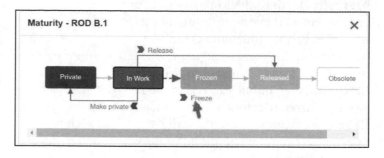

Click **Freeze** to put the ROD component in the Frozen state. This will prevent anyone in your Collaborative space from making a design change to the ROD component. The Frozen status provides the ability to perform a design review, so no one can make changes to the model.

The MySession panel is updated. The ROD component is displayed in a Frozen State.

Change the maturity of the Slot Mate assembly.

Select **Slot Mate** from the MySession panel.

Click **Maturity** from the Lifecycle tab in the Action bar.

The Maturity - Slot Mate B.1 dialog box is displayed. View the Maturity graph. The highlighted box designates the present Maturity state.

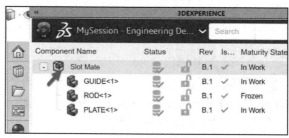

Hover over **Freeze** as illustrated. View the arrow movement from "In Work" to "Freeze".

Click **Include structural objects**. This option changes the Maturity state of the children along with the parent.

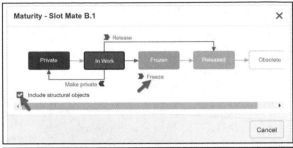

Click **Freeze**.

View the updated MySession panel. The assembly and all reference components are Frozen.

Modify the Maturity State "In Work" of the ROD component.

Select **ROD** from the MySession panel.

Click **Maturity** from the Lifecycle tab. The Maturity dialog box is displayed. View the Maturity graph.

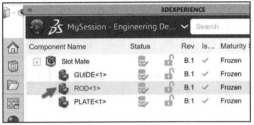

Click **Restart work**.

View the updated MySession panel.

View the Maturity state of two components ROD (In Work) and PLATE (Frozen).

Select **ROD** from the MySession panel.

Hold the **Ctrl key** down.

Select **PLATE**.

Release the **Ctrl key**.

Click **Maturity** from the Lifecycle tab.

The Maturity - 1 Object(s) dialog box is displayed.

View your options. **Note**: For each object, an icon indicates if the transition can be applied to the object: *Applicable Transition*, *Not Applicable Transition*.

Change the ROD Maturity status to Frozen.

Select **Frozen** from the drop-down menu as illustrated. View the icon changes.

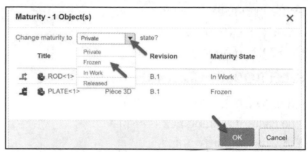

Click **OK** from the Maturity - 1 Object(s) dialog box.

View the results. Close the Report dialog box.

Click **Close**.

Note: To exclude an object and not apply the maturity transition to it, double-click the corresponding Applicable Transition icon. The icon changes to (Not Applied Transition) and the object's maturity state is not modified when the Maturity transition is applied.

Revisions. Displays the revisions, branches, and duplicates of an object.

Select **Slot Mate** from the MySession panel.

Click **Revisions** from the Lifecycle tab in the Action bar. The Revisions dialog box is displayed. **Note**: You can have different revisions at different Maturity States.

Close the Revisions dialog box.

Click **Close**.

New Revision. Creates a new revision of the selected object; you can also select the content of a different object.

Select **Slot Mate** from the MySession panel.

Click **New Revision** from the Lifecycle tab in the Action bar. View the New Revision dialog box.

Create a new revision. Accept the default revision (C.1).

Enter **Revision Comment**.

Click **Revise**.

View the updated MySession panel. You created a new revision. The Collaborative Lifecycle App locked the assembly. No changes can be made by anyone in your Collaborative space. Save the assembly to the platform. The Maturity State displays "In Work".

Right-click the **Slot Mate Status** icon.

Click **Save with Options**.

The Save to 3DEXPERIENCE dialog box is displayed. View the update.

Add a **Revision Comment**. **Note**: The Revision Comment box was removed in the latest **3D**EXPERIENCE hot fix.

Check the **Unlock files** after saving box.

Click **Save** from the Save to 3DEXPERIENCE dialog box.

View the updated MySession panel.

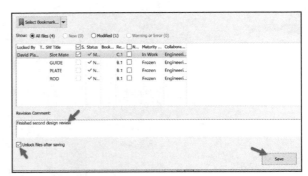

There are two additional options to review.

New Revision From: Create a revision of the selected structure using the content of another structure. In the dialog box, select the source, which can be a revision, a duplicate or a branch. Choose the appropriate option for each child node, and click Revise.

New Branch. Creates a new branch from the selected object. In the dialog box, select the credentials to assign to the new branches. If you select one object, you can enter a title for the new branch. If you select several objects, you can enter a prefix for the new branches. You can enter a comment for the new branches in the Revision Comment box.

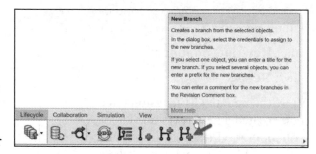

This entire lesson was within the SOLIDWORKS environment connected to the 3DEXPERIENCE platform.

We are finished with this lesson.

Project Summary

In this project, you explored SOLIDWORKS and the **3D**EXPERIENCE platform. The project was written to answer a few key questions for the new user. What is the **3D**EXPERIENCE platform? How does it work within my SOLIDWORKS session?

The platform enables you to collaborate with students and teams on and off campus through a browser in a secure cloud environment. Think of the **3D**EXPERIENCE platform simply as a SOLIDWORKS Add-In on your desktop. Through the Task Pane, students can manage, revise, and share information. Experience a single modeling environment. Access the platform within the SOLIDWORKS Task Pane.

Detail steps were provided to obtain a **3D**EXPERIENCE ID (login credentials) to access the platform.

Nine step-by-step PDF tutorials and models are provided to practice and understand the tools and advantages of SOLIDWORKS and the platform.

- 3DEXPERIENCE Lesson 1: Getting Started with SOLIDWORKS and the Platform

- 3DEXPERIENCE Lesson 2: SOLIDWORKS Save and Revision

- 3DEXPERIENCE Lesson 3: SOLIDWORKS Bookmarks, Share and Delete

- 3DEXPERIENCE Lesson 4: SOLIDWORKS and Lifecycle Maturity States

- 3DEXPERIENCE Lesson 5: SOLIDWORKS, Collaborative Space and Bookmarks

- 3DEXPERIENCE Lesson 6: SOLIDWORKS and Search

- 3DEXPERIENCE Lesson 7: SOLIDWORKS with 3DPlay

- 3DEXPERIENCE Lesson 8: SOLIDWORKS and 3DDrive

- 3DEXPERIENCE Lesson 9: SOLIDWORKS and 3DSwym Communities

Notes:

Appendix

SOLIDWORKS Keyboard Shortcuts

Below are some of the pre-defined keyboard shortcuts in SOLIDWORKS:

Action:	Key Combination:
Model Views	
Rotate the model horizontally or vertically	**Arrow** keys
Rotate the model horizontally or vertically 90 degrees	**Shift** + **Arrow** keys
Rotate the model clockwise or counterclockwise	**Alt** + left of right **Arrow** keys
Pan the model	**Ctrl** + **Arrow** keys
Magnifying glass	**g**
Zoom in	**Shift + z**
Zoom out	**z**
Zoom to fit	**f**
Previous view	**Ctrl + Shift + z**
View Orientation	
View Orientation menu	**Spacebar**
Front view	**Ctrl + 1**
Back view	**Ctrl + 2**
Left view	**Ctrl + 3**
Right view	**Ctrl + 4**
Top view	**Ctrl + 5**
Bottom view	**Ctrl + 6**
Isometric view	**Ctrl + 7**
NormalTo view	**Ctrl + 8**
Selection Filters	
Filter edges	**e**
Filter vertices	**v**
Filter faces	**x**
Toggle Selection Filter toolbar	**F5**
Toggle selection filters on/off	**F6**
File menu items	
New SOLIDWORKS document	**Ctrl + n**
Open document	**Ctrl + o**
Open From Web Folder	**Ctrl + w**
Make Drawing from Part	**Ctrl + d**
Make Assembly from Part	**Ctrl + a**
Save	**Ctrl +s**
Print	**Ctrl + p**
Additional items	
Access online help inside of PropertyManager or dialog box	**F1**
Rename an item in the FeatureManager design tree	**F2**

Action:	Key Combination:
Rebuild the model	**Ctrl + b**
Force rebuild - Rebuild the model and all its features	**Ctrl + q**
Redraw the screen	**Ctrl + r**
Cycle between open SOLIDWORKS document	**Ctrl + Tab**
Line to arc/arc to line in the Sketch	**a**
Undo	**Ctrl + z**
Redo	**Ctrl + y**
Cut	**Ctrl + x**
Copy	**Ctrl + c**
Paste	**Ctrl + v**
Delete	**Delete**
Next window	**Ctrl + F6**
Close window	**Ctrl + F4**
View previous tools	**s**
Selects all text inside an Annotations text box	**Ctrl + a**

In a sketch, the **Esc** key un-selects geometry items currently selected in the Properties box and Add Relations box.

In the model, the **Esc** key closes the PropertyManager and cancels the selections.

Use the **g** key to activate the Magnifying glass tool. Use the Magnifying glass tool to inspect a model and make selections without changing the overall view.

Use the **s** key to view/access previous command tools in the Graphics window.

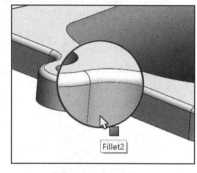

Modeling - Best Practices

Best practices are simply ways of bringing about better results in easier, more reliable ways. The Modeling - Best Practice list is a set of rules helpful for new users and users who are trying to experiment with the limits of the software.

These rules are not inflexible, but conservative starting places; they are concepts that you can default to, but that can be broken if you have good reason. The following is a list of suggested best practices:

- Create a folder structure (parts, drawings, assemblies, simulations, etc.). Organize into project or file folders.

- Construct sound document templates. The document template provides the foundation that all models are built on. This is especially important if working with other SOLIDWORKS users on the same project; it will ensure consistency across the project.

- Generate unique part filenames. SOLIDWORKS assemblies and drawings may pick up incorrect references if you use parts with identical names.

- Apply Custom Properties. Custom Properties is a great way to enter text-based information into the SOLIDWORKS parts. Users can view this information from outside the file by using applications such as Windows Explorer, SOLIDWORKS Explorer, and Product Data Management (PDM) applications.

- Understand part orientation. When you create a new part or assembly, the three default Planes (Front, Right and Top) are aligned with specific views. The plane you select for the Base sketch determines the orientation.

- Learn to sketch using automatic relations.

- Limit your usage of the Fixed constraint.

- Add geometric relations, then dimensions in a 2D sketch. This keeps the part from having too many unnecessary dimensions. This also helps to show the design intent of the model. Dimension what geometry you intend to modify or adjust.

- Fully define all sketches in the model. However, there are times when this is not practical, generally when using the Spline tool to create a freeform shape.

- When possible, make relations to sketches or stable reference geometry, such as the Origin or standard planes, instead of edges or faces. Sketches are far more stable than faces, edges, or model vertices, which change their internal ID at the slightest change and may disappear entirely with fillets, chamfers, split lines, and so on.

- Do not dimension to edges created by fillets or other cosmetic or temporary features.

- Apply names to sketches, features, dimensions, and mates that help to make their function clear.

- When possible, use feature fillets and feature patterns rather than sketch fillets and sketch patterns.

- Apply the Shell feature before the Fillet feature, and the inside corners remain perpendicular.

- Apply cosmetic fillets and chamfers last in the modeling procedure.

- Combine fillets into as few fillet features as possible. This enables you to control fillets that need to be controlled separately, such as fillets to be removed and simplified configurations.

- Create a simplified configuration when building very complex parts or working with large assemblies.

- Use symmetry during the modeling process. Utilize feature patterns and mirroring when possible. Think End Conditions.

- Use global variables and equations to control commonly applied dimensions (design intent).

- Add comments to equations to document your design intent. Place a single quote (') at the end of the equation, then enter the comment. Anything after the single quote is ignored when the equation is evaluated.

- Avoid redundant mates. Although SOLIDWORKS allows some redundant mates (all except distance and angle), these mates take longer to solve and make the mating scheme harder to understand and diagnose if problems occur.

- Fix modeling errors in the part or assembly when they occur. Errors cause rebuild time to increase, and if you wait until additional errors exist, troubleshooting will be more difficult.

- Create a Library of Standardized notes and parts.

- Utilize the Rollback bar. Troubleshoot feature and sketch errors from the top of the design tree.

- Determine the static and dynamic behavior of mates in each sub-assembly before creating the top-level assembly.

- Plan the assembly and sub-assemblies in an assembly layout diagram. Group components together to form smaller sub-assemblies.

- When you create an assembly document, the base component should be fixed, fully defined or mated to an axis about the assembly origin.

- In an assembly, group fasteners into a folder at the bottom of the FeatureManager. Suppress fasteners and their assembly patterns to save rebuild time and file size.

- When comparing mass, volume and other properties with assembly visualization, utilize similar units.

- Use limit mates sparingly because they take longer to solve and whenever possible, mate all components to one or two fixed components or references. Long chains of components take longer to solve and are more prone to mate errors.

Helpful On-line Information

The SOLIDWORKS URL:
http://www.SOLIDWORKS.com
contains information on Local
Resellers, Solution Partners,
Certifications, SOLIDWORKS
users groups and the
3DEXPERIENCE platform.

Use the SOLIDWORKS Task
Pane to obtain access to Customer
Portals, User Groups,
Manufacturers, Solution Partners,
and the **3D**EXPERIENCE
Platform.

Helpful on-line SOLIDWORKS information is available from
the following URLs:

- http://www.swugn.org/

List of all SOLIDWORKS User groups.

- https://www.solidworks.com/sw/edu
  cation/certification-programs-cad-
  students.htm

The SOLIDWORKS Academic
Certification Programs.

- http://www.solidworks.com/sw/in
  dustries/education/engineering-
  education-software.htm

The SOLIDWORKS Education
Program:

- visit
  https://3dexperience.virtualtester.
  com/#home to obtain additional
  SOLIDWORKS Certification
  exam information.

On-line tutorials are for educational
purposes only. Tutorials are
copyrighted by their respective owners.

SOLIDWORKS Document Types

SOLIDWORKS has three main document file types: Part, Assembly and Drawing, but there are many additional supporting types that you may want to know. Below is a brief list of these supporting file types:

Design Documents	Description
.sldprt	SOLIDWORKS Part document
.slddrw	SOLIDWORKS Drawing document
.sldasm	SOLIDWORKS Assembly document

Templates and Formats	Description
.asmdot	Assembly Template
.asmprp	Assembly Template Custom Properties tab
.drwdot	Drawing Template
.drwprp	Drawing Template Custom Properties tab
.prtdot	Part Template
.prtprp	Part Template Custom Properties tab
.sldtbt	General Table Template
.slddrt	Drawing Sheet Template
.sldbombt	Bill of Materials Template (Table-based)
.sldholtbt	Hole Table Template
.sldrevbt	Revision Table Template
.sldwldbt	Weldment Cutlist Template
.xls	Bill of Materials Template (Excel-based)

Library Files	Description
.sldlfp	Library Part file
.sldblk	Blocks

Other	Description
.sldstd	Drafting standard
.sldmat	Material Database
.sldclr	Color Palette File
.xls	Sheet metal gauge table

Project 8: Answer Key

1. What is the Modulus of Elasticity?

- The slope of the Deflection-Stress curve

- **The slope of the Stress-Strain curve in its linear section**

- The slope of the Force-Deflection curve in its linear section

- The first inflection point of a Strain curve

2. What is Stress?

- A measure of power

- A measure of strain

- A measure of material strength

- **A measure of the average amount of force exerted per unit area**

3. Which of the following assumptions are true for a static analysis in SOLIDWORKS Simulation with small displacements?

- **Inertia effects are negligible and loads are applied slowly**

- The model is not fully elastic. If loads are removed, the model will not return to its original position

- **Results are proportional to loads**

- **All the displacements are small relative to the model geometry**

4. What is Yield Stress?

- **The stress level beyond which the material becomes plastic**

- The stress level beyond which the material breaks

- The strain level above the stress level which the material breaks

- The stress level beyond the melting point of the material

5. A high quality Shell element has _____ nodes.

- 4
- 5
- **6**
- 8

6. Stress σ is proportional to _____ in a Linear Elastic Material.

- **Strain**
- Stress
- Force
- Pressure

7. The Elastic Modulus (Young's Modulus) is the slope defined as _____ divided by _____.

- Strain, Stress
- **Stress, Strain**
- Stress, Force
- Force, Area

8. Linear static analysis assumes that the relationship between loads and the induced response is _____.

- Flat
- **Linear**
- Doubles per area
- Translational

9. In SOLIDWORKS Simulation, the Factor of Safety (FOS) calculations are based on one of the following failure criteria.

- **Maximum von Mises Stress**
- **Maximum shear stress (Tresca)**
- **Mohr-Coulomb stress**
- **Maximum Normal stress**

10. The Yield point is the point where the material begins to deform at a faster rate than at the elastic limit. The material behaves _____ in the Plastic Range.

- Flatly
- Linearly
- **Non-Linearly**
- Like a liquid

11. What are the Degrees of Freedom (DOFs) restrained for a Solid?

- None
- **3 Translations**
- 3 Translations and 3 Rotations
- 3 Rotations

12. What are the Degrees of Freedom (DOFs) restrained for Truss joints?

- None
- **3 Translations**
- 3 Translations and 3 Rotations
- 3 Rotations

13. What are the Degrees of Freedom (DOFs) restrained for Shells and Beams?

- None
- 3 Translations
- **3 Translations and 3 Rotations**
- 3 Rotations

14. Which statements are true for Material Properties using SOLIDWORKS Simulation?

- **For solid assemblies, each component can have a different material**
- For shell models, each shell cannot have a different material and thickness
- **For shell models, the material of the part is used for all shells**
- For beam models, each beam cannot have a different material

15. A Beam element has _____nodes (one at each end) with _____degrees of freedom per node plus_____ node to define the orientation of the beam cross section.

- 6, 3, 1
- 3, 3, 1
- **3, 6, 1**
- None of the above

16. A Truss element has _____ nodes with _____ translational degrees of freedom per node.

- **2, 3**
- 3, 3
- 6, 6
- 2, 2

17. In general, the finer the mesh the better the accuracy of the results.

- **True**
- False

18. How does SOLIDWORKS Simulation automatically treat a Sheet metal part with uniform thickness?

- **Shell**
- Solid
- Beam
- Mixed Mesh

19. Use the mesh and displacement plots to calculate the distance between two _____ using SOLIDWORKS Simulation.

- **Nodes**
- Elements
- Bodies
- Surfaces

20. Surface models can only be meshed with _____ elements.

- **Shell**
- Beam
- Mixed Mesh
- Solid

21. The shell mesh is generated on the surface (located at the mid-surface of the shell).

- **True**
- False

22. In general, use Thin shells when the thickness-to-span ratio is less than _____.

- **0.05**
- .5
- 1
- 2

23. The model (a rectangular plate) has a length to thickness ratio of less than 5. You extracted its mid-surface to use it in SOLIDWORKS Simulation. You should use a _____.

- Thin Shell element formulation
- **Thick Shell element formulation**
- Thick or Thin Shell element formulation, it does not matter
- Beam Shell element formulation

24. The model, a rectangular sheet metal part, uses SOLIDWORKS Simulation. You should use a:

- **Thin Shell element formulation**
- Thick Shell element formulation
- Thick or Thin Shell element formulation, it does not matter
- Beam Shell element formulation

25. The Global element size parameter provides the ability to set the global average element size. SOLIDWORKS Simulation suggests a default value based on the model volume and _____ area. This option is only available for a standard mesh.

- Force
- Pressure
- **Surface**
- None of the above

26. A remote load applied on a face with a Force component and no Moment can result in: Note: Remember (DOFs restrain).

- **A Force and Moment of the face**
- A Force on the face only
- A Moment on the face only
- A Pressure and Force on the face

27. There are _____ DOFs restrain for a Solid element.

- **3**
- 1
- 6
- None

28. There are _____ DOFs restrain for a Beam element.

- 3
- 1
- **6**
- None

29. What best describes the difference(s) between a Fixed and Immovable (No translation) boundary condition in SOLIDWORKS Simulation?

- There are no differences
- There are no difference(s) for Shells but it is different for Solids
- **There is no difference(s) for Solids but it is different for Shells and Beams**
- There are only differences(s) for a Static Study

30. Can a non-uniform pressure of force be applied on a face using SOLIDWORKS Simulation?

- No

- Yes, but the variation must be along a single direction only

- **Yes. The non-uniform pressure distribution is defined by a reference coordinate system and the associated coefficients of a second order polynomial**

- Yes, but the variation must be linear

31. You are performing an analysis on your model. You select five faces, 3 edges and 2 vertices and apply a force of 20lbf. What is the total force applied to the model using SOLIDWORKS Simulation?

- 100lbf

- 1600lbf

- 180lbf

- **200lbf**

32. Yield strength is typically determined at _____ strain.
- 0.1%

- **0.2%**

- 0.02%

- 0.002%

33. There are four key assumptions made in Linear Static Analysis: 1. Effects of inertia and damping are neglected, 2. The response of the system is directly proportional to the applied loads, 3. Loads are applied slowly and gradually, and_____ .

- **Displacements are very small. The highest stress is in the linear range of the stress-strain curve**

- There are no loads

- Material is not elastic

- Loads are applied quickly

34. How many degrees of freedom does a physical structure have?

- Zero

- Three - Rotations only

- Three - Translations only

- **Six - Three translations and three rotational**

35. Brittle material has little tendency to deform (or strain) before fracture and does not have a specific yield point. It is not recommended to apply the yield strength analysis as a failure criterion on brittle material. Which of the following failure theories is appropriate for brittle materials?

- **Mohr-Columb stress criterion**

- Maximum shear stress criterion

- Maximum von Mises stress criterion

- Minimum shear stress criterion

36. You are performing an analysis on your model. You select three faces and apply a force of 40lb. What is the total force applied to the model using SOLIDWORKS Simulation?

- 40lb

- 20lb

- **120lb**

- Additional information is required

37. A material is orthotropic if its mechanical or thermal properties are not unique and independent in three mutually perpendicular directions.

- True

- **False**

38. An increase in the number of elements in a mesh for a part will:

- Decrease calculation accuracy and time

- **Increase calculation accuracy and time**

- Have no effect on the calculation

- Change the FOS below 1

39. SOLIDWORKS Simulation uses the von Mises Yield Criterion to calculate the Factor of Safety of many ductile materials. According to the criterion:

- **Material yields when the von Mises stress in the model equals the yield strength of the material**

- Material yields when the von Mises stress in the model is 5 times greater than the minimum tensile strength of the material

- Material yields when the von Mises stress in the model is 3 times greater than the FOS of the material

- None of the above

40. SOLIDWORKS Simulation calculates structural failure on:

- Buckling

- Fatigue

- Creep

- **Material yield**

41. Apply a uniform total force of 200lb on two faces of a model. The two faces have different areas. How do you apply the load using SOLIDWORKS Simulation for a Linear Static Study?

- Select the two faces and input a normal to direction force of 200lb on each face

- Select the two faces and a reference plane. Apply 100lb on each face

- **Apply equal force to the two faces. The force on each face is the total force divided by the total area of the two faces**.

- None of the above

42. Maximum and Minimum value indicators are displayed on Stress and Displacement plots in SOLIDWORKS Simulation for a Linear Static Study.

- **True**

- False

43. What SOLIDWORKS Simulation tool should you use to determine the result values at specific locations (nodes) in a model using SOLIDWORKS Simulation?

- Section tool
- **Probe tool**
- Clipping tool
- Surface tool

44. What criteria are best suited to check the failure of ductile materials in SOLIDWORKS Simulation?

- Maximum von Mises Strain and Maximum Shear Strain criterion
- **Maximum von Misses Stress and Maximum Shear Stress criterion**
- Maximum Mohr-Coulomb Stress and Maximum Mohr-Coulomb Shear Strain criterion
- Mohr-Coulomb Stress and Maximum Normal Stress criterion

45. Set the scale factor for plots_____ to avoid any misinterpretation of the results, after performing a Static analysis with gap/contact elements.

- Equal to 0
- **Equal to 1**
- Less than 1
- To the Maximum displacement value for the model

GLOSSARY

Alphabet of Lines: Each line on a technical drawing has a definite meaning and is drawn in a certain way. The line conventions recommended by the American National Standards Institute (ANSI) are presented in this text.

Alternate Position View: A drawing view superimposed in phantom lines on the original view. Utilized to show range of motion of an assembly.

Anchor Point: The origin of the Bill of Material in a sheet format.

Annotation: An annotation is a text note or a symbol that adds specific information and design intent to a part, assembly, or drawing. Annotations in a drawing include specific note, hole callout, surface finish symbol, datum feature symbol, datum target, geometric tolerance symbol, weld symbol, balloon and stacked balloon, center mark, centerline marks, area hatch and block.

ANSI: American National Standards Institute.

Area Hatch: Apply a crosshatch pattern or solid fill to a model face, to a closed sketch profile, or to a region bounded by a combination of model edges and sketch entities. Area hatch can be applied only in drawings.

ASME: American Society of Mechanical Engineering, publisher of ASME Y14 Engineering Drawing and Documentation Practices that controls drawing, dimensioning and tolerancing.

Assembly: An assembly is a document in which parts, features and other assemblies (sub-assemblies) are put together. A part in an assembly is called a component. Adding a component to an assembly creates a link between the assembly and the component. When SOLIDWORKS opens the assembly, it finds the component file to show it in the assembly. Changes in the component are automatically reflected in the assembly. The filename extension for a SOLIDWORKS assembly file name is *.sldasm.

Attachment Point: An attachment point is the end of a leader that attaches to an edge, vertex, or face in a drawing sheet.

AutoDimension: The Autodimension tool provides the ability to insert reference dimensions into drawing views such as baseline, chain, and ordinate dimensions.

Auxiliary View: An Auxiliary View is similar to a Projected View, but it is unfolded normal to a reference edge in an existing view.

AWS: American Welding Society, publisher of AWS A2.4, Standard Location of Elements of a Welding Symbol.

Axonometric Projection: A type of parallel projection, more specifically a type of orthographic projection, used to create a pictorial drawing of an object, where the object is rotated along one or more of its axes relative to the plane of projection.

Balloon: A balloon labels the parts in the assembly and relates them to item numbers on the bill of materials (BOM) added in the drawing. The balloon item number corresponds to the order in the Feature Tree. The order controls the initial BOM Item Number.

Baseline Dimensions: Dimensions referenced from the same edge or vertex in a drawing view.

Bill of Materials: A table inserted into a drawing to keep a record of the parts and materials used in an assembly.

Block: A symbol in the drawing that combines geometry into a single entity.

BOM: Abbreviation for Bill of Materials.

Broken-out Section: A broken-out section exposes inner details of a drawing view by removing material from a closed profile. In an assembly, the Broken-out Section displays multiple components.

CAD: The use of computer technology for the design of objects, real or virtual. CAD often involves more than just shapes.

Cartesian Coordinate System: Specifies each point uniquely in a plane by a pair of numerical coordinates, which are the signed distances from the point to two fixed perpendicular directed lines, measured in the same unit of length. Each reference line is called a coordinate axis or just axis of the system, and the point where they meet is its origin.

Cell: Area to enter a value in an EXCEL spreadsheet, identified by a Row and Column.

Center Mark: A cross that marks the center of a circle or arc.

Centerline: An axis of symmetry in a sketch or drawing displayed in a phantom font.

CommandManager: The CommandManager is a Context-sensitive toolbar that dynamically updates based on the toolbar you want to access. By default, it has toolbars embedded in it based on the document type. When you click a tab below the Command Manager, it updates to display that toolbar. For example, if you click the Sketch tab, the Sketch toolbar is displayed.

Component: A part or sub-assembly within an assembly.

ConfigurationManager: The ConfigurationManager is located on the left side of the SOLIDWORKS window and provides the means to create, select and view multiple configurations of parts and assemblies in an active document. You can split the ConfigurationManager and either display two ConfigurationManager instances, or combine the ConfigurationManager with the FeatureManager design tree, PropertyManager or third party applications that use the panel.

Configurations: Variations of a part or assembly that control dimensions, display and state of a model.

Coordinate System: SOLIDWORKS uses a coordinate system with origins. A part document contains an original origin. Whenever you select a plane or face and open a sketch, an origin is created in alignment with the plane or face. An origin can be used as an anchor for the sketch entities, and it helps orient perspective of the axes. A three-dimensional reference triad orients you to the X, Y, and Z directions in part and assembly documents.

Copy and Paste: Utilize copy/paste to copy views from one sheet to another sheet in a drawing or between different drawings.

Cosmetic Thread: An annotation that represents threads.

Crosshatch: A pattern (or fill) applied to drawing views such as section views and broken-out sections.

Cursor Feedback: The system feedback symbol indicates what you are selecting or what the system is expecting you to select. As you move the mouse pointer across your model, system feedback is provided.

Datum Feature: An annotation that represents the primary, secondary and other reference planes of a model utilized in manufacturing.

Depth: The horizontal (front to back) distance between two features in frontal planes. Depth is often identified in the shop as the thickness of a part or feature.

Design Table: An Excel spreadsheet that is used to create multiple configurations in a part or assembly document.

Detail View: A portion of a larger view, usually at a larger scale than the original view. Create a detail view in a drawing to display a portion of a view, usually at an enlarged scale. This detail may be of an orthographic view, a non-planar (isometric) view, a section view, a crop view, an exploded assembly view or another detail view.

Detailing: Detailing refers to the SOLIDWORKS module used to insert, add and modify dimensions and notes in an engineering drawing.

Dimension Line: A line that references dimension text to extension lines indicating the feature being measured.

Dimension Tolerance: Controls the dimension tolerance values and the display of non-integer dimensions. The tolerance types are *None, Basic, Bilateral, Limit, Symmetric, MIN, MAX, Fit, Fit with tolerance* or *Fit (tolerance only).*

Dimension: A value indicating the size of the 2D sketch entity or 3D feature. Dimensions in a SOLIDWORKS drawing are associated with the model, and changes in the model are reflected in the drawing, if you DO NOT USE DimXpert.

Dimensioning Standard - Metric: - ASME standards for the use of metric dimensioning require all the dimensions to be expressed in millimeters (mm). The (mm) is not needed on each dimension, but it is used when a dimension is used in a notation. No trailing zeros are used. The Metric or International System of Units (S.I.) unit system in drafting is also known as the Millimeter, Gram Second (MMGS) unit system.

Dimensioning Standard - U.S: - ASME standard for U.S. dimensioning uses the decimal inch value. When the decimal inch system is used, a zero is not used to the left of the decimal point for values less than one inch, and trailing zeros are used. The U.S. unit system is also known as the Inch, Pound, Second (IPS) unit system.

DimXpert for Parts: A set of tools that applies dimensions and tolerances to parts according to the requirements of the ASME Y.14.41-2009 standard.

DimXpertManager: The DimXpertManager lists the tolerance features defined by DimXpert for a part. It also displays DimXpert tools that you use to insert dimensions and tolerances into a part. You can import these dimensions and tolerances into drawings. DimXpert is not associative.

Document: In SOLIDWORKS, each part, assembly, and drawing is referred to as a document, and each document is displayed in a separate window.

Drawing Sheet: A page in a drawing document.

Drawing Template: A document that is the foundation of a new drawing. The drawing template contains document properties and user-defined parameters such as sheet format. The extension for the drawing template filename is .DRWDOT.

Drawing: A 2D representation of a 3D part or assembly. The extension for a SOLIDWORKS drawing file name is .SLDDRW. Drawing refers to the SOLIDWORKS module used to insert, add, and modify views in an engineering drawing.

Edit Sheet Format: The drawing sheet contains two modes. Utilize the Edit Sheet Format command to add or modify notes and Title block information. Edit in the Edit Sheet Format mode.

Edit Sheet: The drawing sheet contains two modes. Utilize the Edit Sheet command to insert views and dimensions.

eDrawing: A compressed document that does not require the referenced part or assembly. eDrawings are animated to display multiple views in a drawing.

Empty View: An Empty View creates a blank view not tied to a part or assembly document.

Engineering Graphics: Translates ideas from design layouts, specifications, rough sketches, and calculations of engineers & architects into working drawings, maps, plans and illustrations which are used in making products.

Equation: Creates a mathematical relation between sketch dimensions, using dimension names as variables, or between feature parameters, such as the depth of an extruded feature or the instance count in a pattern.

Exploded view: A configuration in an assembly that displays its components separated from one another.

Export: The process to save a SOLIDWORKS document in another format for use in other CAD/CAM, rapid prototyping, web or graphics software applications.

Extension Line: The line extending from the profile line indicating the point from which a dimension is measured.

Extruded Cut Feature: Projects a sketch perpendicular to a Sketch plane to remove material from a part.

Face: A selectable area (planar or otherwise) of a model or surface with boundaries that help define the shape of the model or surface. For example, a rectangular solid has six faces.

Family Cell: A named empty cell in a Design Table that indicates the start of the evaluated parameters and configuration names. Locate Comments in a Design Table to the left or above the Family Cell.

Fasteners: Includes Bolts and nuts (threaded), Set screws (threaded), Washers, Keys, and Pins to name a few. Fasteners are not a permanent means of assembly such as welding or adhesives.

Feature: Features are geometry building blocks. Features add or remove material. Features are created from 2D or 3D sketched profiles or from edges and faces of existing geometry.

FeatureManager: The FeatureManager design tree located on the left side of the SOLIDWORKS window provides an outline view of the active part, assembly, or drawing. This makes it easy to see how the model or assembly was constructed or to examine the various sheets and views in a drawing. The FeatureManager and the Graphics window are dynamically linked. You can select features, sketches, drawing views and construction geometry in either pane.

First Angle Projection: In First Angle Projection the Top view is looking at the bottom of the part. First Angle Projection is used in Europe and most of the world. However, America and Australia use a method known as Third Angle Projection.

Fully defined: A sketch where all lines and curves in the sketch, and their positions, are described by dimensions or relations, or both, and cannot be moved. Fully defined sketch entities are shown in black.

Foreshortened radius: Helpful when the centerpoint of a radius is outside of the drawing or interferes with another drawing view: Broken Leader.

Foreshortening: The way things appear to get smaller in both height and depth as they recede into the distance.

French curve: A template made out of plastic, metal or wood composed of many different curves. It is used in manual drafting to draw smooth curves of varying radii.

Fully Defined: A sketch where all lines and curves in the sketch, and their positions, are described by dimensions or relations, or both, and cannot be moved. Fully defined sketch entities are displayed in black.

Geometric Tolerance: A set of standard symbols that specify the geometric characteristics and dimensional requirements of a feature.

Glass Box method: A traditional method of placing an object in an *imaginary glass box* to view the six principle views.

Global Coordinate System: Directional input refers by default to the Global coordinate system (X-, Y- and Z-), which is based on Plane1 with its origin located at the origin of the part or assembly.

Graphics Window: The area in the SOLIDWORKS window where the part, assembly, or drawing is displayed.

Grid: A system of fixed horizontal and vertical divisions.

Handle: An arrow, square or circle that you drag to adjust the size or position of an entity such as a view or dimension.

Heads-up View Toolbar: A transparent toolbar located at the top of the Graphic window.

Height: The vertical distance between two or more lines or surfaces (features) which are in horizontal planes.

Hidden Lines Removed (HLR): A view mode. All edges of the model that are not visible from the current view angle are removed from the display.

Hidden Lines Visible (HLV): A view mode. All edges of the model that are not visible from the current view angle are shown gray or dashed.

Hole Callouts: Hole callouts are available in drawings. If you modify a hole dimension in the model, the callout updates automatically in the drawing if you did not use DimXpert.

Hole Table: A table in a drawing document that displays the positions of selected holes from a specified origin datum. The tool labels each hole with a tag. The tag corresponds to a row in the table.

Import: The ability to open files from other software applications into a SOLIDWORKS document. The A-size sheet format was created as an AutoCAD file and imported into SOLIDWORKS.

Isometric Projection: A form of graphical projection, more specifically, a form of axonometric projection. It is a method of visually representing three-dimensional objects in two dimensions, in which the three coordinate axes appear equally foreshortened and the angles between any two of them are 120º.

Layers: Simplifies a drawing by combining dimensions, annotations, geometry and components. Properties such as display, line style and thickness are assigned to a named layer.

Leader: A solid line created from an annotation to the referenced feature.

Line Format: A series of tools that control Line Thickness, Line Style, Color, Layer and other properties.

Local (Reference) Coordinate System: Coordinate system other than the Global coordinate system. You can specify restraints and loads in any desired direction.

Lock Sheet Focus: Adds sketch entities and annotations to the selected sheet. Double-click the sheet to activate Lock Sheet Focus. To unlock a sheet, right-click and select Unlock Sheet Focus or double click inside the sheet boundary.

Lock View Position: Secures the view at its current position in the sheet. Right-click in the drawing view to Lock View Position. To unlock a view position, right-click and select Unlock View Position.

Mass Properties: The physical properties of a model based upon geometry and material.

Menus: Menus provide access to the commands that the SOLIDWORKS software offers. Menus are Context-sensitive and can be customized through a dialog box.

Model Item: Provides the ability to insert dimensions, annotations, and reference geometry from a model document (part or assembly) into a drawing.

Model View: A specific view of a part or assembly. Standard named views are listed in the view orientation dialog box such as isometric or front. Named views can be user-defined name for a specific view.

Model: 3D solid geometry in a part or assembly document. If a part or assembly document contains multiple configurations, each configuration is a separate model.

Motion Studies: Graphical simulations of motion and visual properties with assembly models. Analogous to a configuration, they do not actually change the original assembly model or its properties. They display the model as it changes based on simulation elements you add.

Mouse Buttons: The left, middle, and right mouse buttons have distinct meanings in SOLIDWORKS. Use the middle mouse button to rotate and Zoom in/out on the part or assembly document.

Oblique Projection: A simple type of graphical projection used for producing pictorial, two-dimensional images of three-dimensional objects.

OLE (Object Linking and Embedding): A Windows file format. A company logo or EXCEL spreadsheet placed inside a SOLIDWORKS document are examples of OLE files.

Ordinate Dimensions: Chain of dimensions referenced from a zero ordinate in a drawing or sketch.

Origin: The model origin is displayed in blue and represents the (0,0,0) coordinate of the model. When a sketch is active, a sketch origin is displayed in red and represents the (0,0,0) coordinate of the sketch. Dimensions and relations can be added to the model origin but not to a sketch origin.

Orthographic Projection: A means of representing a three-dimensional object in two dimensions. It is a form of parallel projection, where the view direction is orthogonal to the projection plane, resulting in every plane of the scene appearing in affine transformation on the viewing surface.

Parametric Note: A Note annotation that links text to a feature dimension or property value.

Parent View: A Parent view is an existing view on which other views are dependent.

Part Dimension: Used in creating a part, they are sometimes called construction dimensions.

Part: A 3D object that consist of one or more features. A part inserted into an assembly is called a component. Insert part views, feature dimensions and annotations into 2D drawing. The extension for a SOLIDWORKS part filename is .SLDPRT.

Perspective Projection: The two most characteristic features of perspective are that objects are drawn smaller as their distance from the observer increases and foreshortened: the size of an object's dimensions along the line of sight are relatively shorter than dimensions across the line of sight.

Plane: To create a sketch, choose a plane. Planes are flat and infinite. Planes are represented on the screen with visible edges.

Precedence of Line Types: When obtaining orthographic views, it is common for one type of line to overlap another type. When this occurs, drawing conventions have established an order of precedence.

Precision: Controls the number of decimal places displayed in a dimension.

Projected View: Projected views are created for Orthogonal views using one of the following tools: Standard 3 View, Model View or the Projected View tool from the View Layout toolbar.

Properties: Variables shared between documents through linked notes.

PropertyManager: Most sketch, feature, and drawing tools in SOLIDWORKS open a PropertyManager located on the left side of the SOLIDWORKS window. The PropertyManager displays the properties of the entity or feature so you specify the properties without a dialog box covering the Graphics window.

RealView: Provides a simplified way to display models in a photo-realistic setting using a library of appearances and scenes. RealView requires graphics card support and is memory intensive.

Rebuild: A tool that updates (or regenerates) the document with any changes made since the last time the model was rebuilt. Rebuild is typically used after changing a model dimension.

Reference Dimension: Dimensions added to a drawing document are called Reference dimensions, and are driven; you cannot edit the value of reference dimensions to modify the model. However, the values of reference dimensions change when the model dimensions change.

Relation: A relation is a geometric constraint between sketch entities or between a sketch entity and a plane, axis, edge or vertex.

Relative view: The Relative View defines an Orthographic view based on two orthogonal faces or places in the model.

Revision Table: The Revision Table lists the Engineering Change Orders (ECO), in a table form, issued over the life of the model and the drawing. The current Revision letter or number is placed in the Title block of the Drawing.

Right-Hand Rule: Is a common mnemonic for understanding notation conventions for vectors in 3 dimensions.

Rollback: Suppresses all items below the rollback bar.

Scale: A relative term meaning "size" in relationship to some system of measurement.

Section Line: A line or centerline sketched in a drawing view to create a section view.

Section Scope: Specifies the components to be left uncut when you create an assembly drawing section view.

Section View: You create a section view in a drawing by cutting the parent view with a cutting, or section line. The section view can be a straight cut section or an offset section defined by a stepped section line. The section line can also include concentric arcs. Create a Section View in a drawing by cutting the Parent view with a section line.

Sheet Format: A document that contains the following: page size and orientation, standard text, borders, logos, and Title block information. Customize the Sheet format to save time. The extension for the Sheet format filename is .SLDDRT.

Sheet Properties: Sheet Properties display properties of the selected sheet. Sheet Properties define the following: Name of the Sheet, Sheet Scale, Type of Projection (First angle or Third angle), Sheet Format, Sheet Size, View label, and Datum label.

Sheet: A page in a drawing document.

Silhouette Edge: A curve representing the extent of a cylindrical or curved face when viewed from the side.

Sketch: The name to describe a 2D profile is called a sketch. 2D sketches are created on flat faces and planes within the model. Typical geometry types are lines, arcs, corner rectangles, circles, polygons, and ellipses.

Spline: A sketched 2D or 3D curve defined by a set of control points.

Stacked Balloon: A group of balloons with only one leader. The balloons can be stacked vertically (up or down) or horizontally (left or right).

Standard views: The three orthographic projection views, Front, Top and Right positioned on the drawing according to First angle or Third angle projection.

Suppress: Removes an entity from the display and from any calculations in which it is involved. You can suppress features, assembly components, and so on. Suppressing an entity does not delete the entity; you can unsuppress the entity to restore it.

Surface Finish: An annotation that represents the texture of a part.

System Feedback: Feedback is provided by a symbol attached to the cursor arrow indicating your selection. As the cursor floats across the model, feedback is provided in the form of symbols riding next to the cursor.

System Options: System Options are stored in the registry of the computer. System Options are not part of the document. Changes to the System Options affect all current and future documents. There are hundreds of Systems Options.

Tangent Edge: The transition edge between rounded or filleted faces in hidden lines visible or hidden lines removed modes in drawings.

Task Pane: The Task Pane is displayed when you open the SOLIDWORKS software. It contains the following tabs: SOLIDWORKS Resources, Design Library, File Explorer, Search, View Palette, Document Recovery and RealView/PhotoWorks.

Templates: Templates are part, drawing and assembly documents that include user-defined parameters and are the basis for new documents.

Third Angle Projection: In Third angle projection the Top View is looking at the Top of the part. First Angle Projection is used in Europe and most of the world. America and Australia use the Third Angle Projection method.

Thread Class or Fit: Classes of fit are tolerance standards; they set a plus or minus figure that is applied to the pitch diameter of bolts or nuts. The classes of fit used with almost all bolts sized in inches are specified by the ANSI/ASME Unified Screw Thread standards (which differ from the previous American National standards).

Thread Lead: The distance advanced parallel to the axis when the screw is turned one revolution. For a single thread, lead is equal to the pitch; for a double thread, lead is twice the pitch.

Tolerance: The permissible range of variation in a dimension of an object. Tolerance may be specified as a factor or percentage of the nominal value, a maximum deviation from a nominal value, an explicit range of allowed values, be specified by a note or published standard with this information, or be implied by the numeric accuracy of the nominal value.

Toolbars: The toolbar menus provide shortcuts enabling you to access the most frequently used commands. Toolbars are Context-sensitive and can be customized through a dialog box.

T-Square: A technical drawing instrument, primarily a guide for drawing horizontal lines on a drafting table. It is used to guide the triangle that draws vertical lines. Its name comes from the general shape of the instrument where the horizontal member of the T slides on the side of the drafting table. Common lengths are 18", 24", 30", 36" and 42".

Under-defined: A sketch is under defined when there are not enough dimensions and relations to prevent entities from moving or changing size.

Units: Used in the measurement of physical quantities. Decimal inch dimensioning and Millimeter dimensioning are the two types of common units specified for engineering parts and drawings.

Vertex: A point at which two or more lines or edges intersect. Vertices can be selected for sketching, dimensioning, and many other operations.

View Palette: Use the View Palette, located in the Task Pane, to insert drawing views. It contains images of standard views, annotation views, section views, and flat patterns (sheet metal parts) of the selected model. You can drag views onto the drawing sheet to create a drawing view.

Weld Bead: An assembly feature that represents a weld between multiple parts.

Weld Finish: A weld symbol representing the parameters you specify.

Weld Symbol: An annotation in the part or drawing that represents the parameters of the weld.

Width: The horizontal distance between surfaces in profile planes. In the machine shop, the terms length and width are used interchangeably.

Zebra Stripes: Simulate the reflection of long strips of light on a very shiny surface. They allow you to see small changes in a surface that may be hard to see with a standard display.

Notes:

Index